Cambridge Archaeological a

PLACE-NAMES

OF

SOUTH-WEST YORKSHIRE

PLACE-NAMES

OF

SOUTH-WEST YORKSHIRE

THAT IS, OF SO MUCH OF THE WEST RIDING AS
LIES SOUTH OF THE AIRE FROM
KEIGHLEY ONWARDS

BY

ARMITAGE GOODALL, M.A.

LATE SCHOLAR OF QUEENS' COLLEGE, CAMBRIDGE

REVISED EDITION

Cambridge :

at the University Press

1914

CAMBRIDGE UNIVERSITY PRESS
Cambridge, New York, Melbourne, Madrid, Cape Town,
Singapore, São Paulo, Delhi, Tokyo, Mexico City

Cambridge University Press
The Edinburgh Building, Cambridge CB2 8RU, UK

Published in the United States of America by Cambridge University Press, New York

www.cambridge.org
Information on this title: www.cambridge.org/9781107669499

First published 1914
First paperback edition 2011

A catalogue record for this publication is available from the British Library

ISBN 978-1-107-66949-9 Paperback

PREFACE

THE material for this work has gradually accumulated during the last seven or eight years, and the work itself owes its existence to the interest aroused during journeyings—almost daily—in and about the northern part of the district dealt with. Names like Thornhill and Langfield made no secret of their origin, but such as Barugh and Chevet, Halifax and Hipperholme, Gildersome and Golcar, raised their voices in continual challenge. The minor names of the district—names of hamlets and farms, and woods and lanes—proved equally provocative. The meaning of Kirkgate and Westgate was obvious, but what were Skeldergate and Cluntergate? There were strange names like Drub and Hades, Backhold and Featherbed; there were others of imposing appearance like Paris and London; there were names obviously Celtic, and others obviously Scandinavian; and behind them all were interesting points in history, both general and ecclesiastical. And so one gradually moved forward, and at last what began in mere curiosity ended in definite purpose.

In regard to the scope of the work the door has been thrown wide open; even rivers have been included, and the result—in an area covering less than half the Riding—is a list of about 1,500 names. In order to secure the advantage arising from comparative methods, names have frequently been considered in groups; and in order to make the work as valuable as possible from the historical point of view an attempt has been made to put on record every existing name where such elements as *by*, *thwaite*, *thorpe*, and *scholes* are involved.

The publications of the Yorkshire Archæological Society, its Journal and its Record Series, have provided the greater part of the material necessary; but a particular debt must be acknowledged in regard to the two volumes of the Wakefield

Court Rolls. It is to the details given in these Court Rolls that one owes the possibility of dealing with large numbers of minor names—some particularly interesting.

As far as our own country is concerned, the scientific study of place-names is quite modern; almost all the really helpful works on the subject have been published during the present century. Among these may be named monographs by the late Professor Skeat on the place-names of Berkshire, Bedford, Cambridge, Hertford, and Huntingdon; Dr Moorman's book on West Riding place-names; Dr Wyld's on the place-names of Lancashire; and the two books by Mr Duignan on those of Staffordshire and Worcestershire. I have made considerable use of all these works, and wish to acknowledge my great indebtedness to them.

Among those to whom I owe gratitude for personal assistance, chief of all is the late Professor Skeat, whose unrivalled stores of knowledge and experience were willingly placed at my disposal on several occasions. My heartiest acknowledgements are also due to Mr E. C. Quiggin, Fellow of Gonville and Caius College, Cambridge, for invaluable help in connection with names of Celtic origin; and in addition I am greatly indebted to Mr C. M. Drennan of Christ's College, Cambridge, and to the Rev. H. Dewhurst of St Andrew's, Leytonstone, for many helpful suggestions and criticisms. Lastly, I must express my appreciation of the great care with which the task of printing has been carried out, and my indebtedness to the Staff of the University Press for many evidences of kindness and consideration.

A. G.

May 1913

A number of corrections and improvements have been made in the body of the work, and others are given at the end of the volume.

A. G.

June 1914

CONTENTS

CHAPTER I

INTRODUCTION

Chief Sources of our Place-names—Need of Early Forms—The Story of York—
Continuity yet change—The Story of Whitby—Births and Deaths—The Rivalries of
Place-names—Their Mimicries—Their Deceptions—Their Limitations—A Difference
between Ancient and Modern—Customary Forms—Secondary Forms—Importance of
Minor Place-names—Limits of the Enquiry.

The place-names of South-west Yorkshire are largely of Anglian origin, but there is a considerable section which is Scandinavian and due to the Vikings. Among sources less important—indeed, far less important—four may be named: Celtic, Roman, Norman, and Modern English. There are thus six sources, of which two require sub-division, for the Celtic names have come to us at different periods, and the Scandinavian by different avenues. *Chief Sources of our Place-names.*

Our enquiry must, of course, be based on historical methods, and its first step must be to discover as far as possible early records of the names to be considered. Two Berkshire place-names, Courage and Seacourt, show this need conspicuously. In pre-Conquest times the former appears as Cusan-ricge, that is, 'Cusa's ridge,' while the latter appears as Seofecan-wyrthe, 'Seofeca's worth,' that is 'Seofeca's farm.' What the explanations might have been if history had been ignored can be readily imagined. In our own district, there are many examples which show just as clearly the need of early spellings, but perhaps most striking of all is the name Golcar, which in former days rejoiced in such forms as Gouthelaghcharthes, Gouthlacharwes, and Goullakarres. *Need of Early Forms.*

The story of gradual development disclosed by early forms is by no means the least attractive source of interest opened out *The Story of York.*

by the enquiry.   Take York as an example.   In the *Historia Britonum* the Celtic name is given as Ebrauc, which corresponds to an earlier form like the stem in Eburacum[1], the name given by the Antonine Itinerary.   But to the Anglian settlers in the 6th century such a name was meaningless, and later on—in the A.-S. Chronicle for 867 for example—we find it in a new guise, Eoforwic, 'boar place,' a name which in course of years became Everwic.   Meanwhile the Danes came upon the scene and impressed upon the word a new pronunciation, of which the written form was Jorvik; and it is from this we get the modern name, for Jorvik was pronounced Yorwick, which later became Yorick, and finally York.   Thus the original Celtic name has been handled by Roman, Anglian, and Dane; it has been changed, but never discarded.   The Roman did no more than change the ending; the Anglian and Dane while refraining from the creation of a new name, twisted that which came down to them into a form they could understand. Centuries passed by, and the word suffered further transformation; yet it is still the true descendant of the Celtic form, a remarkable example of unbroken continuity, bearing witness to corresponding continuity in the history of the city itself[2].

Continuity yet Change.   This continuity is one of the chief characteristics of place-names.   It is well to remember, however, that there is another side to the picture, for we are dealing not so much with counters as with living things.   Our place-names, like the men who use them, change; they have their evolutions and their revolutions. Yet, in the midst of all, they possess a persistence quite remarkable, and often fulfil their duty as proper names long after they have lost the meaning they were originally intended to convey.

The Story of Whitby.   In direct contrast with the story of York is that of another historic town, namely, Whitby.   In the 7th century Bede records its name as Streanæshalch.   But in the opening words of a 12th century document dealing with the foundation of the Abbey we find its situation described as

'in loco qui olim Streoneshalc vocabatur, deinde
Prestebi appellabatur, nunc vero Witebi vocatur.'

[1] The *b* probably represents *v*.
[2] Freeman, *English Towns and Districts*, pp. 275–6.

Thus the Angles described the site of the Abbey as Streoneshalc, while under the Danes it was called Prestebi, the opposite bank of the Esk being Witebi. At a later date Prestebi became subordinate to Witebi, and finally was altogether superseded by it; and so to-day Whitby reigns supreme.

Thus, while York provides an example of continuity, Whitby *Births and* gives one of entire change—the death of one and the birth of *Deaths.* another. Doubtless every century has seen examples of this character, for our place-names are in no small degree a record of the activities of the ages. The 19th century, which saw the upspringing of many centres of population, gave us such names as Savile Town and Ripleyville. The 18th century gave us Fulneck; and perhaps it was the 17th which provided such Bible names as Egypt and Machpelah and Padan-aram. Much earlier came Roche and Grange and Abdy, the product of religious movements in the 12th and 13th centuries. On the other hand Hethewalley in Flockton and Rameldhagh in Huddersfield, with many others known only to ancient charters and deeds, are entirely lost. Every century has seen the departure of some, and the arrival of others. There has been, in fact, living development. A living people has impressed its needs upon its local names, and they perforce have assumed the character of living things.

Of this there is further evidence in the rivalry sometimes *The* shown. Pontefract and Pomfret, after the overthrow of Tate- *Rivalries of Place-* shale and Kirkebi, have maintained for centuries a struggle for *names.* the mastery, which is not yet fully decided. Further south the rivalry between Greseburg and Gresebroke has been decided in favour of the former, now Greasborough; and in the same way the rivalry between Wridelesford and Wodelesford has resulted in victory for the latter, now Woodlesford. Occasionally the usurper fails, as in the case of Emley, where for a time Elmley won much favour; but sometimes it succeeds, as at Queensbury and Norristhorpe.

There are instances not a few where names show mimicry, *Their* changing their form in sympathy with their neighbours. Many *Mimicries.* names in which the original ending was -um have now the ending -holme, as in Hipperholme and the various Mytholms.

The present form of Almondbury is due to similar influence, and so are those of Falhouse and Endcliffe, as well as Ashday, Bailiffe Bridge, Skelmanthorpe, and Thornhills.

Their Deceptions. Not only do place-names show something of mimicry, they may be very deceptive in other ways. Very frequently, in spite of appearances, such names as Newsholme and Newton have long histories; excellent examples are Newhall near Wath and Newsholme near Keighley which appear as early as the Domesday record, and Newton near Wakefield which was in existence in 1190. In many cases the Marshes are no longer swampy, the Lunds no longer groves, and the Tons no longer farmsteads.

Their Limitations. In a word, place-names have their limitations. Among our Yorkshire hills and dales—to mention limitations of another character—we must not expect such examples of poetic appropriateness as are occasionally found among the Celtic peoples. We shall not find, as in Ireland, a brook called 'little silver.' We shall discover little of the heroic, the romantic, or the legendary. Indeed, there will be much that is frankly 'pedestrian,' for the chief characteristic of our English place-names is to describe the simplest facts in the simplest way.

A Difference between Ancient and Modern. There is yet another characteristic to be noted, namely, the profound difference between names of modern creation and those which come down from ancient times. The latter were never merely conventional, like our modern Bellevues and Claremonts; they were the offspring of the automatic operation of the human mind, and possessed in every case a meaning at once simple, appropriate, and well-defined. Allerton was the farm beside the alders; Thurgoland was the land of Thorgeir; Micklebring the great slope; Bradfield the broad field; and Wooldale the valley of the wolf.

Customary Forms. The question may fairly be asked : What are the customary forms taken by ancient place-names? And the answer is the more important because of its bearing on some of our investigations.

A reference to early charters shows that our ancient place-names almost invariably consisted either of one element or two. We find such single-element names as Bury, Cliffe, Dean, Elm, Chart, and Thorn, each the designation of some simple feature

in the landscape. But along with these there is a large body
of names which add to the substantival element a word of
adjectival force describing in the simplest way the general
appearance or situation of the place,—names like Ashwood,
Easton, Highfield, and Lowmoor. Then, further, there are
those compound names which show ownership. The Teutonic
settlers 'called the lands after their own names.' In this way
our modern village names form the memorials—often, indeed,
the sole existing memorials—of many an unknown adventurer
who settled upon some waste, or occupied the lands of some
predecessor whom he had dislodged[1]. Instances where the
person can be historically identified must of necessity be rare,
but with the aid of ancient records we are able in many cases
to recognize the name itself, however disguised in the modern
spelling. The part taken by personal names in the building-up
of place-names is so important that a subsequent note is given
up entirely to their history and formation.

In the names of fields, woods, and streams, and among other Secondary
minor place-names, it is not uncommon, however, to find in- Forms.
stances where there are three or even four elements. Among
such instances—early spellings being given in each case—are
Presterodestihel, Priestroyd stile, *Asschewellerode*, Ashwell royd,
Wlveleyheud, Woolley head. A glance is sufficient to show
that all these are secondary formations, the earlier forms being
Presterode, the priest's clearing, *Asschewelle*, the well beside the
ash-tree, and *Wlveley*, wolf-lea. Indeed, we may take it as
a fixed law that names consisting of three elements are never
primary. It is on this ground that the series of names in -stall
found west of Halifax—Cruttonstall, Heptonstall, Nettletonstall, Rawtonstall, Saltonstall, Shackletonstall, Wittonstall—may
be declared derivatives from earlier forms like Crumton and
Hepton. The latter is actually found in another secondary
form, viz., *Hepton Brige*, that is, Hebden Bridge. For the same
reason the names Earlsheaton, Cleckheaton and Kirkburton,
may without hesitation be claimed as secondary forms, and
historical records fully bear out the claim.

[1] Taylor, *Names and their Histories*, p. 350.

Import-
ance of
Minor
Place-
names.

The reference to minor place-names must not be supposed to be merely casual. When comparative methods are to be applied the help of minor names becomes very valuable. With their assistance many puzzles will find solution, while without it the examples necessary for analysis and comparison will often fail. In other words, the method of comparison is of great value, and in its application minor names play an important part.

Limits
of the
Enquiry.

The present enquiry is limited to that part of Yorkshire which lies south of the Aire, keeping as the northern boundary— where the Aire fails—a line drawn east and west a little north of Keighley. But the enquiry is limited in other ways. It is quite impossible to deal with more than a small proportion of the minor names of the area; indeed a single parish like Bradfield or Ecclesfield or Saddleworth would itself provide materials for a considerable work. And there are still further limits, for the results of the enquiry must of necessity be limited in their success. In many cases it is quite impossible to speak with assurance, the facts are so meagre and inconclusive. Such success as is finally possible can only come after continued effort; and many of the chief puzzles will remain unsolved until other districts bring their own contribution to our assistance. All that can be hoped for in the present attempt is that it may prove sound in its general principles, and that, in spite of short-comings, it may show elements of solid value.

CHAPTER II

THE ANGLIAN ELEMENT

I. HISTORICAL AND GENERAL.

The Coming of the Angles—Lands named after Settlers—Britons not exterminated—Importance of Minor Names—Common Anglian Terminations—Woods and Forests—Wild Animals—Domestic Animals—Cultivation of the Land—Other Occupations—Religious Beliefs.

It is impossible to date the settlement of Yorkshire by the Anglians; their coming is shrouded in mist[1]. Of Bernicia we know that in 547 Ida the Flamebearer succeeded to the kingdom—' Her Ida feng to rice' is the expression of the Chronicle; but there is no account of his coming, and no description of the resistance offered by the Brigantes. Indeed, it seems certain that the Romano-British kingdoms on the eastern side of Britain had come to an end about the middle of the 5th century, and that the first Anglian settlements in the north must have taken place quite early in the story of the Saxon invasion.

The Coming of the Angles.

We learn from the Chronicle that in 560 Ælla became King of Deira, and we know that on the death of Ælla Deira was attached to Bernicia by Æthelfrith the Destroyer. Of this warlike king it is said that he conquered more lands from the Britons than any other king; yet, notwithstanding this, the kingdom of Elmete still survived, and it was not until the days of Edwin early in the 7th century that it was finally absorbed into the Anglian dominions.

Until the coming of the Danes 250 years later the Angles held the upper hand. They settled as lords of the soil, and gave their names to their possessions. South of the Aire we

Lands named after Settlers.

[1] Freeman, *English Towns and Districts*, p. 276.

find 150 places which bear the name of an Anglian settler.   In two instances, Bowling and Cridling, the name is a simple patronymic.   In twenty-four instances it is a patronymic joined to some common termination like -ham or -worth or -ton.

Britons not exterminated.   Yet, deep in the Pennines on the western border, there were doubtless many hills and valleys left in the hands of the conquered Celts.   Anglian place-names witness to this with striking emphasis.   There is, for example, Wales, which means foreigners, that is Welshmen, and represents a community of Britons living side by side with Anglian settlers.   There are also Walsh, Walshaw, and Walton, all pointing in the same direction.   The two Brettons each appear to represent the farm of a man called Bret, that is, Briton.   Kimberworth and Cumberworth have a similar significance—the farm of a man called Cymbra or Cumbra, the Welshman, one of the Cymri.   Further, there is Hewenden, the valley of the servants.   And beyond all these there are the Celtic place-names which still survive.

Importance of Minor Names.   As we shall see later, the Scandinavian contribution to our place-names is very considerable, yet, as we should expect, the Angles have provided the greater number, more especially of our township-names.   Indeed, no point stands out more clearly than that township-names are much more Anglian than the ancient minor names ; it is almost wholly in the latter that Celtic survivals are to be found, and among them the proportion of Scandinavian names is undoubtedly greater than among township-names.   In regard to the name-list given later it should be remembered, however, that many Anglian minor names are omitted as requiring no explanation, where corresponding names from other sources are inserted.

Common Anglian Terminations.   Quite the most common termination is *ton*, an enclosure ; but this, like many others, such as *cliffe* and *moor* and *land*, may be either Anglian or Scandinavian.   Of the distinctively Anglian terminations the most common is *ley*, a lea or meadow ; and others which occur with frequency are *bridge, croft, den, field, ford, ham, hill, wood*, and *worth*.

Woods and Forests.   Apart from the triangle between Goole and Ferrybridge and Doncaster, the country seems to have been fairly rich in copse and woodland.   Among the names which refer to thickets, woods, or copses, there are a large number ending with the word 'shaw,' which means a small wood; such are Bradshaw and Crawshaw, Kershaw and Birkenshaw. A smaller number end with 'hurst,' such as Ashhurst, Elmhirst, and Hazlehurst. In addition there are a few names in -greave, like Hesslegreave and Hollingreave, and one in -holt, namely, Gledholt. Last of all there are names ending with -wood, like Blackwood and Whitwood, Sowood and Eastwood and Outwood.

But there is further evidence which because of the number of names involved is even more striking. I refer to the various words which denote a clearing. The most common of these, 'royd,' occurs in field-names many hundreds of times. Less common is the word 'riding' or 'ridding,' which comes from OE *hryding*, a patch of cleared land. Still less common, but far more interesting, is the Norse word 'thwaite' of which the surviving examples number twenty-five.

Among the trees mentioned we find most frequently the oak, the thorn, the holly, and the hazel, while under the form 'aller,' which is Anglian, and 'owler,' which is Scandinavian, the alder also is very common. Other trees which occur occasionally are the elm, yew, birch, willow, maple, poplar, and aspen.

It is interesting to note that quite a number of places are designated by a simple tree-name. In south-west Yorkshire the examples include Crabtree, Ewes (yews), Hessle, Lighthazels, Oakes, Popples (poplars), Thickhollins, Thorne, and Thornes.

Chief among the wild animals was the wolf, referred to in at least eight names, such as Woolley, Wooldale, Woolrow, and Woolgreaves. The hart also has given rise to several names, among them Hartcliff, Hartley, and Harthill; but in the case of the various villages called Hartshead, the first element is doubtless a personal name. The two places called Earnshaw bear witness to the former existence of eagles, and the two called Brockholes to the presence of the badger, which formerly was called the brock. Wild Animals.

The rearing of cattle had a very important place in the rural economy. Shepley, Shipley, and Shibden are so called from the rearing of sheep; Swinden, Swinton, and Swinnow, from the keeping of swine; Horsfall and Horsehold from the Domestic Animals.

keeping of horses. We are reminded by Hardwick and Hard-castle of the herds once sheltered there, and by Stotfold and Stoodley of important stud-farms. Near Halifax there is evidence of the cattle-rearing energies of our forefathers in the place-names Cruttonstall, Heptonstall, Rawtonstall, Saltonstall and Wittonstall, while such a name as Bellhouse may perhaps point to the means by which a great army of farm servants was controlled.

Cultiva-tion of the Land.
Scattered about the riding there are many names like Shutts, Doles and Fordoles, Eastfield and Westfield, which recall the system of cultivation formerly in existence.

Each township had its 'Common fields,' and, as the rotation of crops was triennial, the fields were three in number. Each of these fields was divided into smaller portions called shotts or shutts, and these in their turn were cut up into acre or half-acre strips separated by green banks of unploughed turf. The shape of these strips was governed by the needs of the ploughman and his team of oxen : the length was that of a normal furrow, a furlong ; the breadth was two or four rods, eleven or twenty-two yards. In each of the great open fields there were hundreds of these strips or selions or doles, and the most striking mark of the system was the way in which one individual held isolated strips scattered throughout the whole area. The 'bundle' of strips held by one person was called a virgate or yardland, and the number might reach fifty or sixty.

The method of cultivation was co-operative. Instead of each man ploughing his own strips—widely separated as they were from each other—one man provided the plough, another the harness, and another the labour, while several lent each an ox to make up the full team of eight. The ploughman passed from strip to strip until he had ploughed one for every owner ; passing on, he continued to plough until he had again done service for every man ; and so he went forward until his task was done.

This system, which broke down as feudalism passed away, carries us back to the earliest days of our national history, yet its remnants are still to be found, sometimes in the balks or linches by which the strips were separated, and sometimes in the local place-names.

But there were other occupations. Orgreave near Rotherham, Other Oc-
a name found in Domesday, and Orpyttes near Sheffield—now cupations.
Pitsmoor—point to iron-mining ; and such names as Kilnshaw,
Kilnhurst, and Cowley bear witness to the important industry of
charcoal-burning.

Then, further, there are several names which give hints
about the dwelling-houses of our forefathers. Lofthouse—a
Scandinavian name—which occurs in Domesday Book as *Loftose*
and *Locthuse*, bears witness to the existence of the two-storied
house in pre-Conquest days. Such houses could not have been
common seeing they were sufficiently noteworthy to become the
distinguishing feature of a district.

Hints are not wanting even in such matters as the building of bridges. In the Domesday Survey only one bridge is
mentioned, and that of the most modest dimensions, namely,
Agbrigg. Further, we find in Domesday only one ferry, that
over the river Aire at *Fereia* ; and it is not until 1199 that the
name *Ferribrig* appears to prove the ferry superseded. Later,
in 1275, we have evidence that the Calder was spanned by a
bridge at Brighouse, clear sign that there was a considerable
body of traffic along the important road which here crossed
the river.

There are several names which within the compass of three or
four syllables present a brief synopsis of history : Ferrybridge—
ferry first, and later bridge ; Dunford Bridge and Cooper Bridge
—now a bridge, but once a ford. A name like Stainforth may
perhaps show that in olden days fords were sometimes paved ;
and not unfrequently the name of the ford contained a warning
as in the case of Rufford, where the river bed was uneven, and
Strangford, where the current was strong.

Some of the place-names carry us back to the religious Religious
beliefs of remote ages. Ramsden, for example, the valley of Beliefs.
Ram or Hramn, that is, of Raven, has its link with totemism,
the primitive animal worship which looked upon each tribe as
descended from some animal, bird, or tree. On the other hand,
the ancient British Christians have not handed down to us so
much as a single place-name derived from the Church; there are
no such names as the Cornish St Germans or St Keyne, and

none like the Welsh Llanberis or Eglwysfair[1]. Indeed, traces of the Christianity of the early Anglians are not easy to find. It is true that Bede speaks of a church built in Campodunum by Paulinus, but he also speaks of its destruction, burnt by the Pagans who had slain the King. It is true that Domesday Book records a 'priest town,' namely, Preston or Purston near Pontefract, but that may be either Anglian or Scandinavian. Indeed, it is not until the appearance of the Vikings that we find definite signs of Christian influence.

2. PERSONAL NAMES.

Ancient Names of Two Kinds—Those of One Element—Those of Two Elements —Effects of the Norman Conquest—Tribal Names in Place-names—Action and Reaction—Light on Meaning of Surnames.

Ancient Names of Two Kinds. The personal names in use among our Anglian forefathers were of two kinds, those of one stem, and those of two. It will be helpful to consider the former in three classes: (1) original names, (2) names varied by means of diminutive or other terminations, (3) names obtained by the shortening of double-stem names.

Those of One Element. The names of the first class, that is, original names of one stem, are of peculiar interest. They are of the earliest period. Some are names of animals and natural objects; others are untranslateable, bearing no obvious meaning in the language as we know it. Among them are such names as Aba, Aca, Cuda, Dud, Dun, names which may appear in various guises, as for example, Abba, Acca, Cudda, Dudde, and Dunne.

The names of the second class, single-stem names varied by the addition of diminutive or other terminations, form a very large group. The terminations most largely used are the patronymic suffix -ing; the endings -ac -ec -ic -oc -uc, -er and -re, -et and -ot ; and the diminutive -el.

---

[1] Llanberis means 'the church dedicated to St Peris,' and Eglwysfair 'the church dedicated to St Mary.'

A long list of patronymics ending in -ing might be made. It will be sufficient to give a few examples: Ading, Bridling, Busling, Colling, Cnotting, Cridling, Golding, Loding, Manning, Upping, Willing. With the help of these it was possible to have a twofold system of personal names not wholly unlike our modern plan of Christian name and surname; indeed, such a method was in existence in the earliest days, when a man might be described as Gamel Golding, the patronymic being added to the personal name.

Single-stem names formed by the addition of the terminations -ac -ec -ic -oc -uc are probably diminutives, and account for many ancient names. From Dudd is obtained Duddac, from Puda Pudec, from Willo Willoc; and such modern names as Coppock, Silcock, Pinnock and Puttock, have doubtless arisen in the same way. The terminations -er and -re, -et and -ot, are also of considerable importance. From Azo we get Azer, from Ota Oter, from Bar Baret, from Lufa Lufet. It is from single-stem names such as those already enumerated that a very large number of the monosyllabic or dissyllabic surnames at present in use owe their origin—Black, Dodd, Dunn, Tate, Hick and Sadd; Blacker, Berner, Abbott, Barrett, and others in great number, as well as diminutives like Abel, Brunel, Cuttell, Lovell, and Riddell.

The third class consists of short forms of double-stem names. These were used as pet-names, as names of friendship and endearment. They were formed from the first element of the original name, the final consonant if single being in most cases doubled, and the vowel -a or -e added. In this way Eadbald was shortened to Eadda, Ælfwine to Ælla, Cuthwulf to Cutha, Hygebald to Hygga.

When we come to the historical period the use of a patronymic as an additional name was passing away, and each person bore, as a rule, but one name. These names, at any rate so far as historical personages are concerned, were almost invariably formed of two elements joined together according to the rules of composition; we might have, for example, adjective and noun, or noun in apposition with noun, or, occasionally, adjective in apposition with adjective. The elements employed were comparatively few in number, but the changes rung upon them were very numerous. Taking the following nine roots, *æthel* noble, *ēad* rich, *sige* victory, *cēol* ship, *wil* desirable, *wulf* wolf, *beorht* bright, *mund* protector, *bald* bold, we can build up at once twenty well-known names :—

Æthelwulf	Æthelbeorht	Æthelmund	Æthelbald
Eadwulf	Eadbeorht	Eadmund	Eadbald
Sigewulf	Sigebeorht	Sigemund	Sigebald
Ceolwulf	Ceolbeorht	Ceolmund	Ceolbald
Wilwulf	Wilbeorht	Wilmund	Wilbald

It is not a little astonishing to find that examples of this class provide so few of our modern names. Such as remain to-day are chiefly in use as Christian names, and owe their vitality in the first instance to the fame of some great king like Alfred or Edward or Edmund, or to some other adventitious **Effects of the Norman Conquest.** circumstance. A complete transformation in the names of the people was in fact one of the results of the Conquest. With the Anglians and Saxons almost infinite variety had been possible, but as the native names yielded to the Norman this variety passed away. The husbandman doubtless held fast to his Hic and Dodd and Dunn, and so provided us with a considerable number of our modern surnames, but for the knight or squire no name would serve but one of Norman ancestry. The result was extreme impoverishment—half the men called John or William, a quarter called Richard, Robert, or Thomas—and so the need of surnames.

Tribal Names in Place-names. But there is a further question, namely, How far do tribal names enter into the formation of place-names? It has already been pointed out that Cumberworth, Kimberworth, and the Brettons, appear to have as their first element names which refer to the nationality of the persons named. And it seems clear as a result of our enquiry that the name of a tribe may become the name of an individual belonging to the tribe, and the personal name thus obtained may then become the first element of a place-name. Hunshelf is 'Hun's ledge of land,' and Hunster is 'Hun's place,' the individual in each case being so named he was one of the tribe of the Huns.

Friezland was in the possession of Frese, a Frisian, and so also was Fryston; Wales was a settlement of Welshmen; Denby of a Dane; and in the name Normanton we have permanent record that the place was settled by a Northman.

In the course of centuries the names of persons and places have acted and reacted upon each other. During the time of the Anglian settlement places received their names from their owners; later, when surnames became a necessity, we find them borrowed from place-names. But in more recent times place-names have once more been formed from personal names—from the names of sailors, soldiers, statesmen, explorers, and pioneers. There is perhaps no more striking example than that of Wellington. First there was the Saxon patronymic Welling; from that was derived the ancient place-name Wellington; from this came the modern title and the modern surname; and finally from the soldier who bore the title, the place-name Wellington in every British colony. Action and Reaction.

One of the most interesting of the secondary results which spring from our enquiry is the light thrown on the origin of surnames. Take three examples, Armitage, Hallows, and Wormald. Light on Meaning of Surnames.

The first is exceedingly common in the neighbourhood of Huddersfield, and it is not unusual to find the theory put forward that Armitage Bridge received its name from some person called Armitage. It is quite certain, however, that the name of the place is derived from an ancient hermitage which existed there as early as the 13th century; and it follows that the surname springs from the place-name, not the place-name from the surname.

The surname Hallows is duly recorded by Bardsley, but without explanation; on the other hand, Hallas, though very common in the West Riding, is left unrecorded. It will be seen, however, from the note on Hale, that Hallas and Hallows are the same word, and that the locality from which they originate may be either Hallas near Bingley, or Hallas in Kirkburton, the source of the name being OE *healh*, a corner or meadow.

Wormald is no less interesting. Its chief habitat, according to Bardsley, is the West Riding, while its meaning is 'son of

Wormbald.' But, further, Bardsley gives the name Wormall, linking it with a place-name recorded in 1379 as Wormwall ; and under the same head he gives the alternative form Wormell. A reference to the place-name Wormald, explained near the end of this work, will show that the origin of all these names is a place formerly called Wlfrunwell and Wulfrunwall, a name which passed through variations like Wollerenwalle, and Wolronwall, to Wornewall, Wormewall, Wormall, and finally Wormald.

CHAPTER III

THE SCANDINAVIAN ELEMENT

1. HISTORICAL SUMMARY.

The Coming of the Vikings—Period of Plunder—Period of Colonization—The Kingdom of York—Danish and Norse Kings—Supremacy of Wessex—Period of Political Conquest—Norway and Harold Fairhair—The Kingdom of Dublin—Settlement of Iceland—Character of the Norsemen.

It was in the year 787, according to the Saxon Chronicle, The that the first of the Vikings reached the shores of England; and Coming it was in January 793 that the monastic house of Lindisfarne Vikings. was 'laid waste with dreadful havoc,' its treasures carried away, its altars desecrated, its monks slaughtered, scattered, or enslaved. In after years the Vikings preferred the summer for their excursions, but their methods were none the less barbarous, and the terror they inspired may be gathered from the prayer of the Litany, 'A furore Normannorum, libera nos, Domine.'

For many years the strangers made their raids in small Period of. parties, disappearing as soon as they had gained their object; Plunder.. but gradually the petty squadrons which harassed the coast made way for larger hosts. In 867 York fell before them, and their armies 'rode over Deira.' In 869, after seizing Nottingham, they returned to York and stayed there a year. In 876 they invaded Yorkshire once more, but with a new purpose. Hitherto their object had been plunder—gold and slaves; but now they came to colonize. 'After the sons of Lodbrok[1] had conquered the Period of country,' says the Saga of Olaf Tryggvason[2], 'Northumberland Coloniza-tion.

---

[1] Ragnar Lodbrok, two of whose sons were Halfdan and Ivar 'the Boneless.'
[2] See p. 16 of the translation by Sephton.

G.   2

was largely colonized by Northmen'; and a notable passage in the Chronicle tells us that 'Halfdan divided out the lands of Northumbria, and henceforth they continued ploughing and tilling them.'

It is to this period—the time of Alfred the Great—that our country owes its first great instalment of Viking blood. How important was this instalment and what was its character may be gathered in some degree from such statements as that in Egil's Saga, where we are told that in the reign of Athelstan, two generations later, 'almost every family of note in Northern England was Danish by the father's or the mother's side.' Anglian and Viking were of nearly related blood; their customs and speech were largely the same; they could well understand each other; and it is not surprising that fusion between the two races readily took place.

The Kingdom of York. From the time of Halfdan there existed in the north something like a regular monarchy, York being for several generations the centre of the Scandinavian interest in England[1], and Yorkshire 'as much a Scandinavian province as Scania or Zealand.' In 876, as we have seen, chief power was in the hands of Halfdan; from 880 to 894 Cuthred ruled; and in 911 a second Halfdan together with Eowils[2].

Danish and Norse Kings. Up to this point we are concerned entirely with incursions from the east—with Danish settlers and Danish kings. But, after this time, the kings came from the west—Norsemen from the kingdom of Dublin—and among them we find Ragnald in 919, Sihtric from 921 to 927, Olaf in 940, Olaf Cuaran and a second Ragnald from 941 to 944, Olaf Cuaran again in 949, and Yric from 952 to 954[2]. Two of these, Olaf Cuaran and the first Ragnald, are figures of great interest. Both had romantic careers, and both were known as kings of 'the Dubhgaill and the Fingaill,' of the dark foreigners and the fair, that is, of Danes and Norsemen. In fame, however, Olaf Cuaran has far outstripped his predecessor, for he is the Havelock Cuheran

---

[1] See p. 16 of the Saga of Olaf Tryggvason, Sephton's translation, where we are told that 'Eric fixed his residence at York, where the sons of Lodbrok are said to have dwelt aforetime.'

[2] Collingwood, *Scandinavian Britain*, pp. 119–144.

whose story has given to the world no less a personage than Hamlet, prince of Denmark.

In former years the city of York had twice been supreme. Under the Romans it was the dwelling-place of Cæsars and the seat of empire; under Edwin and Oswald it was once again the centre of power; and now under its Viking rulers supremacy seemed a third time within its reach. From the throne of Edwin, says Freeman, 'a new Penda threatened England[1].' But in 937 the outlook was altogether changed, and, though seventeen years *Supremacy* elapsed before the final submission, Athelstan's great victory at *of Wesesx.* Brunanburh sealed the fate of the Viking sovereignty in the north.

And so for a generation there was peace. The period of plunder had passed away, and the period of colonization was bearing its fruit. The descendants of the Vikings came more and more under the influence of Christianity; as early as the middle of the 10th century, indeed, we find ecclesiastics whose names are Scandinavian. Year by year fresh links were forged to bind the races more closely together.

But in the last decade of the 10th century the predatory attacks were renewed. After the battle of Maldon the Vikings were bought off. Then Northumbria was attacked and the shores of the Humber were ravaged. And at last, elated by success, and touched to the quick by the massacre of 1002, the *Period of* Danes decided to attempt the conquest of the whole country, *Political Conquest.* the result being that from 1013 to 1042 the realm was governed by Danish kings, Cnut and Harold and Harthacnut. Doubtless the Scandinavian settlements increased—though in a peaceful way. We know, for instance, that a large Danish colony settled in London, and the memory of its burial-place still lingers on in the name of the Church of Saint Clement Danes[2]. We shall scarcely be at fault if we assume that the Viking population of the north received at this time many similar additions.

If we turn back for two centuries we shall find the country *Norway* from whence many of the Vikings came—mountainous Norway *and Harold* —full to overflowing with a vigorous and high-spirited people. *Fairhair.*

---

[1] *English Towns and Districts*, p. 289.

[2] Freeman, *Norman Conquest*, I, pp. 538 and 572.

A continual stream of adventurers poured forth from its shores. At that time the country was divided and sub-divided among petty kings or chieftains. There were tribes, indeed, but no nation; and it required the strong hand of Harold Fairhair and the stern struggle of a lifetime to weld the people together into one united state. In the process the stream of adventurers increased. 'Because of the unpeace,' says the story of the settlement of Thorsness, 'many well-born men fled from their heritage out of Norway, some eastward over Keel, some westward over the sea[1].' And the account goes on to say that there were some that used to keep themselves 'of a winter in the Southreys or Orkneys,' while 'of a summer they would harry in Norway and do much harm in Harold's kingdom[1].' In consequence, Harold fitted out an expedition and reduced to subjection all the islands north and west of Scotland and even as far south as the Isle of Man. Many of his opponents were slain; many fled to Ireland or to Iceland; and from that time forward the Orkneys and Shetlands, the Hebrides and Man, continued under the power of Norway for many centuries.

In this story of the stubborn clash of will with will there are two points of contact with the present subject—Ireland and Iceland.

The Kingdom of Dublin. Ireland had long been the prey of the Viking hosts, the earliest forays taking place near the close of the 8th century, and the earliest comers being Danes. But in 852 a Norse kingdom of Dublin was founded by Olaf the White, and this kingdom was maintained with varying fortunes until the middle of the 10th century. In the meantime, as we have seen, the Danish kingdom of York had been founded, and soon there sprang up the closest relationship between the two kingdoms. Members of the same house were kings in Dublin and in York. There was constant intercourse between them. The Irish Sea was their common highway. And so, as the east had given us Danes, the west now gave us Norsemen.

Settlement of Iceland. But there is a connection also with Iceland. The settlement of that country was largely due to the despotism of Harold

---

[1] *Origines Islandicæ*, I, p. 253.

Fairhair. Of many of its settlers we are told that they 'fled before King Harold,' or 'were at odds' with him. The period of settlement extended from 874 to 934; it took place in fact at the very time when Yorkshire was under the power of the Vikings. Partly because of the isolation of the people, and partly by reason of their literary power, the language spoken by the settlers has continued almost unchanged down to modern times, and modern Icelandic differs but little from the language of the Viking hosts who invaded England. Still more interesting is the fact that the literature which sprang up has preserved to us the elements of the ancient tongue, and has provided us with a mine of information in all matters connected with the Northmen. Thus, the enquirer who desires to understand the place-names of modern Yorkshire must needs have recourse to Icelandic chronicles and sagas, and that not only for sidelights, but also for information of the most direct importance.

What manner of man the Norseman of early days proved himself to be has been vividly pourtrayed by Professor York Powell. 'The character of the people of the west coast of Norway about the end of the eighth century,' he says, 'is illustrated in some measure by certain poems in the Eddic collection, which we take to be of earlier date than the rest, and which, unlike the rest, bear pretty plain marks of Norwegian origin. From these it is possible to get a picture of the population whence the Wicking emigrant came; it is of a type which we pride ourselves upon as essentially British—a sturdy, thrifty, hardworking, law-loving people, fond of good cheer and strong drink, of shrewd, blunt speech, and a stubborn reticence when speech would be useless or foolish; a people clean-living, faithful to friend and kinsman, truthful, hospitable, liking to make a fair show, but not vain or boastful; a people with perhaps little play of fancy or great range of thought, but cool-thinking, resolute, determined, able to realise the plainer facts of life clearly and even deeply. Of course some of these characteristics are common to other nations in their rank or development, but taken together they show a character such as no other race of that day could probably claim, and enable us to understand how that quiet storage of force had gone on which, when released, was

Character of the Norsemen.

capable of such results as the succeeding three centuries witnessed with amazement[1].'

2. General Character of Scandinavian Place-names.

Icelandic Place-names—Viking Names in Yorkshire—Maritime Terms brought Inland—Divisions: Wapentakes and Ridings—Tingley and Husting—Religious Beliefs—Burial Customs.

Icelandic Place-names.
An examination of the place-names which occur in the various sagas and other Icelandic literature enables us to obtain a very clear insight into the methods adopted by the settlers. The names may first be divided into two classes, (1) Simple, (2) Compound.

The former class consists of those place-names which consist of but one element, that element being as we should expect descriptive of some simple topographical feature, e.g. *Berg* a rock, *Borg* a castle, *Hvammr* a grassy slope, *Lundr* a grove, *Tunga* a tongue—common nouns elevated to the dignity of proper names.

Far more numerous is the class of compound names, a class which may be sub-divided into three groups. There are first those names which add to some word of topographical meaning the name of the owner. Such names as these form a very large proportion of the whole. We find, for example, *Grims·dalr*, *Grims·ey*, *Grims·nes*, that is, Grim's dale, Grim's island, Grim's ness; *Steins·holt*, Stein's wood; *Hana·tūn*, Hane's enclosure; *Thororms·tunga*, Thororm's tongue of land.

Secondly, there are those compound names where the purpose is not to show ownership but to give a simple natural description—names where the substantive is qualified by an adjective, or by a noun used adjectivally. This group also contains a very large number of names. The descriptive word is usually of the simplest character, specifying the points of the compass, the colour, the dimensions, the soil, the position. We find *Vest·fold*, Westfold; *Rauða·sandr*, Redsand; *Breiða·vík*, Broadwick; *Lang·dalr*, Longdale. Occasionally the names of trees and

[1] *Scandinavian Britain*, p. 21.

animals are utilised, as in the case of *Espe·höll*, Aspenhill;
Svīna·vatn, Swinemere; *Sauða·nes*, Sheepness.

The third group consists of those compound names which
refer to some historic event, social custom, or religious rite.
This is a comparatively small division, but one of great interest.
Several such names bear witness to the tragedies then almost a
commonplace of daily life. Thore and Ref had a quarrel about
forty cattle which were claimed by both; and when they fought
Thore fell and with him eight men, and the hillocks near which
they fought were afterwards called *Thores·hólar*, Thore's hillocks[1].
Some Irish thralls belonging to the early settlers, after a treacherous murder were captured and slain, and the islands on which
they were put to death were afterwards called *Vestmanna·eyjar*,
that is, islands of the Westmen[2].

There are several places named from crosses set up for
Christian worship. Of Jarl Torf Einarr and his companions we
are told that having previously set up an axe in one place, and
an eagle in another, 'in the third place they set up a cross and
called it *Kross·áss*,' Crossridge[3]. In addition to such names as
these, there are others connected with the government of the
country, *Thing·völlr*, parliament-field, and *Lög·berg*, the rock of
laws; names of peculiar interest because of their connection
with the development of national life.

Place-names derived from the Vikings, like those of Anglian
origin, will usually, therefore, be of one or two elements; and if
of two elements the former will be of an adjectival character
and the latter substantival. In native Celtic place-names the
order is usually reversed, the substantival being first, the adjectival last. Instances where names possess three elements are,
of course, to be found, but they may in every case be declared
secondary formations.

The broad principles governing the question being thus laid
down, it will be interesting to see what is the actual contribution
made by the Vikings, whether Danes or Norsemen. *(margin: Viking Names in Yorkshire.)*

It must first be noted that, just as in the case of Anglian and
Scandinavian, so in that of Dane and Norseman, many words

---

[1] *Origines Islandicæ*, I, p. 30.　　　[2] *Ibid.* I, p. 23.
[3] *Ibid.* I, p. 170.

were possessed in common.   In regard to the greater number of
Scandinavian words we are unable to distinguish whether their
origin is Norse or Danish.   Among words of this character the
following are of frequent occurrence:—beck, biggin, by, carr,
crook, garth, gate, holme, howe, lathe, lund, mire, nab, rake, raw
or row, scar, scoe or skew, scout, storth, wath, with, and such
tree-names as ask, birk, busk, hessle, and owler.

Maritime Terms brought Inland.   It is not surprising to find that the Vikings, having given
up their seafaring life and settled down to a career of peaceful
industry, still retained some of their old habits of thought.   In
describing the features of the country in which they had settled
they not unfrequently made use of terms connected with their
former occupations.   Something of the same kind had already
taken place in the mother-country.   One of the great mountain
ranges of Norway was called Kjolen from its resemblance to a
ship's keel, and a deep cleft between two Norwegian mountains
is to-day called Kjepen because of its likeness to a gigantic
rowlock.

   Kjölr, a ship's keel, appears to have given us the four names
Keelam or Keelham which doubtless mean 'the ridges.'

   Vik, a bay, seems to have given the Cumbrian word 'wike,'
which denotes 'a narrow opening between rising grounds.'   From
this word we probably get the name Wyke, which occurs near
Bradford and Horbury.

Divisions: Wapen-takes and Ridings.   The part played by the Vikings in the government of the
country is indicated in a striking way by the names of its chief
divisions.   Though the formation of townships was in the main
due to the Anglians, the grouping of townships into Wapentakes, and of Wapentakes into Ridings appears to have been
the work of the Vikings.

   The word Wapentake, from ON *vapna·tak*, means literally
'weapon-touching.'   In its original sense it appears to have
been derived from an ancient method of expressing approval
adopted by the Northmen in their assemblies.   Later, the word
took up new senses.   It meant a vote or resolution; it also
meant the breaking-up of parliament when the men resumed
the weapons they had laid aside during the session.   But in that
part of England which formed the Danelagh the word came to

mean a portion of a county corresponding to the 'hundred' of purely English shires. The wapentakes were often named from some conspicuous object near the place of meeting—a cross as at Staincross and Osgoldcross, a ford as at Strafford, a hill as at Tickhill, a bridge as at Agbrigg; and the place of meeting was usually in the open country, at some distance from the chief town, lest its inhabitants should unduly influence the proceedings.

The word Riding comes from an earlier form *thriding*, which is to be connected with the ON word *thriŏ·jungr*, a third part. In DB we find such spellings as *Nort Treding, Est Treding, West Treding*, forms which at an early date settled down into the more smooth and euphonious North Riding, East Riding, and West Riding.

No less interesting are the names Tingley and Husting Knowl, as well as the name Bierlow, for they carry us back to the very centre of the public life of the Scandinavian settlers.

Tingley and Husting.

In Tingley, formerly *Thing·lawe*, we have the survival of the ON word *thing*, an assembly, meeting. This word is found in the Icelandic *Thing·völlr*, the field where the parliament of the island held its annual assemblies. Six places in Great Britain show the exact equivalent of the Icelandic name—Thingwall in Cheshire, Thingwall in South-west Lancashire, Tynwald in the Isle of Man, Tinwald in Dumfries, Dingwall in Caithness, and Tingwall in the Shetlands.

At one time Norway had three assemblies of this kind, one for each of its three great districts, Frosta, Gula, and Eidsifia. The annual meetings of the 'things' were held at midsummer, and lasted for two weeks, those present being accommodated in booths set up near the place of meeting.

It seems clear that Tingley, *Thing·lawe*, assembly-hill, was just such a place of meeting. Here the Viking settlers met together annually to transact public business, to decide cases of dispute, and to promulgate their decrees. The 'lowe' is still to be seen, and near at hand a well-known horse-fair is held which probably owes its origin to the meetings of the 'thing' and the buying and selling which accompanied them.

An interesting question arises at this point, namely, whether

Tingley was the meeting-place for one wapentake only—that of Morley—or for several wapentakes combined? In other words, Did the 'thing' consist of a federation of smaller districts as in Norway, and were these smaller districts our present wapentakes?

There is a striking piece of evidence in favour of the suggestion that Tingley was the united meeting-place for several wapentakes. It is this. All the wapentakes in the neighbourhood receive their names from the place of meeting—Agbrigg, Osgoldcross, Skyrack, Staincross—and if the wapentake of Morley had its meeting at Tingley, we should expect it to be called the wapentake of Tingley. If, however, the wapentake of Morley met at a definite spot then called Morley, while the annual united meeting was held at Tingley, any incongruity in the system of names would be removed.

Still another word connected with the Viking methods of government is Knowler Hill, Liversedge, 1560 *Hustin Knowll*. Here the prefix is from the ON *hūs·thing*, a word denoting a smaller assembly than either of those just discussed. To such a meeting a king, earl, or captain would summon the people connected with his *hūs*, his guardsmen or the men of his estate.

Religious
Beliefs.

Passing on to the religion of the Vikings, we must remember that during their sojourn in Ireland the Norsemen had been brought into contact with Celtic Christianity. There were many, doubtless, who held the old beliefs, and there were others who, side by side with something of Christianity, retained much that was distinctly heathen. Among the records of the contemporary settlements in Iceland there are indications of just such a state of things. The Landnama Book, speaking of Aud, widow of Olaf the White, tells us that she spent her later years in Iceland, and had her prayer-place at Kross-holar, that is, Cross-hillocks; 'there she caused crosses to be set up, for she was baptized and of the true faith.' But the account goes on to say that 'her kinsmen afterwards used to hold these hillocks holy, and a high-place was made there, and sacrifices offered[1].' We are also told of a certain Helge that he put his trust in Christ and after Him named his homestead Krist·nes, 'but yet

[1] *Origines Islandica*, I, p. 79.

he would pray to Thor when at sea, and in hard stresses, and in all things that he thought of most account[1].'

Among evidences of the old heathenism are the names Lund and London, from ON *lund*, a grove. Vigfusson tells us that in Iceland places called Lund were connected with the worship of groves, and the Landnama Book relates of a man called Geat that he dwelt at Lund and sacrificed to the grove—'ok biō at Lunde; hann blōtaðe lundenn[2].'

But if there are relics of Scandinavian heathenism there are also evidences of Scandinavian Christianity, and, strangely enough, these evidences are more distinct than those of either Anglian or Celtic Christianity.

The chief signs are the words *cross* and *kirk*. The DB references do not, however, include more examples than Crosland, Staincross, Osgoldcross, and South Kirkby. But Dobcross in Saddleworth and Kirkby in Pontefract seem clearly of early date, while some of the Crossleys may also be early. On the other hand the prefix in Kirkburton, Kirkheaton, and Kirk Bramwith does not appear until late; Kershaw (*Kirkeschawe*) cannot be traced beyond the 14th century, and Woodkirk does not appear before the 12th century.

The ON *haugr*, a word used to describe the artificial burial-mounds of the Vikings, may fitly be mentioned at this point. It has given us the word 'how' or 'howe,' and appears under various guises, as in Carlinghow, Flanshaw, Clitheroe, and Wincobank. Though frequently joined to a personal name—doubtless that of the person there interred—the word is to be found under other circumstances, as in the case of Howley, Slitheroe, Grenoside, and Stenocliffe. Burial Customs.

In Icelandic literature there are many references to these burial-mounds. We read of a chapman that as he voyaged along the coast of Norway he related the story of Vatnarr, and described him as a noble man. And 'when they lay off Vatnarr's howe he dreamed that King Vatnarr came to him and spoke to him: "Thou hast told my story, therefore I will reward thee; seek thou treasure in my howe and thou shalt find." He sought, and found there much treasure[3].'

---

[1] *Origines Islandicæ*, I, p. 149.   [2] *Ibid.* I, p. 162.   [3] *Ibid.* I, p. 272.

In the Laxdala Saga is the following account: 'Hoskuld
died, and his death was much grieved for, first by his sons, and
next by all his relations and friends. His sons had a worthy
howe made for him, but with him, in the howe, was put little
money. And, when this was over, the brothers began to talk
over the matter of preparing a burial-feast after their father, for
at that time such was the custom[1].'

Another passage from the Laxdala Saga reads as follows:
'So now they drank together Olaf's bridal feast and the funeral
honours of Unn. And on the last day of the feast Unn was
carried to the howe that was prepared for her. She was laid in
a ship in the howe, and in the howe much treasure was laid with
her[2].'

3. NORSE OR DANISH.

Norse Test-words: *schole, gill, thwaite*—Celtic Loan-words: *cross, ergh*—Distinctive
Vowels and Consonants—Danish Test-word: *thorpe*—Importance of Minor Names—
Distribution of *by* and *thorpe*—Distribution of Norse Test-words—The Domesday
Survey—The Settlement largely Peaceful—The Conqueror's Vengeance—The Re-
peopling—Strong Norse Settlements—Strong Danish Settlements.

When we come to the task of. distinguishing between the
Norse and Danish elements we must place in the front rank
two words found in Norse but not in Danish—in West Scandi-

*Norse
Test-
words:
schole, gill,
thwaite.*

navian but not in East Scandinavian—namely, 'schole' and
'gill.' The first represents ON *skáli*, a shieling, log-hut, shed[3],
and occurs in the form Schole or Scholes eighteen times. The
second comes from ON *gil*, a valley or ravine[3], and is found
fourteen times.

In the second rank comes 'thwaite,' from ON *thveit*, a
clearing, a word which may be claimed as Norse for geographical
reasons. The West Riding examples of this name number
seventy-two, of which twenty-six are found south of the Aire;
but the East Riding provides no more than a single example.
It appears therefore that though the word is found in Denmark
in the form *tved*, the Danish settlers in Yorkshire made little use
of it.

[1] *Origines Islandicæ*, II, p. 179.   [2] *Ibid.* II, p. 150.
[3] Björkman, *Scandinavian Loan-Words in Middle English*, p. 283.

In the third rank we must place two words neither Danish Celtic
nor Norse, the words 'cross' and 'ergh.' These are Celtic loan- Loan-
words which have come to us from the other side of the Irish *cross, ergh.*
Sea[1]. The list of early place-names where 'cross' is the first
element is particularly instructive. In Ireland and the South of
Scotland there are many such names, and in England there are
more than thirty. An analysis of the English examples brings
out two points with great clearness.

In the first place English examples occur almost wholly in
the north-west—in Cumberland, Westmorland, Lancashire, and
the West Riding. Here we find Crosby nine times, while Crosby-
thwaite occurs once, and Crosthwaite thrice. Other examples
of similar character—Crosscanonby, Crossrigg, Crossens, the
Crosdales, Croslands, and Crostons—occur only in the same
area. Against these, however, we must set Crosby in Lincolnshire and the two Norfolk names Crostwick and Crostwight.

In the second place the yoke-fellow of 'cross' is invariably a
word of Scandinavian origin. This yoke-fellow, though frequently
a word which cannot be Danish, is never one which cannot be
Norse. Among Scotch examples we find Crosaig (= Crosvik),
Crosbost (= Crosbolstaðr[2]), Crosby, Crosgills, Croskirk, Crosspol,
and Crosston; and among English examples—not to repeat the
list already given—it is interesting to find the Norfolk examples,
Crostwick and Crostwight, recorded in the Domesday Survey as
Crostueit and Crostwit, where the terminal corresponds to the
name 'thwaite.'

Seeing that the word is associated with Norse terminations
and with districts settled by Norsemen, we may fairly claim it
as a Norse test-word, provided always that the names dealt with
are of early date.

Passing now to 'ergh,' which represents ON *erg*, a shieling
or summer farm, and is derived through *erg* from OIr. *airge*, we
find the conditions just described almost exactly repeated. In
a district of North-west England stretching in a crescent from
the Solway to the Mersey there are (or were) twenty-six names

---

[1] For 'cross' see the *New English Dictionary*.
[2] Compare the Norw. place-names Myklebost, Helgebost (Aasen).

with this terminal. Below is the full list, names without modern equivalents being starred:

Cumberland :

Cleator,	St Bees,	early	*Cletergh*
Salter,	„	„	*Saltergh*
Winder,	„	„	*Windergh*

Westmorland :

Potter,	Kendal,	1301	*Pottergh*
Skelsmergh,	„	1301	*Skelmesergh*
Docker,	„	1294	*Docherga*
Mozergh,	„	early	*Mozergh*
Sizergh,	„	1301	*Siresdergh*
Ninezergh,	„	1254	*Nissandesergh*
Mansergh,	Kirkby Lonsd.	1226	*Manesarghe*

Lancashire :

Torver,	Coniston,	1202	*Thorwerghe*
*Cabbanarghe,	Wennington,	1247	*Cabbanarghe*
Medlar,	Kirkham,	1235	*Midelergh*
Kellamergh,	„	1246	*Kelgrimesarge*
Goosnargh,	Preston,	1086	*Gusanarghe*
Grimsargh,	„	1086	*Grimsarge*
*Siuritharghe,	Bretherton,	1250	*Siuritharghe*
Anglezarke,	Chorley,	1208	*Anlauesargh*
*Oddisherhe,	Formby,	1213	*Oddisherhe*
Bretargh,	Liverpool,	1358	*Bretargh*

Yorkshire :

*Snelleshargh,	Bentham,	1260	*Snellesherg*
Feizor,	Settle,	1299	*Feghesargh*
Battrix,	Bowland,	1342	*Bathirarghes*
*Gamellesarges,	„	†1232	*Gamellesarges*
*Stratesergh,	Gisburn,	1086	*Stratesergh*
Golcar,	Huddersfield,	1086	*Gudlagesargo*

In thirteen of these the first element is undoubtedly a Scandinavian personal name: Skelm, Man, Kabbi, Snel, Boðvar, Kolgrim, Grim, Sigrið, Anlaf, Odd, Bret, Gamel, Guðlaug. And in four others it is a Scandinavian common noun: *klettr* a rock, *salt* salt, *vind* wind, and *dokk* a swampy place. The word is, therefore, curiously similar to 'cross'; but perhaps most noteworthy is the correspondence existing between the habitat of the two words.

At this point an appeal must be made to two well-known distinctions between Norse and Danish.

In the West Scandinavian dialects (Icelandic and Norse), at a period probably before 1000, a noteworthy assimilation of consonants was developed by which *nk* became *kk*, *nt* became *tt*, and *rs* became *ss*[1]. That this assimilation took effect before the end of the Viking settlements in Yorkshire seems clear from the word 'drucken,' a common dialect-form equivalent to 'drunken.' In consequence of this change we find such pairs of words as the following:

> Dan. *klint*,   Norw. *klett*,   a rock
> „   *brink*,   „   *brekka*, a slope
> „   *slank*,   „   *slakke*, a hollow.

In South-west Yorkshire, however, the Domesday record presents no assured example of any of these words, and I have found no modern representative of either 'klint' or 'klett.' On the other hand 'brink' and 'breck' occur with some frequency, but the latter may be simply English, and, further, a Swedish dialect-word *brakka* quite prevents us from claiming it as distinctly Norse. Lastly 'slack' is frequently found, especially on the western border, but there is no companion-word 'slank,' and a Swedish dialect-word *slakk* raises the same doubts as in the case of 'breck.'

The second distinction relates to a vowel change by which Icelandic *ei* is represented in East Scandinavian (Danish and Swedish) by *e*, the diphthong remaining uncontracted in West Scandinavian. This change began to show itself soon after 800, and was completed in Denmark before 1050[2]. As a result we get the following forms:

> Icel. *steinn*, Dan. *sten*,   a stone
> „   *thveit*,   „   *tved*,   a clearing
> „   *grein*,   „   *gren*,   a branch
> „   *beit*,   „   *bed*,   pasturage.

---

[1] Björkman, *Scandinavian Loan-words in Middle English*, pp. 168–176. Flom, *Scandinavian Influence on Southern Lowland Scotch*, p. 7.

[2] Björkman, *Scandinavian Loan-words in Middle English*, p. 36. Flom, *Scandinavian Influence on Southern Lowland Scotch*, p. 6.

An examination of the Domesday Survey shows five names in South-west Yorkshire where the first element is connected with ON *steinn* or Dan. *sten*:

Stainforth,	Doncaster,	DB *Steinford, Stenforde,*	CR 1232 *Steinford*
Stainton,	„	DB *Staintone, Stantone,*	PF 1166 *Steinton*
Stainborough,	Barnsley,	DB *Stainburg,*	CR 1252 *Steinborg*
Staincross,	„	DB *Staincros,*	PF 1166 *Steincros*
Stainland,	Halifax,	DB *Stanland,*	PT 1379 *Stayneland.*

Further, there are three examples where the Scandinavian *ei* or *e* is involved in the personal names Steinn, Thorgeir, and Thorsteinn:

Stancil,	near Doncaster,	DB *Steineshale,*	RC 1232 *Stansale*
Thurgoland,	near Barnsley,	DB *Turgesland,*	PF 1202 *Turgarland*
Thurstonland,	Huddersfield,	DB *Tostenland,*	PF 1202 *Thurstanland.*

Lastly, there is a single name where *thveit* or *thvet* is involved:

Langthwaite, Doncaster, DB *Langetouet,* PF 1167 *Langethwaite.*

Although *sten* in Stenforde and *touet* (= *thwet*) in Langetouet seem clearly Danish, it would scarcely be wise, in view of the alternative and later forms, to predicate more than Danish influence on words originally Anglian or Norse. In regard to Tostenland and Turgesland (= Turgerland) it will be observed that the change from Thorsteinn and Thorgeir to Thorsten and Thorger is due to the weak stress on the second syllable; and in regard to Stainton and Stainland it seems clear that Anglian influence has been at work.

To sum up, we may take it as certain that the Viking settlers, whether Danes or Norsemen, usually brought with them the uncontracted *ei*.

Danish Test-word: *thorpe.*

Passing now to Danish test-words we are immediately met by a difficulty, for the only word of serious importance, 'thorpe,' may be either English or Scandinavian; compare OE *thorp*, and ON *thorp*. An examination of the Domesday record shows that seven of our South-west Yorkshire 'thorpes' are to be found there, viz.:

Armthorpe, DB *Ernulvestorp, Einulvestorp*
Goldthorpe, DB *Guldetorp, Goldetorp, Godetorp*
Hexthorpe, DB *Hestorp, Estorp*

Rogerthorpe,	DB *Rogartorp*
Skelmanthorpe,	DB *Scelmertorp, Scemeltorp*
Thorpe (Leeds),	DB *Torp*
Throapham,	DB *Trapun*

And, if we enquire what is the origin of the yoke-fellow in each case, we find that Ernulf, Einulf, Gulde, Hegg, Rogar, Skelmer, may all be Scandinavian, while Ernulf and Einulf may possibly be English, and Golde also, though the last is probably nothing more than a variant of Gulde. It appears, therefore, that our Domesday 'thorpes' are most probably Scandinavian, a conclusion greatly strengthened by what is known of the East Riding examples. Doubtless many of our South-west Yorkshire 'thorpes' are of late origin and possess yoke-fellows which are not Scandinavian; yet even these may be claimed as lineal descendants of Scandinavian names and rightly described by the same term.

But, though they are Scandinavian, can our 'thorpes' be definitely ascribed to the Danes? To find an answer we must look at the geographical distribution of the word. Counties like Berkshire, Bedford,. Cambridge, Hertford, and Huntingdon, almost purely English in their place-names, do not count a dozen thorpes among them. Lancashire, which though predominantly Anglian is partly Norse, has only three. But the counties of York, Lincoln, and Norfolk, well-known for their Danish connections, possess at least three hundred. Yorkshire alone has a hundred and eighty, of which fifty-five are in the East Riding and sixty-three in the southern part of the West Riding. It appears, therefore, that those parts of Yorkshire known to be more Norse than Danish contain but a small proportion of 'thorpes,' while the remaining districts, those in the south-east and south-west, have a far greater proportion.

To sum up, it seems clear that while 'thorpe' may be accepted as distinctively Danish, we may claim 'cross,' 'ergh,' 'gill,' 'schole,' and 'thwaite' as distinctively Norse.

About one-fourth of the names entered in the Domesday record are of Scandinavian origin. When minor names are examined, however, there are districts where the proportion is considerably higher. The neighbourhood of Wakefield provides

Importance of Minor Names.

a striking illustration, for while the township names are largely
Anglian—Wakefield, Stanley, Warmfield, Horbury, Crigglestone—the number of 'thorpes' is quite remarkable.

Perhaps it will be most helpful if we take the names 'by,'
'thorpe,' 'thwaite,' 'schole,' 'gill,' 'cross,' and 'ergh,' and see in
what localities each occurs; and for this purpose the whole
area may be divided into three parts, (*a*) Western, (*b*) Central,
(*c*) Eastern, taking as the lines of demarcation first a line
running north and south through Bradford, Huddersfield, and
Holmfirth, and next a similar line running through Pontefract
and east of Rotherham. Of the three divisions the Western is
smallest, the others being approximately equal.

Distribu-
tion of *by*
and *thorpe*.

Taking first the word 'by,' the best of all tests for Scandinavian settlements, we find the thirty examples divided between
the three districts as follows:

	Western	Central	Eastern
By	9	12	9

The ending occurs, indeed, from Keighley in the north to Maltby
in the south, and from Sowerby in the west to Fockerby in the
east; thus the influence of the Vikings has been felt throughout
the whole area.

Taking next the Danish test-word 'thorpe,' which occurs
sixty-three times, we find its distribution is as follows:

	Western	Central	Eastern
Thorpe	4	41	18

Hence, it would seem that the Danes settled more frequently in
the east than the west, though to some extent they appear to
have penetrated the whole area.

Distribu-
tion of
Norse
Test-
words.

Passing on to the tests for Norse settlements—'schole,' 'gill,'
'thwaite,' 'cross,' and 'ergh'—we get the following results:

	Western	Central	Eastern
Schole	9	9	o
Gill	2	9	o
Cross	2	7	o
Ergh	1	o	o
Thwaite	6	16	4

It is doubtless true that 'gill' could only occur along the western
border, and it is equally true that 'thwaite' could only occur
where woodland formerly existed, yet, taken together, these tests

prove conclusively that there was an immigration of Norsemen from the west.

Hitherto we have taken every instance of the seven tests, even though some are of modern origin. It must of course be remembered that while some of these words may have produced no names beyond those given by the Viking settlers, as in the case of 'ergh' and 'by,' others have become living elements in the language, 'cross' and 'thorpe' and 'gill' for example. Under these circumstances it will be interesting to take for examination only those words which occur in the Domesday record : *The Domesday Survey.*

	Western	Central	Eastern
By	3	2	6
Thorpe	0	3	6
Schole	0	0	0
Gill	0	1	0
Cross	1	2	0
Ergh	1	0	0
Thwaite	0	0	1

Thus DB gives 11 out of 29 'bys,' but only 9 out of 64 'thorpes,' and only 5 out of 75 Norse test-words. This is very different from the state of things found among the 'thorpes' and 'bys' of the East Riding, where almost all the names now existing are to be found in DB. Possibly the difference between the two Ridings is due to the fact that a great part of the settlement in the West Riding was later than that in the East and conse- quently of a peaceful character. That it was indeed largely of a peaceful character is shown by the fact that not a few townships with Anglian names possess important members where the name is Scandinavian; among the rest there are Gawthorpe in Ossett, Rawthorpe in Dalton, Scholes in Cleck- heaton, Barnby in Cawthorne, Staincross in Darton, Wilby in Cantley, and Dirtcar in Crigglestone. *The Settlement largely Peaceful.*

At this point another side of the picture must be noted, namely, the effect of the devastation wrought in 1069 by William the Norman. In 1086, when the Domesday Survey was made, the results of the Conqueror's campaign of fire and sword were still to be seen. Township after township was unable to raise even a single shilling for the king's tax-gatherers, and time after time the pitiful entry appears, 'It is waste.' From Penistone *The Con- queror's Venge- ance.*

to Bradford, and from Meltham to Beeston—with here and there an oasis, as at Denby and High Hoyland, or at Thornhill, Mirfield, Hartshead and Liversedge—the country was devastated and in a great measure depopulated. But this is not all. Not only were many townships thus recorded as waste; others were entirely omitted. On the western borders no mention is made of the townships of Oxenhope, Heptonstall, Erringden, Soyland, Norland, Barkisland, Skircoat, Halifax, Ovenden, Rishworth, Scammonden, Marsden, Slaithwaite, Linthwaite, Lingards, and —after an interval filled in by Meltham, Holme, Penistone and Thurlstone—Langsett and Bradfield. Whatever may be said in explanation of some of these omissions, it seems clear that in 1086 much of the borderland was almost devoid of inhabitants.

The Re-peopling.
Thus a problem of great interest arises, namely, How was this tract of country afterwards peopled or repeopled? An answer has already been given by Professor Collingwood who suggests that to a great extent it must have been repeopled by immigrants from Cumberland and Westmorland[1]. If this be the correct answer, it will do much to account for the fewness of the Scandinavian names in Domesday Book as compared with those now existing.

An interesting piece of evidence is provided by the names Erringden and Cruttonstall near Hebden Bridge.

Cruttonstall is the name of a farm situate in Erringden, formerly *Ayrykedene*, which is now a township. This farm appears in WCR 1308 as *Crumtonstall*, and there can be little hesitation in associating it with the DB *Crumbetonestun*. But while *Crumbetonestun* is never found in any record later than DB, no record of *Ayrykedene* is found earlier than the 13th century. It does not seem unreasonable to suppose that the area called *Crumbetonestun* in DB was afterwards called *Ayrykedene*, Eric's Valley. This is a point of considerable importance, giving as it does some cause for believing there was —at any rate in this district—a Scandinavian immigration after the date of the Domesday Survey.

A remarkable series of 'thwaites,' all of them wanting in DB, is to be found in the district around Barnsley and Penistone.

[1] *Scandinavian Britain*, p. 178.

Among existing names there are Alderthwaite, Birthwaite, Strong Norse Settlements.
Butterthwaite, Falthwaite, Gilthwaite, Gunthwaite, Hornthwaite,
Huthwaite, Linthwaite, and Ouselthwaite; but there are also
several obsolete examples, for instance, Micklethwaite, Oggethwaite, and Thunnethwaite. This district has, indeed, quite the
strongest body of thwaites in South-west Yorkshire, and at the
same time it was one of those which suffered most severely under
the Conqueror. It seems not at all improbable that the repeopling
was of the kind suggested by Professor Collingwood; and yet near
at hand is Staincross, a Norse name found in the Domesday record.

There is evidence of another Norse settlement near Huddersfield[1]. But in this case it is certain that, at least in part, the
date was pre-Conquest, for DB has two decisive names, Crosland and Golcar. Thus, although Linthwaite and Lingards and
Slaithwaite do not occur until later, it is extremely probable
that the whole series is of pre-Conquest origin.

A similar, though much smaller, group of Norse names occurs
near Keighley, and in this case the DB name Micklethwaite—
situate quite near though outside our area—seems once more to
point to a pre-Conquest settlement.

Passing next to the 'thorpes,' we are at once met by the fact Strong Danish Settlements.
that there is a cluster of twelve in the immediate neighbour-
hood of Wakefield[2]. The names are Alverthorpe, Chapelthorpe,
Gawthorpe, Hollingthorpe, Kettlethorpe, Kirkthorpe, Milnthorpe,
Ouchthorpe, Painthorpe, Snapethorpe, Woodthorpe, Wrenthorpe.
This is a very remarkable series, and though none of the names
appear in DB and some may be comparatively modern, the
inference that Wakefield was a strong Danish centre is irre-
sistible. Indeed, the valley of the Calder from Castleford to
Sowerby shows twenty-three out of the sixty-three thorpes in
South-west Yorkshire, more than a third of the whole number.

Other districts where the thorpes are numerous are the valley
of the Don from Hatfield to Sheffield, and the district which
lies between Doncaster and Wakefield. But in the neighbourhood of Sheffield there is at the same time evidence of Norse
settlements, just as there is in the district around Halifax.

---

[1] See the note on Huddersfield.   [2] See the note on Wakefield.

CHAPTER IV

THE CELTIC ELEMENT

The Celts and their Tongue—Goidelic and Brythonic—Documentary Evidence—
Names of Rivers—Names of Hills and Valleys—Anglian Borrowings—Norse Borrowings.

The Celts and their Tongue. Before the landing of Julius Cæsar on the coast of Kent the British Islands were for centuries occupied by various tribes of people who reached our shores from the countries now known as France and the Netherlands. These tribes, though by no means homogeneous in race, are usually described as Celts, their language being, indeed, substantially the same.

Goidelic and Brythonic. In Ireland, as years passed by, the particular form of the Celtic tongue now known as Goidelic was evolved, and this was in all probability carried overseas by Irish colonists to Scotland and the Isle of Man during the early centuries of the Christian era. There are thus three modern dialects representing ancient Goidelic, namely, Irish, the Gaelic of Scotland, and the Manx of the Isle of Man.

In Great Britain, on the other hand, the form of speech prevalent when the Romans first reached our shores was that known as the Brythonic branch of the Celtic family. This is represented to-day by modern Welsh and Breton; but a Brythonic dialect survived in Strathclyde up to the 12th century, and Cornish was a living tongue up to the 18th. We may conclude, therefore, that until South-west Yorkshire was conquered by the Angles the language spoken there was a dialect of Brythonic—in other words, a speech resembling early Welsh.

Beginning in the mists of pre-historic days the Celtic period extended right through the Roman occupation to the early years of the 6th century—possibly even for a century beyond, when, as we learn from Nennius, Edwin of Northumbria ' seized Elmet and expelled Cerdic its king.' But doubtless the speech lingered, especially in remote districts, for centuries—occasionally, in a word here and there, even down to the present time. It is true that to-day few townships possess a name even in part Celtic, yet among minor place-names—among the names of hills and valleys, woods and lanes, rivers and hamlets—there are Celtic survivals in considerable numbers.

Concerning some of the early Celtic names we possess documentary evidence. The Antonine Itinerary contains three that are assured in this way, two definitely identified with particular places, the third less certain. These three are : *Documentary Evidence.*

> *Danum*, Doncaster.
> *Lagecium* or *Legeolium*, Castleford.
> *Cambodunum* or *Camulodunum*, probably Slack, near Huddersfield.

We have also two names given in early chronicles : the Celtic name of the battle fought in 633 between Edwin of Northumbria and Penda as given by Nennius and the Welsh Chronicle, and the name of Conisborough recorded in Geoffrey of Monmouth and Pierre de Langtoft :

> *Meicen*, *Meiceren*, *Meigen*, Hatfield.
> *Kaerconan*, *Conane*, Conisborough.

In addition Haigh has suggested (YAS Journal, IV, 61–65), the identification of four places mentioned by the Ravenna Geographer with sites in our district :

> *Alunna* with Castleshaw in Saddleworth.
> *Caluuium*, with a place near the confluence of the Colne and Calder.
> *Medibogdum* with Methley.
> *Rerigonium* with a place near Ripponden.

These identifications, however, are unsupported. Lastly, Bede has the name *Campodonum* which has sometimes been thought to be Doncaster, Alfred's version giving *Donafeld* in its stead ; but here again the matter must be left unsettled.

In only one instance, that of Doncaster, can we be sure that the modern name is a lineal descendant of the recorded Celtic name, though it is possible that in the case of Conisborough we have a second example.

Names of Rivers. Passing from the names guaranteed by documentary evidence we find ourselves treading on very treacherous ground. Among river-names we may enumerate as survivals from pre-Anglian days the Ouse, Aire, Calder, and Don, with their tributaries the Colne and Dove, as well as the Tame and Chew on the Lancashire side and the Derwent on the Derbyshire border. Possibly in addition to these we may count Lud- in Ludwell, Ludden- in Luddenden and Rib- in Ribble and Ribbleden.

Names of Hills and Valleys. Among the names of valleys and hollows we find four examples of Combes or Cowmes ; and among names of hills and rocks we may count as Celtic the Tors on the Derbyshire border, the first element in Chevinedge, and perhaps, in addition, the Rose Hills of which there are three.

Anglian Borrowings. At this point it will be helpful to consider the conditions under which the Angles made use of names borrowed from their Celtic predecessors. There were three possible methods.

1. The Celtic name might be taken over unchanged, with or without a knowledge of its meaning, and without the addition of any Anglian term. This appears to have taken place in such instances as Balne, Cowmes, Howcans, and Krumlin.

2. The Celtic name might be taken over with a full knowledge of its meaning and joined to some Anglian term. The word would thus become a true loan-word and enter fully into the language of its adoption.

3. The Celtic name might be taken over as a true proper noun, and joined, irrespective of its meaning, to an Anglian word. In that case it would assume the position and function of an adjective, being placed before, not after, the Anglian term to which it was attached.

Experience shows that it is chiefly in the third of these divisions that Celtic survivals are to be found. Place-names made from river-names assume this form quite regularly, as in

the case of Airmyn, Colnebridge, Dovecliffe, and **Ousefleet.**
There is quite a series of names in -den which may have arisen
in this way, including Alcomden, Luddenden, Ribbleden, the
two Bogdens and the two Sugdens. Other examples possibly
of a similar character are Cartworth, Catbeeston, Conisborough,
Crigglestone, Crimsworth, Featherbed, Featherstone, Mountain,
and Sugworth.

The list is, however, not yet complete, for there are quite a
number of places where the two elements are less closely linked
together, as in the case of Sude Hill. Words probably Celtic
and used in this way include Allen or Allan (6), Anna (3),
Crumack (3), Mankin, Sude, and Pennant.

A second group of Celtic names consists of loan-words Norse Bor-
introduced by the Norse immigrants in the 10th century, rowings.
namely, the words 'cross' and 'ergh'; but the reader must be
referred for these to the chapter on the Scandinavian element.

CHAPTER V

THE ROMAN, NORMAN, AND MODERN ELEMENTS

1. ROMAN.

Erming Street, the great Roman road to the north, passed from Doncaster to Castleford and so by way of Tadcaster to Aldborough or York. Another great Roman road, Riknild Street, came from Derby; entering the county near Beighton it proceeded northward by Swinton and Thurnscoe, and assuming a course almost parallel to Erming Street crossed the Calder at Normanton and came to Aldborough by way of Woodlesford. To the east of these roads lay a wilderness of swamp and morass; to the west forbidding hills. But, just as South-west Yorkshire was crossed by two almost parallel roads from the south making for Aldborough, two similar roads came from the north-west and with Manchester as their objective traversed some of the wildest of the hills on the Lancashire border. All the survivals from the Roman occupation are connected, directly or indirectly, with these roads. They consist of two Anglian words: Caster, borrowed from Lat. *castra*, a camp, and Street borrowed from Lat. *strata*. These are found in Castleford, Doncaster, and perhaps Castleshaw; in Adwick-le-Street, Strafford, Streethouse, Ossett Streetside, and Tong Street.

2. NORMAN.

In the Norman period there is greater variety, though even here the number of names is small. Pontefract, with its doublet Pomfret, is quite the most interesting example. The former is Latin, and due to the lawyers and chroniclers of the 11th and

12th centuries; the latter is French, and doubtless comes to us from the lips of Norman knights and squires. Strangely enough, the Latin and Norman names have lived side by side for eight centuries, and have driven the native names out of the field. The loss of these native names is very significant; it shows how under the influence of the de Lacy family Pontefract became the rendezvous for crowds of Norman retainers and adventurers.

In some parts of England it is quite common to find the name of a Norman family attached to a Saxon place-name. Well-known instances like Stoke Mandeville and Berry Pommeroy will be at once recalled. In South-west Yorkshire the only examples are Burghwallis, Farnley Tyas, Newton Wallis, Stubbs Lacy, and Whitley Beaumont. Early members of the two families connected with Whitley and Farnley were called William de Bellomonte, and Baldwin le Teys or Baldwinus Teutonicus. Another Norman family name occurs in Lascelles Hall.

A little group of names, including Grange, Roche, Spital, Friarmere, and Abdy, serves as a memorial of the ancient religious houses and of their work during many centuries. Other words connected with the religious life of the past are Armitage, found in the parish of South Crosland, and Chapel, found in Chapelthorpe and Chapeltown.

Two other survivals come to us from the Law, the French particles enclosed in the words Laughton-en-le-Morthen and Adwick-le-Street, and the name Purprise given to a farm in the neighbourhood of Heptonstall. Encroachments upon the property of the community or of the crown were described by the French term 'purpresture' and the word 'purprise' came to mean enclosed land.

Only three other examples remain. One is the curious name Hitchells which is found near Doncaster and appears to be derived from OFr *escheles*, ladders. The second is the word Grice, which meant a flight of steps, an ascent or slope, and comes from OFr *greis*, a derivative of Lat *gradus*. And the third is the name Richmond which occurs near Sheffield.

But there is a second way in which the coming of the

Norman has exercised an influence on our Yorkshire place-names, for the spellings in DB and other ancient documents show peculiarities obviously due to Norman scribes. These peculiarities arose from the fact that the Normans were foreigners, accustomed, on the one hand, to the orthographical methods of the French, and unaccustomed, on the other hand, to many of the sounds used by the English. A few of the more noteworthy points may now be enumerated.

1. Unfamiliar to the Norman were the two sounds of *th*. In the initial position he usually wrote *t* instead, as in *Torp* for Thorp and *Torn* for Thorn; in the medial position he often wrote *d*, as in *Medeltone* now Melton; and occasionally he left the consonant altogether unrepresented, as in *Ferestane* for Featherstone. When at last he began to use the sign *th*, it was frequently misplaced, as in *Thofthagh* for Toftshaw.

2. Such initial consonant-groups as *sn* and *st* were often changed to *esn* and *est*, as in *Esneid* for Snaith; and, occasionally, initial *s* was written where it had no rightful place, as in *Scroftune* and *Scusceuurde* for Crofton and Cusworth.

3. Other consonant-groups which gave him difficulty were *kn* for which he sometimes wrote *n*, and *ks* for which he gave *z* or *s*; compare *Notingeleia* for Knottingley and *Chizeburg* for Kexborough.

4. The guttural in Drighlington and Laughton he represented by an *s*; compare DB *Dreslintone* and *Lastone*. And quite frequently, especially before *n* and *m*, he wrote *o* instead of *u*. An interesting illustration of this is provided by the name Dudmanstone—or, rather, by the personal name which provides the first element. This name appears in an ancient document of the year 824 as Dudeman (Searle), but in the 13th century, in WCR 1296, we find it in the form Dodeman, while in the 17th century, in RE 1634 Dudmanston, the form is Dudman. There can be little doubt that during this long period the pronunciation of the first syllable had always been the same.

3. MODERN NAMES.

A few words must be said about the names which have arisen during recent centuries.

Perhaps the most interesting are those derived from Biblical sources. As our Parish Registers show, there was a considerable period during which Christian names drawn from Holy Writ were held in great favour. During that period such names as Faith and Mercy, Abel and Seth, Rachel and Jemima, were very common. And, apparently during the same period, place-names from the same source were held in equal esteem. Examples are to be found like Padan-aram and Machpelah, Egypt and Mount Tabor, Bethany and Jericho and Paradise, while in the neighbourhood of Halifax we find farms called Noah's Ark and Solomon's Temple.

Another series of modern names has arisen from the desire to substitute for the ancient name some more high-sounding designation. During the 19th century Queensbury took the place of Queenshead, a name which had itself previously supplanted the earlier name Causewayend. In the same period Norristhorpe was substituted for Doghouse as the name of a hamlet in Liversedge.

Still another series of names is connected with great captains of industry; such are Akroydon, Ripleyville, and Saltaire. Others due to the industrial expansion of the 19th century, often strikingly inappropriate, include New Brighton and High Scarborough, Mount Pleasant and Bellevue and Claremont.

The great events of modern history have also had their influence. To this we owe Waterloo and Odessa, Portobello and Alma, as well as, perhaps, the Dunkirks and Quebecs.

It may seem at first sight that Paris and London should be explained by similar methods, but a closer examination reveals the probability, strange as it may appear, that they are indigenous to the soil, the former of Anglian origin and the latter Scandinavian.

CHAPTER VI

LIST OF ABBREVIATIONS

1. LANGUAGES OR DIALECTS

AFr	=Anglo-French	ME	=Middle English
AS	=Anglo-Saxon	MHG	=Middle High German
Bret	=Breton	Norw	=Norwegian
Dan	=Danish	ODan	=Old Danish
Du	=Dutch	OE	=Old English
Fr	=French	OFr	=Old French
Fris	=Frisian	OHG	=Old High German
Gael	=Gaelic	OIr	=Old Irish
Germ	=German	ON	=Old Norse
Icel	=Icelandic	OS	=Old Saxon
Ir	=Irish	OW	=Old Welsh
Lat	=Latin	Sw	=Swedish
LL	=Late Latin	W	=Welsh

2. EARLY RECORDS OF PLACE-NAMES

AR =Yorkshire Assize Rolls; YAS Record Series
BCS =Birch's Cartularium Saxonicum
BD =Bosville Deeds; YAJ, Vol. XIII
BM =Burton's Monasticon Eboracense
BPR =Bingley Parish Register
CC =Calverley Charters; Thoresby Society, Vol. VI
CH =Charter—unspecified
CR =Calendar of Charter Rolls; Rolls Series
DB =Domesday Book for Yorkshire; Skaife
DC =Dewsbury Church and Manor; YAJ, Vols. XX–XXI
DN =Dodsworth's Notes; YAJ, Vols. VI, VII, VIII, X, XI, XII
FC =Memorials of Fountains Abbey; Surtees, Vols. XLII and LXVII
GC =Cartæ et Munimenta de Glamorgan; Cardiff, 1910
HH =Hunter's Hallamshire (Gatty), 1869
HPR =Halifax Parish Register
HR =Hundreds Rolls
HS =Harrison's Survey of Sheffield

HW   = Halifax Wills ; Messrs Clay and Crossley
IL   = Index Locorum to Charters and Wills in the British Museum
IN   = Inquisition
KC   = Kirkstall Coucher Book ; Thoresby Society, Vol. VIII
KCD   = Kemble's Codex Diplomaticus
KCR   = Knaresborough Court Rolls
KF   = Knights' Fees ; Surtees, Vol. XLIX
KI   = Kirkby's Inquest ; Surtees, Vol. XLIX
KP   = Kirklees Priory ; YAJ, Vol. XVI
LAR   = Lancashire Assize Rolls ; Lanc. and Ches. Hist. Soc.
LC   = Lacy Compoti ; YAJ, Vol. VIII
LF   = Lancashire Fines ; Lanc. and Ches. Hist. Soc.
LI   = Lancashire Inquisitions ; Lanc. and Ches. Hist. Soc.
LN   = Landnama Book, included in Origines Islandicæ ; Vigfusson
        and Powell, 1905
MPR   = Methley Parish Register
NV   = Nomina Villarum ; Surtees, Vol. XLIX
PC   = Pontefract Chartulary ; YAS Record Series
PF   = Pedes Finium
PM   = Calendarium Inquisitionum Post Mortem
PR   = Pipe Rolls
PT   = Poll Tax Return ; YAJ, Vols. V and VI
RC   = Rievaulx Chartulary; Surtees, Vol. LXXXIII
RE   = Ramsden Estate Maps
RPR   = Rothwell Parish Register
SC   = Selby Coucher Book ; YAS Record Series
SE   = Savile Estate Maps
SM   = Speed's Map of the West Riding
TPR   = Thornhill Parish Register
VE   = Valor Ecclesiasticus ; temp. Henry VIII
WC   = Whalley Coucher Book ; Lanc. and Ches. Hist. Soc.
WCR   = Wakefield Court Rolls ; YAS Record Series
WH   = Watson's History of Halifax
WHS   = Stevenson's notes on Yorkshire Surveys ; English Historical
        Review, Jan. 1912
WPR   = Wath on Dearne Parish Register
WRM   = Wakefield Rectory Manor ; Taylor
YAJ   = Yorkshire Archæological Journal
YAS   = Yorkshire Archæological Society
YD   = Yorkshire Deeds, YAS Record Series ; and YAJ, Vols. XII,
        XIII, XVI, XVII
YF   = Yorkshire Fines ; YAS Record Series
YR   = Registers of Archbishops Gray 1225–1255, Giffard 1266–1279,
        and Wickwane 1279–1285 ; Surtees, Vol. LVI, CIX, CXIV
YS   = Yorkshire Lay Subsidies ; YAS Record Series

3. OTHER ABBREVIATIONS—DICTIONARIES, ETC.

EDD   = English Dialect Dictionary ; Wright, Oxford, 1898-1905.
NED   = New English Dictionary; Murray, Oxford, 1888 etc.
NGN   = Nomina Geographica Neerlandica ; Leyden, 1892-1901.
Aasen   = Norsk Ordbog ; Aasen, Christiania, 1900.
Björkman   = Nordische Personennamen in England ; Björkman, Halle, 1910.
Brons   = Friesische Namen ; Brons, Emden, 1878.
Clarke   = Clarke's Yorkshire Gazetteer ; London, 1828.
Dineen   = Dineen's Irish Dictionary; Dublin, 1904.
Förstemann = Altdeutsches Namenbuch ; Förstemann, Nordhausen, 1859.
Falk   = Norwegisch-Danisches Etymologisches Wörterbuch ; Falk
             und Torp, Heidelberg, 1911.
Falkman   = Ortnamnen i Skane ; Falkman, Lund, 1877.
Gazetteer = Bartholomew's Gazetteer of the British Isles.
Gonidec   = Dictionnaire de la Langue Celto-Bretonne ; Le Gonidec,
             Angoulême, 1821.
Hatzfeld   = Hatzfeld's French Dictionary; Paris, 1871 etc.
Hogan   = Hogan's Onomasticon Goedelicum ; Dublin, 1910.
Holder   = Alt-Celtischer Sprachschatz ; Holder, Leipzic.
Jamieson   = Jamieson's Scottish Dictionary; Paisley, 1879.
Jellinghaus = Westfälischen Ortsnamen ; Jellinghaus, Kiel und Leipzig,
             1902.
Kelly   = Kelly's West Riding Directory; 1912.
Larsen   = Larsen's Dano-Norwegian Dictionary; Copenhagen, 1910.
Leithaeuser = Bergische Ortsnamen ; Leithaeuser, Elberfeld, 1901.
Littré   = Dictionnaire de la Langue Française ; Littré, Paris, 1883.
Macbain   = Macbain's Gaelic Etymological Dictionary; Stirling, 1911.
Madsen   = Sjælandske Stednavne ; Madsen, Copenhagen, 1863.
Middendorff = Altenglisches Flurnamenbuch ; Middendorff, Halle, 1902.
Naumann   = Altnordische Namenstudien ; Naumann, Berlin, 1912.
Nielsen   = Olddanske Personnavne ; Nielsen, Copenhagen, 1883.
Oman   = Oman's Swedish Dictionary.
O'Reilly   = O'Reilly's Irish Dictionary; Dublin, 1864.
Peiffer   = Noms de Lieux (France, Corse, et Algérie); Peiffer, Nice,
             1894.
Pughe   = Pughe's Welsh Dictionary; Denbigh, 1891.
Richthofen = Altfriesisches Wörterbuch ; Richthofen, Göttingen, 1840.
Rietstap   = Aardrijkskundig Woordenboek van Nederland ; Rietstap,
             Groningen, 1892.
Robinson   = Gazetteer of France ; Robinson, London, 1793.
Rygh   = Gamle Personnavne i Norske Stedsnavne, and Norske Gaard-
             navne ; Rygh, Christiania, 1898 etc.
Searle   = Searle's Onomasticon Anglo-Saxonicum ; Cambridge, 1897.

Scheler   = Dictionnaire d'Étyṃologie Française ; Scheler, Bruxelles,
             1888.
Schönfeld   = Wörterbuch der altgermanischen Personen- und Völkernamen ;
             Schönfeld, Heidelberg, 1911.
Skeat   = Skeat's Etymological Dictionary ; Oxford, 1910.
Spurrell   = Spurrell's Welsh Dictionary ; Carmarthen, 1905.
Stokes   = Urkeltischer Sprachschatz ; Stokes, Göttingen, 1894.
Stratmann   = Stratmann's Middle-English Dictionary ; Oxford, 1891.
Torp   = Wortschatz der Germanischen Spracheinheit ; Torp, Göttin-
             gen, 1909.
Vigfusson   = Cleasby and Vigfusson's Icelandic Dictionary ; Oxford, 1874.
Williams   = Williams' Cornish Dictionary ; Llandovery and London, 1865.
Zoëga   = Zoëga's Old Icelandic Dictionary ; Oxford, 1910.

For minor names reference has been made to the six-inch maps of the
Ordnance Survey.

CHAPTER VII

LIST OF ADDITIONAL WORKS QUOTED OR USED

Anglo-Norman Influence on English Place-names ; Zachrisson, Lund, 1909.
Anglo-Saxon Britain ; Grant Allen, 1904.
Blandinger til Oplysning om Dansk Sprog i ældre og nyere Tid ; Copenhagen.
British Family Names ; Barber, 1903.
British Place-names in their Historical Setting ; McClure, 1910.
Celtic Britain ; Rhys, 1884.
Celtic Researches ; Nicholson, 1904.
Cornish Language, Handbook of the ; Jenner, 1904.
Crawford Charters ; Napier and Stevenson, Oxford, 1895.
Deutscher Flussnamen ; Lohmeyer, Göttingen, 1881.
Domesday Inquest, The ; Ballard, 1906.
England before the Norman Conquest ; Oman, 1910.
Englische Ortsnamen im Altfranzösischen ; Westphal, Strasburg, 1891.
Englische und Niederdeutsche Ortsnamen ; Anglia, Vol. xx, 257–334, Björkman, Halle, 1898.
English Dialect Grammar ; Wright, Oxford, 1905.
English Towns and Districts ; Freeman, 1883.
English Village Community, The ; Seebohm, 1884.
Études Étymologiques sur les noms des villes (etc.) de la Province du Brabant ; Chotin, Paris, 1859.
Französischen Ortsnamen, Keltischer Abkunft ; Williams, Strasburg, 1891.
Irish Names of Places ; Joyce, 1901–2, two series.
Lincolnshire and the Danes ; Streatfeild, 1884.
Manx Names ; Moore, 1903.
Names and their Histories ; Taylor, 1898.
Place-names of Argyll ; Gillies, 1906.
      „   „ Bedfordshire ; Skeat, 1906.
      „   „ Berkshire ; Skeat, 1911.
      „   „ Cambridgeshire ; Skeat, 1901.
      „   „ Decies ; Power, 1907.
      „   „ Derbyshire ; Davis, 1880.

Place-names of Hertfordshire ; Skeat, 1904.
    „   „ Huntingdonshire ; Skeat, 1904.
    „   „ Lancashire ; Wyld and Hirst, 1911.
    „   „ Liverpool and District ; Harrison, 1898.
    „   „ Norfolk ; Munford, 1870.
    „   „ Nottinghamshire ; Mutschmann, 1913.
    „   „ Ross and Cromarty ; Watson, 1904.
    „   „ Scotland ; Johnston, 1903.
    „   „ Shetland ; Jakobsen, 1897.
    „   „ Staffordshire ; Duignan, 1902.
    „   „ Suffolk ; Skeat, 1913.
    „   „ Warwickshire ; Duignan, 1912.
    „   „ West Aberdeenshire ; Macdonald, 1899.
    „   „ Worcestershire ; Duignan, 1905.
Roman Roads in Britain ; Codrington, 1905.
Saga Book of the Viking Club.
Saxon Chronicles, Two ; Earle and Plummer, 1892.
Scandinavian Britain ; Collingwood, 1908.
Scandinavian Element in the English Dialects ; Wall, 1897.
Scandinavian Influence in Lowland Scotch ; Flom, 1900.
Scandinavian Loan-words in Middle English ; Björkman, 1900–2.
Scottish Land-names ; Maxwell, 1894.
Sprache der Urkunden aus Yorkshire im 15 Jahrhundert, Die ; Baumann, 1902.
West Riding Place-names ; Moorman, 1911.
Words and Places ; Taylor, 1902.
Yorkshire Archæological Society's Journal and Record Series.
Zur Lautlehre der altenglischen Ortsnamen im Domesday Book ; Stolze, Berlin, 1902.

CHAPTER VII

LIST OF ADDITIONAL WORKS QUOTED OR USED

Anglo-Norman Influence on English Place-names; Zachrisson, Lund, 1909.
Anglo-Saxon Britain; Grant Allen, 1904.
Blandinger til Oplysning om Dansk Sprog i ældre og nyere Tid; Copenhagen.
British Family Names; Barber, 1903.
British Place-names in their Historical Setting; McClure, 1910.
Celtic Britain; Rhys, 1884.
Celtic Researches; Nicholson, 1904.
Cornish Language, Handbook of the; Jenner, 1904.
Crawford Charters; Napier and Stevenson, Oxford, 1895.
Deutscher Flussnamen; Lohmeyer, Göttingen, 1881.
Domesday Inquest, The; Ballard, 1906.
England before the Norman Conquest; Oman, 1910.
Englische Ortsnamen im Altfranzösischen; Westphal, Strasburg, 1891.
Englische und Niederdeutsche Ortsnamen; Anglia, Vol. XX, 257–334, Björkman, Halle, 1898.
English Dialect Grammar; Wright, Oxford, 1905.
English Towns and Districts; Freeman, 1883.
English Village Community, The; Seebohm, 1884.
Études Étymologiques sur les noms des villes (etc.) de la Province du Brabant; Chotin, Paris, 1859.
Französischen Ortsnamen, Keltischer Abkunft; Williams, Strasburg, 1891.
Irish Names of Places; Joyce, 1901–2, two series.
Lincolnshire and the Danes; Streatfeild, 1884.
Manx Names; Moore, 1903.
Names and their Histories; Taylor, 1898.
Place-names of Argyll; Gillies, 1906.
     ,,   ,, Bedfordshire; Skeat, 1906.
     ,,   ,, Berkshire; Skeat, 1911.
     ,,   ,, Cambridgeshire; Skeat, 1901.
     ,,   ,, Decies; Power, 1907.
     ,,   ,, Derbyshire; Davis, 1880.

ADDINGFORD, Horbury, is the name of a ford across the Calder, now almost or entirely disused. The name should be compared with Addingham, † 1130 *Addingeham*, where the first element is the gen. pl. of the patronymic Adding recorded in LV.

ADLINGFLEET, EDLINGTON.—Under the date 763 a new translation of the AS Chronicle has the following statement : 'Then was Petwin consecrated Bishop of Whitern at Adlingfleet.' A reference to the original, however, shows that the name is *Ælfet·ee*, which presents no points of contact with the name Adlingfleet, though it may be connected with Durham, for the name Elvet occurs in that city. Early spellings of Adlingfleet and Edlington are as follows :

DB	1086	*Adelingesfluet*	DB	1086	*Ellintone, Eilintone*
DN	1220	*Adlingflet*	KI	1285	*Edelington*
YR	1245	*Adelingflet*	YS	1297	*Edelington*
DN	1292	*Athelingflete*	NV	1316	*Edelyngton*
PT	1379	*Adlyngflete*	PT	1379	*Edlyngton*

Personal names derived from OE *æðel*, noble, illustrious, often appear in the Domesday record under the form Adel- or Edel- ; hence *æðeling*, a prince, and the personal name derived therefrom, might appear as Adeling or Edeling. Occasionally *æðel* is reduced to Al- or El-, and, apparently under Norman influence, to Ail- or Eil- (Zachrisson).

ADLINGFLEET, Goole, is plainly either 'Atheling's channel,' or 'the prince's channel,' from OE *flēot*, a running stream, channel, estuary.

EDLINGTON, Conisborough, shows no sign of the genitive, and must be compared with such names as Darrington, Drighlington, and Alverthorpe. The termination is from OE *tūn*, an enclosure, farmstead.

ADWALTON, Bradford, was the scene of an engagement between the Roundheads and Cavaliers in 1643. The earliest record is *Athelwaldon* in PF 1202, and later we get YF 1504 *Adwalton*, SM 1610 *Adwalton*. The meaning is probably 'Athelwald's farmstead,' from OE *tūn*, and the personal name found in DB as Athelwold and in Searle as Æthelweald.

ADWICK-LE-STREET, ADWICK-ON-DEARNE.—

The ancient spellings warrant the explanation 'Ada's habitation,' from the recorded personal name Ada and OE *wīc*, an enclosure, habitation, village.

ADWICK-LE-STREET, Doncaster.   ADWICK-ON-DEARNE, Wath.

DB	1086	*Adewic*	DB	1086	*Hadeuuic*
YI	1304	*Adewike, Athewyke*	CH	1330	*Addewyk*
KC	1325	*Adewyk*	PT	1379	*Addewyk*

The full title Adwick-le-Street shows a curiously mixed origin, 'Adwick' being Anglian, 'le' French, and 'Street' Latin; and it owes its distinctive affix to the fact that a Roman road passed through the village. Where these roads have been obliterated we may often trace their direction by means of such names as this. The great road which passed through Adwick—Erming Street as it is called—is, however, by no means obliterated. In its course from London to Carlisle it entered Yorkshire at Bawtry, crossed the Don at Doncaster, and so by Adwick and Castleford came to Tadcaster, its course being often along the present highway and for many miles along parish boundaries.

AGBRIGG, AGDEN.—

The former is the name of the wapentake in which Wakefield and Huddersfield are situate; it is also the name of a hamlet in Sandal, Wakefield. The latter occurs in Bradfield and near Keighley.

The first syllable in the two names has quite the appearance of being derived from OE *āc*, an oak, for, when followed by *b* and *d*, 'ac' would according to rule become 'ag.' Early spellings of Agbrigg and Agden (Bradfield) show how mistaken any such surmise would be:

DB	1086	*Agebruge, Hagebrige*	HH	1329	*Aykeden*
PF	1166	*Aggebrigge*	CH	1337	*Aykeden*
WCR	1277	*Aggebrigg*			
KF	1303	*Aggebrigg*			

Obviously Agden is from ON *eik*, an oak, and OE *denu*, a valley; compare Ackton. But Agbrigg comes from quite a different source, its first element being either (1) a personal name Aggi recorded by Nielsen, or (2) a stream-name Agge which appears in the early Norwegian place-name Aggedal recorded by Rygh.

In agreement with these we may explain the terminal as from ON *bryggja*, a bridge.

This bridge stood at the point where the Roman road from Pontefract to Wakefield crossed a small side-stream of the Calder. Here, long before the Conquest, and for many generations after, the franklins of the wapentake met together to transact the business of the district and to settle disputes. The main road by which they travelled was doubtless that which came from the neighbourhood of Huddersfield by way of Lepton, Flockton, Overton, and Horbury.

AINLEY, Elland, BM 1199 *Aghenlay*, WCR 1297 *Aveneley* (*v* = *u*), WCR 1314 *Anneley*, WCR 1389 *Anelay*, has developed on similar lines to Hainworth which in 1230 was *Hagenwrthe*. The meaning is 'the lea of Agena,' where *Agena is a short form of some such personal name as Agenulf; compare the ON personal name Agni (Rygh).

AIRE, AIRMYN.—The village of Airmyn or Armin stands near the confluence of the Aire and Ouse. Early records of the name include the following:

DB	1086 *Ermenie, Ermenia*	YI	1295 *Ayremyn*
CR	†1108 *Eyreminne*	SC	1319 *Ayremine*
YI	†1272 *Eyreminne*	PT	1379 *Harmyn*

AIRE appears in PC 1218 as *Air*, but in two records of Airmyn the river-name appears as *Er-* and in others as *Ar-* or *Har-*. The name is related to the Swiss Aar and German Ahr, formerly *Ara*. Lohmeyer describes the word as Teutonic and connects it with OE *earu*, ON *òrr*, swift; but Holder claims it as Celtic. Possibly the change from Er- to Ayre- or Air- is due to the influence of the ON *eyrr*, a sandbank; but see Bairstow.

AIRMYN means simply 'Aire-mouth,' from ON *minne*, the confluence of two streams; compare the early Icelandic name *Dals-minne* (LN), and note also the ancient Yorkshire names *Nidderminne* (CR), at the confluence of the Ouse and Nidd, and *Burmyne* (PT), mentioned under Hoghton (Glass Houghton).

AKROYDON is the name of a portion of Halifax built by Colonel Akroyd, a great benefactor to the town.

ALCOMDEN, Wadsworth, is recorded by Clarke in 1828 as *Alecomden.* See Allan and Combe.

ALDERTHWAITE, ALVERLEY, ALVERTHORPE. It would be difficult to find a better object-lesson than is presented by the early records of these names, which occur respectively in or near Hoyland Nether, Doncaster, and Wakefield.   Early forms are :

PC 1239 *Alwardethuait*   PR 1190 *Alwardeslea*   WCR 1274 *Alvirthorpe*
PC 1239 *Alverdethuait*   CH †1254 *Alwardley*   WCR 1285 *Alverthorp*
YS 1297 *Allurthauyt*   BM 1277 *Alvarlay*   WCR 1291 *Aluerthorp*
CH 1302 *Allertwayte*   PM 1337 *Alverley*   WCR 1300 *Alverthorp*

In each case the first element is a personal name.   In Alderthwaite it appears to be a weak form, probably ON Alfvarði ; in Alverley it is a strong form, probably OE Ælfweard ; and in Alverthorpe it may be ON Alfarr, but in that case we must assume that the -s- of the genitive has been lost, as in the later form of Alverley.   The terminals come from ON *thveit*, a clearing, OE *lēah*, a lea, and ON *thorp*, a village.

ALDWARK, Rotherham, is the site of an ancient fortified post intended to control the passage of the Don.   It was here in all probability that the Roman road from Derby to Aldborough, Riknild Street as it is called, crossed the river.   In YS 1297 the spelling was *Aldewerke*, and in YF 1532 *Aldwark*; and the meaning is ' the old work,' that is, ' the old fortification,' from OE *eald*, old, and *weorc*, a work or fortification, or more accurately from Anglian *ald* and *werc.*   In the North Riding there is another Aldwark, spelt *Aldewerc* in DB ; compare also Newark, ' new work'; Southwark, 'south work '; and bulwark, ' a fortification constructed of tree-trunks,' Dan. *bulværk.*

ALLAN, ALLEN.—In Pudsey there is Allan Brigg ; in Warley, Allan Gate ; in Saddleworth, Allen Bank ; in Wilsden, Allen Moor ; in Norland and Shelley, Allen Wood.

Passing to other parts of England we find places called Allenford in Wiltshire and Hants ; others called Alford in Lincoln and Somerset ; and streams called Allen in Northumberland, Dorset, and Cornwall.   In Scotland the name Allan is applied to three rivers, tributaries of the Forth, Teviot and Tweed; there is a stream called Allander in Berwick; and there are others called Ale Water in Roxburgh and Berwick. Still further, there are Welsh streams called Allen, Aled, and Alwen. Among early spellings we find the following:

> Ptol *Alauna*, a stream called Allan in Stirling.
> Ptol *Alaunos*, a river called Alne in Northumberland.
> BCS *Alleburne, Albourne*, a village in Sussex.
> BCS *Alneceastre*, Alcester, on the Warwick Alne.
> DB *Eleburne*, 1285 *Alburne*, Auburn, East Riding.
> CR *Alebrok* 1267, a Devonshire stream.
> CR *Alan* 1285, 'the water of Alan,' in Cornwall.

Compare these with such German river-names as Ahlbeck and Elbach, Alpe and Elpe, Alster and Elster, and note that among Celtic river-names Holder records *Alana, Alara, Alantia*, and *Alanion*. The latter forms are possibly extensions of the stem **pal-* found in Lat. *palus*, a marsh, the initial *p* being according to rule dropped in Celtic.

ALLERTON, Bradford, DB *Alretone*, PC † 1246 *Alretona* KI 1285 *Allerton*, NV 1316 *Allerton*, is derived from OE *alr, aler, alor*, an alder, and *tūn*, an enclosure or farmstead. The meaning is 'alder-farm'; compare BCS *Alar·sceat*, now Aldershot.

ALMA occurs as the name of a farm in Meltham, and obviously gets its name from the battle fought in 1854.

ALMHOLME, Doncaster.—See Holme.

ALMONDBURY, Huddersfield, occupies a site of very great interest, and the name has given rise to the most varied interpretations. There are two local pronunciations of the name to be recorded, Aimbry (eimbri) and Awmbry (ɔmbri).

DB	1086 *Almaneberie*	NV	1316 *Almanbury*
YR	1230 *Almannebire*	PT	1379 *Almanbery*
DN	1250 *Alemanbir*	HW	1545 *Ambry*
CR	1251 *Alemanebiri*	YF	1549 *Almonbury*
WCR	1274 *Almanbiry*	RE	1634 *Almanburie*

The ending is from OE *byrig*, the dative singular of *burh*, a town or fortified place. Dr Moorman thinks the first element refers to the Alemanni, a South German tribe, and he indicates two historic statements in support of the possibility of an Alemannic settlement during the Roman period.

(1) In his *Historiæ Novæ* the Greek historian Zosimus speaks of a great victory over the Alemanni gained by the Emperor Probus, after which many of the conquered were deported to Britain ; these, he says, 'when settled in that island were serviceable to the Emperor as often as anyone thenceforward revolted.'

(2) Another historian, Aurelius Victor, says that among those present at York who in 310 used their influence to persuade Constantine to assume the imperial power there was a certain Erocus who is described as a King of the Alemanni.

It is plainly not impossible that Almondbury should be the centre of such a settlement.   The value of a strong outpost at such a point to keep in order the tribes in the western hill-country cannot be denied ; neither can we challenge the fitness of Almondbury for such a duty.   On the other hand the analogy of such names as Dewsbury and Barnborough would lead us to look for a personal name as the first element ; and Bardsley records the name *Aleman* as occurring in 1216 and *Alman* in 1379, while the Wakefield Court Rolls mention *Richard Alman* in 1308 and *Richard Aleman* in 1309.   Further, Searle records the names *Ælmanus* and *Ællmann*, and a corresponding weak form would fully explain such early forms as *Almaneberie* and *Almannebire*.

'British coins of the Brigantine type have been found in hoards, in association with Roman imperial and other coins,' both at Almondbury, where sixteen Brigantian coins were found, and at Lightcliffe where the number was four (VCH).

ALTOFTS, Wakefield, PC†1090 *Altoftes*, PC†1140 *Altofts*, PF 1207 'in bosco de *Altoftes*,' KC 1332 *Altoftes*, YF 1509 *Altoftys*.   The second element is from ON *topt*, a green knoll ; and the first comes most probably from a Scandinavian word meaning alder.   Falk, in addition to ON *elri* and *elrir*, alder, gives ON *alr*, and Falkman has the form *al* in Allò with the same meaning.

ALVERLEY, Doncaster.—See Alderthwaite.

ALVERTHORPE, Wakefield.—See Alderthwaite.

AMBLERTHORNE, Northowram, is mentioned in WCR 1546 in the phrase ' William Awmbler of Awmbler Thorn.'

ANGRAM.—In the northern half of the West Riding this name occurs several times, early spellings being HR 1276 *Angrum* for a hamlet in Nidderdale, and BM 1325 *Angrum* for another in Wharfedale. In our own area it occurs in Ecclesfield, HS 1637 *Angerum*, and as a field-name in Mirfield, SE 1708 *Angram*. The name is a dat. plur., and it goes back to the Germanic **angra*, which has given on the one hand Germ. *anger*, Du. *anger*, a meadow or pasture-land, and on the other hand ON *angr*, a bay or firth ; compare the Dutch place-name Angeren, and the Norwegian Hardanger and Stavanger. No example is given by Middendorff; it appears, indeed, to be a Northern word, and, if Scandinavian, it will be applied to a bay-like valley.

ANNA LANE, HANNA WOOD, HANNA MOOR, situated respectively in Thurlstone, Wyke, and Wortley.—If these names are Celtic in origin they may perhaps be related to Irish *an*, water. This word represents an older stem **(p) ana*, a swamp, bog, and is connected with our own word ' fen.' In Hampshire we find the name Anna, KCD 903 *Anna*, and the name Andover, PR 1170 *Andeura*, while in Ayrshire there is a stream called the Ann.

ANSTON, NORTH and **SOUTH**, near Sheffield, DB *Anstan, Anestan, Litelanstan*, CR 1200 *Anestane*, PT 1379 *Anstane*, is probably ' the solitary stone,' from OE *ān*, one, and *stān*, a stone—not from *tūn*, a farmstead.

APPERLEY BRIDGE, Eccleshill, YR 1279 *Apperley*, CC 1351 *Apperlaybrig*, must be compared with the Dutch place-name Aperlo which according to NGN, III, 322 derives its first element from *apa*, a word meaning water, and the common terminal -lo which corresponds to our English -ley, a lea or meadow.

APPLEHAIGH, APPLEYARD, Royston and Thurlstone.—Early records of the latter are YS 1297 *Apelyard,* 1372 *Apilyerd*; of the former, YF 1560 *Appleday,* the *d* being intrusive as in Backhold and Wormald. Both names are from OE *æppel,* apple, and, as the terminations, OE *haga* and *geard,* both signify an enclosure, we may give the meaning in each case as 'the apple orchard.'

ARBOUR, ARBOURTHORNE.—The former occurs as a field-name in Elland, SE *Arbour Closes,* and the latter is found near Sheffield, HS 1637 *Arbor Thorne.*

Amblerthorne seems to be derived from the personal name Ambler, and Arbourthorne may have a similar origin ; compare the surnames Arber and Harbour. But a more likely interpretation of Arbour is 'earth cottage,' from OE *eorðe,* earth, and OE *būr,* ME *bour,* a cottage, chamber, bower.

ARDRON, HORDRON.—A dialect-word *ron* or *rone,* used in the North of England and also in Scotland and Ireland, is explained in EDD as a thick growth of weeds, a tangle of thorns and brushwood. A similar word occurs in the place-names of Shetland—for example, in Longaroni, Queedaronis, and Hoorun —and Jakobsen explains it as a wilderness, a rough hill, from ON *hraun,* a rough place, a wilderness. On the other hand, Aasen has a Norw. word *ron* meaning a corner.

ARDRON occurs in Kirkheaton.

HORDRON, Penistone, spelt *Horderon* in a Chapel-en-le-Frith charter of 1323, appears in PT 1379 in the personal name *Johannes Horderon.* The first element probably comes from the ON personal name *Haurði; compare the strong form Haurð found in LN.

ARDSLEY.—South-west Yorkshire has two places of this name, but early records show they are derived in part from different sources :

ARDSLEY, Wakefield.	ARDSLEY, Barnsley.
DB 1086 *Erdeslawe, Erdeslauue*	PF 1202 *Erdeslegh*
PF 1208 *Erdeslawe*	IN 1320 *Erdesley*
YI 1249 *Erdeslawe*	CR 1371 *Erdesley*
KI 1285 *Ardeslawe*	PT 1379 *Erdeslay*

The former is 'Eard's burial-mound' from OE *hlāw, hlǣw*, a mound or hill, and the latter is ' Eard's lea,' from OE *lēah*, the personal name being equivalent to the first element in such names as Eardhelm and Eardwulf.

Duignan tells us that a forge is known to have existed at Aston near Birmingham as early as 1329, but according to WCR the present ironworks at Ardsley, Wakefield, may claim as their predecessor a 'forg apud Erdeslawe' which existed even earlier, namely, in 1326.

ARKSEY, Doncaster, DB *Archeseia*, YR 1250 *Arkesay*, HR 1276 *Arkeseye*, YS 1297 *Arkessey*, is formed from OE *ēg*, an island, watery land, and a personal name. We may explain DB *Archeseia*, that is *Arkesei*, which represents the latter forms quite fairly, as ' Ark's water-meadow.'

ARMITAGE BRIDGE, Huddersfield, is one of the few West Riding names of French origin. PC † 1212 has the phrase ' *Heremitagie* que jacet juxta Caldwenedenebroc,' the hermitage which lies beside the Caldwenedene brook ; YD 1352 gives *Ermitage* ; PT 1379 speaks of *William del Ermytache* ; and in YF 1514 we find the form *Armitage*. The derivation is from the OF *hermitage* ; and the surname Armitage, so common in the neighbourhood of Huddersfield, owes its origin to this ancient cell. Staffordshire has a parish of the same name, and South-west Yorkshire has two other references to hermits in the names *Armit Hole*, Bingley Register 1653, and *Armetroyde*, Bradfield Register 1708.

ARMLEY, ARMTHORPE.—These names show very plainly the importance of early spellings :

DB 1086 *Ermelai*	DB 1086 *Ernulfestorp, Einulvestorp*	
PC 1155 *Armeslie*	RC 1231 *Arnelthorpe*	
PM 1287 *Armeley*	YR 1237 *Armethorp*	
KC 1300 *Castel Armelay*	YI 1256 *Arnethorpe*	

ARMLEY, Leeds, is 'the lea of Erm,' from OE *lēah*, and a personal name ; note that Forstemann records the name Ermo, and Brons the Frisian name Erme.

ARMTHORPE, Doncaster, is ' the village of Arnulf,' from ON *thorp*, and the ON personal name Arnulfr (Rygh).

ARRUNDEN, Holmfirth.—In WCR 1308 we find *Aundene* which is probably a scribal error for *Arundene*. The meaning is 'Arun's valley,' from OE *denu*, a valley, and the personal name Arun recorded by Barber ; compare Alverley and Alverthorpe.

ASHDAY, Southowram, was thought by Watson to be corrupted from Ashdale ; but such a derivation is entirely negatived by the early forms of the word : 1275 *Astey*, 1277 *Astaye*, 1284 *Astey*, 1308 *Astay*, 1370 *Astay*, all from WCR. The meaning is either ' Asti's island,' or 'east island,' from ON *ey*, or OE *ēg*, ME *ey*, an island or water-meadow. Compare Aston, Pudsey, Wibsey, and see the note on the termination -ey.

ASHURST, Ecclesall, recorded as *Hassherst* in 1318, *Asse-hirst* in 1347, and *Asshehirst* in 1374, is the ' ash-wood ' from OE *æsc*, ash-tree, and *hyrst*, a copse or wood.

ASKERN, Doncaster, is recorded as *Askerne* in PC † 1170, KC 1218, and DN 1318, and as *Askarne* in PT 1379. The meaning is either ' Aski's house,' or ' ash-tree house,' from ON *askr*, and OE *ærn*, a habitation, house.

ASPLEY, Huddersfield, is 'the poplar lea,' from OE *æspe*, the aspen or white poplar, and *lēah*, a lea or meadow ; compare Icel. *ösp*, Dan. and Sw. *asp*.

ASTON IN MORTHEN, Rotherham, DB *Estone*, KI 1285 *Aston*, CR 1335 *Aston in Morthing*, comes from OE *ēast*, east, and *tūn*, an enclosure, homestead. There are in the British Isles as many as sixty Astons, and thirty Eastons, all from OE *Easttun*. See Morthen.

ATTERCLIFFE, Sheffield, which provides a difficult problem, is represented in the following early records :

DB	1086 *Ateclive*	HH	1382 *Attercliff*
YI	1296 *Atterclive*	HS	1637 *Attercliffe*
HH	1366 *Attercliff*	HH	1647 *Attercliffe*

Before double consonants the Domesday scribes often dropped *r* ; OE *hyrst*, for example, was written *hest*, and probably the Domesday record of Attercliffe is imperfect for the same reason. Assuming that *Aterclive* is the correct Domesday form, we may

put forward the explanation ' Atter's cliff', for Nielsen gives an
ODan. name Attær, and in CR 1308 we find the name William
Atter. If this be the correct explanation we have to assume
that the -s- of the genitive was lost in very early days, as in the
case of Alverthorpe, Skelmanthorpe, and Thurstonland. The
name should be compared with the Scandinavian Attermire near
Skipton and Atterby in North Lincolnshire.

AUCKLEY, Doncaster, DB *Alcheslei, Alchelei*, PM 1294
Alkelay, NV 1316 *Alkeley*, IN 1327 *Alkesleye*, PT 1379 *Alkelay*,
YF 1567 *Awkley*, shows the loss of the sign of the genitive. We
may fairly explain DB *Alcheslei*, that is *Alkeslei*, as ' Alk's lea,'
noting that Searle gives the weak form Alca, Brons a Frisian
name Alke, and Nielsen the ODan. Alkæ ; compare ON *ālka*,
a sea-bird, the auk, which may be the source of the personal
name.

AUGHTON, Sheffield, DB *Actone*, PF 1202 *Acton*, 1324
Aghton, YF 1532 *Awghton*, is ' the farmstead beside the oak-
tree', from OE *āc*, oak, and *tūn*, a farmstead. Compare
Deighton, formerly *Dīcton*, and Broughton, formerly *Brōctun*.

AUSTERFIELD, AUSTERLANDS.—The former, near
Bawtry, is doubtless the *Ouestraefelda* or *Estrefeld*, where a
Council of the English Church met in 702 (Eddi). Post-
Conquest records of Austerfield include the following :

DB 1086 *Oustrefeld*	YI 1293 *Oysterfeld*
PM 1237 *Oystrefeud*	CR 1333 *Austerfeld*
YF 1247 *Westerfeud*	PT 1379 *Austerfeld*
HR 1276 *Ousterfeld*	YD 1465 *Austrefeld*

In the post-Conquest forms there are obviously three types :
(1) *Westerfeud*, the field more to the west, from ON *vestri*;
(2) *Oysterfeld*, the field more to the east, from ON *ŷystri*;
(3) *Austerfeld*, east field, from ON *austr*. The terminal comes
from OE *feld*, a field or plain. For the initial diphthong in
Oysterfeld compare the early spellings of Hoyland, DB 1086
Hoiland, PR 1176 *Hoiland*.

AUSTERLANDS, Saddleworth, appears in the Saddleworth
Parish Registers during the 18th century as *Osterlands*.

AUSTONLEY, Holmfirth, DB *Alstaneslei*, WCR 1274 *Alstanley*, WCR 1286 *Alstanley*, may be explained as 'the lea of Alstan.' The personal name appears in DB as Alstan, and in OE as Ealhstan.

BACKHOLD, Southowram, like Wormald, shows an intrusive *d*, early forms being YD 1277 *Bachale*, WCR 1369 *Bakhale*, PT 1379 *Backhall*. The meaning appears to be either 'ridge tongue,' from ON *bak*, a ridge, back, and ON *hali*, a tail, Dan. *hale*, a tongue of land, or 'the corner of land on the ridge,' from OE *bæc* and *healh* ; see Hale.

BADSWORTH, Pontefract, DB *Badesuuorde*, *Badesuurde*, YS 1297 *Baddeswurd*, PT 1379 *Badesworth*, YD 1548 *Baddesworth*, has for its first element a personal name Bad or Badd, while the ending is the OE *weorth*, a holding, farm.

BAGDEN, BAGHILL, BAGLEY, BAGSHAW.—There are at least forty names in the British Isles which show the prefix Bag ; but, while some have Scandinavian terminations like Bagby and Bagthorpe, in others the ending is English as in the names now under discussion.   Early records are as follows :

BAGDEN, Denby, YF 1552 *Bagden*, YF 1560 *Bagden*.
BAGHILL, Pontefract, KC 1284 *Baghill*, PC 1222 *Baggehil*.
BAGLEY, Calverley, CC 1344 *Bagley*, CC 1346 *Bagley*.
BAGLEY, Tickhill, YF 1539 *Bagley*.
BAGSHAW, Sheffield, PT 1379 *Bagschaghe*.

For the last-named the ON *bæki·skógr*, beech-wood, has been suggested ; but *skógr* would give -scoe, -skow, or -skew.   On the other hand we find in KCD *Bacganleah* for Bagley and *Bacgan-broc* for Bagbrook, both in Berkshire.   These suggest the personal name Bacga as the prefix ; compare the Norwegian place-names Baggetorp and Baggerud which come from the ON personal name Baggi (Rygh).   But there is another possibility, for Dr Skeat shows that Bagshot in Berkshire is derived from OE *bæc*, the back, and *sceat*, an angle, nook, corner.   Thus Bagshaw may mean 'back wood,' from OE *sceaga*, a copse or wood, while Bagden may be explained as 'back valley', from OE *denu*, a valley.

BAILDON DIKE, Skelmanthorpe.—See Beldon.

BAILIFFE BRIDGE, BALLIFIELD, Brighouse and Sheffield, have the following early records :

WCR	1374 *Bailibrigge*	CH	1277 *Balifeld*
WCR	1427 *Balybrigg*	YS	1297 *Balifeld*
WH	1775 *Bailey Brigg*	YD	1618 *Ballifield*

In addition, the Hartshead Parish Register has 1698 *Belly-bridge* and 1779 *Belleybridge*. The first element in these names may perhaps come from OFr. *baili*, ME *baili*, a steward or bailiff; compare ME *bali-schepe*, the office of bailiff, and *bally-wycke*, now '*bailiwick*,' the jurisdiction of a bailiff. Less probable is a connection with OFr. *baille*, a barrier, ME *baile, baili, bali* ; compare *balle*, a barrier, used in the place-names of Northern France (Peiffer).

BAIRSTOW, Warley, WCR 1277 *Bayrestowe*, WCR 1285 *Bayrstowe*, WCR 1308 *Bairstowe*, gives some difficulty. OE *bere*, barley, and *stow*, a place, should have given Barstow ; compare Barton, barley-enclosure. Apparently dialectal influence has been at work, and, just as 'rode' became 'royd,' so 'bere' became 'beyre' or 'bayre.'

BAITINGS is close to the county boundary on the road between Halifax and Littleborough. The name is given by WCR as *Baytinges* in 1285, and *Baitings* in 1413. It is derived from ON *beit*, pasturage, and *eng*, a meadow.

BALBY, Doncaster, DB *Ballebi*, CR 1269 *Balleby*, HR 1276 *Balleby*, YI 1279 *Balleby*, is 'the farm of Balli,' from ON *býr*, a farm, and the ODan. personal name Balli. Brons gives a Frisian name Balle.

BALLIFIELD, Sheffield.—See Bailiffe Bridge.

BALNE, BALME, BAWN.—Balme occurs in Kirkheaton as Little Balm and Great Balm, in Liversedge as Balme Ing, and in Cleckheaton as Balme Mill. Bawn is found in Farnléy near Leeds. In Wakefield there is Balne Lane, pron. Bawn (bɔn), and in Manningham Balne Closes. The only township-name,

however, is Balne near Snaith, of which we have the following
early spellings :

PF	1167	*Baune*	LC	1296	*Balne*
SC	1197	*Balnehale*	DN	1317	*Balne*
AR	†1216	*Beln'*	DN	1336	*Balnehecke*
HR	1276	*Balnehal*	PT	1379	*Balne*

In the 14th century the forms *Baulne* and *Bawne* appear
occasionally ; and in the 16th century we find YD 1530 and
YF 1565 *Balme*. It seems clear, therefore, that Balme and
Bawn may be simply variations of Balne.

No Anglian or Scandinavian explanation presents itself, and
we must perforce ask whether any Celtic explanation is possible.

In Irish we find *bail*, a place, and *baile*, a homestead, words
which represent Prim. Celt. **balis* and **baljos* respectively.
From *baile*, which appears in modern names as Bally, about
six thousand Irish place-names spring—one-tenth of the entire
list. Perhaps Balne comes from this Celtic source, the termination being one of the Celtic diminutive endings containing *n*.
In that case the meaning would be 'little farm.' It should be
noted that Hogan records several early names of the form
Balna, and that among our Yorkshire river-names a similar
ending is shown by Colne, Dearne, and Torne, words probably
themselves of Celtic origin.

BANK, BANKFOOT, BANKSIDE, BANKTOP.—The
word 'bank,' a mound or ridge of earth, comes from ON **banke*,
from which come also Icel. *bakki*, Dan. *bakke*, Sw. *backe*. It
occurs in Bankfoot, Bradford ; Bankside, Thorne ; and Banktop,
Southowram and Worsborough. These names may, of course,
be of comparatively recent origin.

BANNER CROSS, Sheffield.—HH 1494 has *Bannerfield*,
but *Bannercross* does not appear until the 17th century. Yet
the name may be early, and Addy suggests a Scandinavian
etymology, *bœna·cross*, prayer-cross, formed after the pattern of
bœna·hūs, house of prayer. It will be seen from the notes on
Gildersome and Kinsley that the development of *-er-* in the
second syllable would not be without precedent. Compare the
Cumbrian name Bannerdale.

BANNISTER.—A deed of the time of Henry VI mentions *Bannesterdike* in connection with Erringden ; a map of Wakefield dated 1728 shows a field called *Bannister Ing*; and in the 14th century the surname *Banastre* was quite common in the neighbourhood of Wakefield. Of Bannister Edge, Meltham, there are no early spellings. The meaning is probably ' Bani's abode,' from the personal name Bani given by Nielsen and ON *staðr*, a stead, place, abode.

BARCROFT, Bingley, HR 1276 *Bercroft*, PT 1379 *Bercroft*, means ' barley croft,' from OE *bere*, barley, and *croft*, a small field. See Barwick, Hertfordshire (Skeat).

BARKISLAND, BARSEY GREEN, Halifax.—The recorded forms include WCR 1275 *Barkesland*, PT 1379 *Barkesland*, WCR 1389 *Barsland*, HW 1515 *Barslande*, HPR 1586 *Barsland*, SM 1610 *Barseland*. In the natural order of things the township should now be called Barsland ; but someone has thought it better to put back the clock, and in doing so has given us the name in its least suitable form. The first element is a personal name, connected doubtless with DB Barch, which in its turn is connected with ODan. Barki. Among other significations, the OE and ON *land* means an estate or country.

BARSEY GREEN, a hamlet of Barkisland, is recorded in WCR 1277 as *Barkeshey*, 1286 *Barkeshay*, 1297 *Barkeseye*. Its prefix is the same as that of Barkisland, and its terminal is from OE *hege*, ME *heye*, a hedge, enclosed place.

BARNBOROUGH.—See Barnsley.

BARNBY, BARNBY DUN.—These are of Scandinavian origin, and the first element is a personal name ; compare ODan. *Barni, a name recorded by Nielsen.

BARNBY, Cawthorne, PC † 1090 *Barneby*, PT 1379 *Barmebe*, is ' the farm of Barni,' from ON *býr*; compare the Norwegian place-name Björneby, formerly *Biærnaby* (Rygh).

BARNBY DUN, Doncaster, DB *Barnebi*, PF 1202 *Barneby*, CR 1232 *Barneby*, has a similar origin and meaning. This name is particularly interesting because with others it appears in the list of festermen who stood sponsor to Archbishop Ælfric early in 1023; the spelling at that date was *Barnabi* and *Bærnabi*.

BARNES, Ecclesfield, YD 1279 *Bernis*, YD 1290 *Bernes*, YS 1297 *Bernis*, YD 1302 *Bernis*, is simply 'the barns,' from OE *bere·ærn*, barley-house, which first became *berern*, afterwards *bern*, and lastly *barn*.

BARNSIDE, Penistone.—See Chevet.

BARNSLEY, BARNBOROUGH, BARNSDALE.— We may fairly classify these as Anglian, the first element being the name Beorn or Beorna recorded by Searle; compare DB Bern, Berne. Early forms are as follows:

DB	1086 *Berneslai*		DB	1086 *Berneburg, Barneburg*
PC	†1090 *Bernesleia*		CR	1215 *Barneburge*
HR	1276 *Bernesley*		HR	1276 *Barneburg*
NV	1316 *Berneslai*		KI	1285 *Barneburg*
PT	1379 *Berneslay*		PT	1379 *Barniburgh*

BARNSLEY may therefore be interpreted 'Beorn's lea,' from OE *lēah*, a lea or meadow.

BARNSDALE, Doncaster, appears to be 'the dale of Beorn,' from OE *dæl*, a valley.

BARNBOROUGH, Doncaster, is 'Beorna's fortified post,' from OE *burg, burh*, ME *burgh*.

BARROW, BARROWCLOUGH, BARROWSTEAD. —The common word 'barrow' comes from ME *berw, barw*, which goes back to OE *beorg, beorh*, a hill, burial-mound.

BARROW, Wentworth, YI 1284 *Barwe*, YI 1515 *Barrowe*, YI 1566 *Barowe*, means simply 'grave-mound.'

BARROWCLOUGH, Northowram, means 'grave-mound valley,' from OE *clōh*, a valley.

BARROWSTEAD, Skelmanthorpe, means 'grave-mound place,' from OE *stede*, a place.

BARUGH, Darton, pronounced 'bark,' DB 1086 *Berg*, PC 1122 *Berx*, YI 1304 *Bergh*, PT 1379 *Bargh*, YI 1523 *Bargh*, like Barrow means 'grave-mound,' from OE *beorg*.

BARSEY.—See Barkisland.

BASSINGTHORPE, Rotherham, must be compared with Bassingthorpe in Lincolnshire and with Bessingby near Bridlington, DB *Basingebi*. The first element is a patronymic formed from ODan. Bassi, Bessi, a name recorded by Nielsen.

BATLEY has the following early forms: DB *Bateleia*, *Bathelie*, PC 1195 *Batelaia*, PF 1202 *Batteleg*, YI 1249 *Batelay*. The meaning is 'Bata's lea,' from OE *lēah*, a lea or meadow.

BATTYEFORD, Mirfield, SE 1708 *Batty Ford*, is pronounced locally with a strong stress on 'ford,' sufficient proof that the name is not ancient.

BAWN.—See Balne.

BAWTRY stands on the Idle at the point where the Great North Road enters the county. The name *Baltrytheleage*, which occurs in the will of Wulfric Spot, dated 1004, is the modern Balterley in Staffordshire (Duignan), and we must therefore make the best of the following early forms:

CR 1232 *Baltry*	PM 1279 *Bautre*	CR 1293 *Baltrie*
PM 1247 *Bautre*	YR 1281 *Bautre*	PT 1379 *Bautre*
YR 1268 *Bautre*	BD 1293 *Baltrey*	YD 1408 *Bautre*
YR 1273 *Bautre*	BD 1293 *Bawtrey*	YD 1567 *Bawtrye*

I take *Baltrey* to be the most primitive of these spellings, and, if so, the interpretation would seem to be 'Balthere's island,' from OE *ēg*, an island or water-meadow. I assume the loss of the sign of the genitive, and I accept the name Balthere (for Baldhere) on the authority of LV. Under these circumstances *Bautre* must be ascribed to popular etymology.

BEAL, Pontefract, DB *Begale*, *Beghale*, PC 1159 *Bekhala*, CR 1215 *Becchehale*, CR 1230 *Begehal*, PM 1311 *Beghale*, PT 1379 *Beghall*, means 'the corner of Bega,' from OE *healh*, a corner. The name St Bees is derived from an Irish saint called Bega.

BECK, BECKFOOT.—The common Northern word 'beck' is of Scandinavian origin; it is derived from ON *bekkr*, Dan. *bæk*, a stream or brook. As a termination it is found in Firbeck and Sandbeck. BECKFOOT, BPR 1634 *Beckfoote*, is a hamlet in the township of Bingley.

BEESTON, BEESTONLEY.—The former, in the city of Leeds, was spelt *Bestone* in DB, while PF 1202 gives *Beston*, KI 1285 *Bestone*, and PT 1379 *Beeston*. In explaining Beeston, Bedfordshire, which has similar early spellings, Professor Skeat says 'The corresponding AS form would be Bēos·tūn where Bēos is the genitive of Bēo, used as a personal name. Thus the sense is " Bee's farm." The name of John Bee occurs in 1428.'

BEESTONLEY lies in the valley which divides Stainland from Barkisland ; the name is a secondary formation, and means ' Beeston lea.'

BEGGARINGTON is the name of a hamlet in Hartshead, 1804 *Begerington*, and another in Queensbury.

BELDON, BELL GREAVE, BELL HAGG, BELL HOUSE, BELL SCOUT, BAILDON.—In Kirkburton we find Beldon Brook ; in Wibsey Beldon Hill ; in Fulstone Bell Greave ; in Erringden Bellhouse Moor and Bell Scout ; in Hallam Bell Hagg ; and in Skelmanthorpe Baildon Dike.

BELLHOUSE, Erringden, WCR 1307 *Bellehus*, WCR 1308 *Bellehouse*, comes from OE *belle*, a bell, and *hūs*, a house. The name could be applied to other structures as well as to the belfries of churches.

BELDON is to be found at least three times in Yorkshire. The terminal represents OE *dūn*, a hill, and it is noteworthy that quite a considerable number of Yorkshire hill-names have as their first element the word Bel-, as for examples Bella, Beldi, Beldoo, Beldow, Bellow, Bell Hill, and Bell Howe, all in the North Riding. Two of these, Beldoo and Beldow, appear to have a Celtic terminal representing Welsh *du*, black, dark, and it seems very probable the first element also is Celtic. Note in this connection Ill Bell, a hill in Westmorland.

BELLE VUE, a district in the city of Wakefield, appears to have received its name from a residence called Belle Vue which was in existence in 1828 (Clarke).

BENT, BENTLEY.—The OE for 'bent' was *beonet*, coarse grass of a reedy character, and the ME was *bent*, which in addition to the original sense signified an open grassy place or moor. Thus the name Bentley means 'the lea covered with coarse grass,' from OE *lēah*, a lea or meadow.

BENTLEY, Doncaster, in addition to DB *Benedleia*, KI 1285 *Benteley*, YS 1297 *Bentelay*, which agree with the above interpretation, has the following DB forms: *Beneslai, Beneslei*, where the sense is 'the lea of *Benn,' and *Benelei*, where the genitival *s* has been lost.

BENTLEY, Emley, DN 1365 *Bentley grange*.

BENTLEY ROYD, Sowerby, WCR 1300 *Benteleyrode, Bentelayrode*, WCR 1326 *Benteleyroide*.

GOOD BENT occurs as the name of a small district on the borders of Meltham.

-BER, -BERGH.—These terminations occur in Gawber, Hoober, and Thrybergh. They represent OE *beorg, beorh*, a hill, grave-mound, barrow.

BESSACAR, Doncaster, was formerly in the possession of Kirkstall Abbey and early records are therefore not to seek; but difficulties arise from the fact that in ancient times the name had two different forms.

PF	1202 *Besacre*	KC	1187 *Besacle*
KC	1209 *Besacre*	KC	1199 *Besacle*
YS	1279 *Besaker*	KC	1202 *Besacle*
PM	1281 *Besakre*	PT	1379 *Besakell, Besakill*

Bessi was an ON personal name representing an earlier form Bersi, and Bessacar may be explained as 'the field of Bessi,' from ON *akr*, a field; compare the Norwegian place-names Besserud and Besseby, the clearing and the farmstead of Bessi, that is, Bersi. For the ending *-acle* see Eccles.

BIERLEY, IDLE, NOSTELL.—A trio of interesting and difficult names of which we have early records as follows:

DB	1086 *Birle*	CC	†1190 *Idla*	DB	1086 *Osele*
CC	†1250 *Birle*	PF	1212 *Hidel*	YR	1216 *Nostell*
WCR	1275 *Byrle*	CC	†1230 *Ydele*	CR	1228 *Nostle*
KF	1303 *Byrill*	CC	1246 *Ydel*	YR	1251 *Nostel*
NV	1316 *Byrell*	KI	1285 *Idell*	CR	1280 *Nostell*
CC	1344 *Birille*	NV	1316 *Idele*	BM	1282 *Nostel*
CC	1424 *Estbirle*	PT	1379 *Idyll*	DC	1313 *Nostella*
YF	1573 *Bierle*	IN	1572 *Idle*	CR	1380 *Nostell*

All these early forms present an -*l* termination which is quite common in the Netherlands and the adjacent parts of Germany. Among the examples provided by Förstemann and Jellinghaus we find Hedel which comes from *Hedilla*, Bürgel which comes from *Burgila*, the Briels and Breuls from OHG *brogil*, marshy land, the Böhls and Bühls from OHG *buhil*, Prim. Germ. *buhila, a hill. Among Yorkshire names which show the same termination there are the North Riding Burrell, DB *Borel*, and the East Riding Nuttle, DB *Notele*; and among South-west Yorkshire names there are Diggle and Ickles.

BIERLEY, Bradford, should rather be Byrell, the ending -ley being obviously a modern addition due to the influence of such near neighbours as Calverley, Rodley, Thackley, and Stanningley. It goes back, I imagine, to an early form *būrila derived from the Germanic *bura, a storehouse. Compare OE *būr*, a storehouse, dwelling, OE *bȳre*, a shed, hut; and note that Förstemann gives an early German place-name of corresponding form, viz. *Burela*.

IDLE, Bradford, lies within a bend of the Aire. The name must be compared with that of the Nottinghamshire river: Bede *Idla*, YD 1567 *Idyll*, YD 1575 *Idell*, YD 1576 *Idle*, now Idle. I suggest that the Yorkshire Idle was originally a stream-name, perhaps cognate in origin with OE *ȳð*, flowing water.

NOSTELL, Pontefract, has a picturesque sheet of water, and the name is doubtless connected with OFris. *nost*, MLG *nòste*, a watering-place for cattle, a horse-pond; compare the North Riding name Nosterfield, KI 1285 *Nosterfeld*, NV 1316 *Nosterfeld*. In view of the unbroken series of forms equivalent to Nostell the Domesday spelling *Osele* must be rejected.

BIERLOW.—This name appears in Brampton Bierlow, Brightside Bierlow, and Ecclesall Bierlow, all near Sheffield. Early spellings are IN 1307 *Brantonbirlagh*, 1412 *Brampton Birlagh*, 1483 *Brampton Birlagh*, YE 1535 *Brampton Byerlawe*. The word signifies a district having its own byrlaw court; it signifies also, as explained by NED, 'the local custom or law of a township, manor, or rural district, whereby disputes as to boundaries, trespass of cattle, etc., were settled without going into the law courts.' It is derived from ON *bȳjar-lög*, the law of the byr, that is, of the village or community. In addition to the matters already mentioned NED tells us that the byrlaw court regulated such points as the date of ploughing, the number of cattle to be turned out upon the common land, the fines for trespass and for damage to fences.

Interesting references to these local customs occur in the Wakefield Court Rolls (II, xxiv). At Alverthorpe in 1298 a man was charged with making an unlawful distress, and in reply he pleaded that the debt for which he was distraining had been declared his due by the judgement of the whole 'Byrrelaghe'; and at Brighouse in 1330 the local court found that Thomas son of Julian had allowed his cattle to graze in the herbage of the Birefield 'contrary to the custom of the Bireleghe.'

BILCLIFFE, BILHAM, BILLEY, BILLINGLEY.— From the OE personal name Bill or Billa came the patronymic Billing, and from this the names of many English villages like Billingford and Billington. In France the patronymic appears in the place-name Billanges, and BCS has such early names as *Billingabyrig* and *Billanoran*. Early records of Bilham and Billingley are as follows:

DB	1086	*Bileham, Bilha'*	DB 1086	*Bilingelei, Bilingelie*
DB	1086	*Bilam, Bilan*	PF 1167	*Billinglea*
PC	†1180	*Bilham, Bilam*	PR 1190	*Billingeleya*
KI	1285	*Billeham, Bylleham*	KI 1285	*Bilingley*
YS	1297	*Billeham*	PT 1379	*Billyngley*

BILHAM, Doncaster, is either 'the home of Billa,' from OE *hām*, or 'the enclosure of Billa,' from OE *hamm*.

BILHAM occurs also in Clayton West.

BILLINGLEY, Doncaster, is 'the lea of the Billings,' from OE *lēah*, a lea or meadow.

BILCLIFFE, Penistone, CH 1329 *Bylcliffe*, YD 1358 *Bilclif*, PT 1379 *Bilclyf*, is 'the cliff of Billa,' from OE *clif*.

BILLEY, Ecclesfield, HH 1366 *Bilhagh*, is 'the enclosure of Billa,' from OE *haga*, an enclosure or small farm.

BINGLEY, Bradford, DB *Bingelei, Bingheleia*, PF 1209 *Bingeleia*, KI 1285 *Byngeley*, KF 1303 *Byngeley*, is 'the lea of

Binga,' from OE *lēah*, a lea or meadow. Support for an OE personal name *Binga is to be found in the Frisian name Binge recorded by Brons. Compare Bingley with Bingham, Notts, DB *Bingeham*.

BIRCHENCLIFFE, Huddersfield, is 'the birch-tree declivity,' from the OE adjectival form *beorcen*, birchen, and *clif*, a cliff, steep hill.

BIRDWELL occurs in Worsborough and Swinton ; the name should be compared with Spinkwell and Ouzelwell.

BIRKBY, BIRKENSHAW.—Though NED describes 'birk' simply as a northern form, there is no doubt that as a rule it is Scandinavian; compare such examples as Birkby, Birkwith, Birkholme, and Briscoe, where the whole name is undoubtedly Scandinavian, with such as Birkenshaw and Birket which have Anglian terminations.

BIRKBY occurs near Brighouse, in Huddersfield, and in Morley, and probably means 'the birch-tree farm,' from Dan. *birk*, the birch-tree. Birkby near Leeds and Birkby near Northallerton are, however, both 'the farm of the Britons,' being recorded in DB as *Bretebi*, and having similar forms of later date.

BIRKENSHAW, Bradford, WCR 1274 *Birkenschawe*, WCR 1307 *Birkynschawe*, PT 1379 *Kirkyngschawe* (the *k* a scribal error), is 'the birch copse,' from OE *sceaga*, a copse.

BIRLEY, BIRSTALL, Ecclesfield and Leeds.—Early spellings of these names are as follows :

DN	1161 *Burleya*	PF	1202 *Burstall*
YS	1297 *Byrley*	BM	1273 *Burstall*
YI	1298 *Birley*	YR	1281 *Byrstalle*
YD	1323 *Birley*	WCR	1286 *Byrstall*
PT	1379 *Byrlay*	WCR	1296 *Birstall*

Birstall corresponds to the German place-name Borstel which is found in Holstein, Hanover, and Westphalia. Jellinghaus records an early form *Burstalle*, and this goes back to an earlier

Burgstall, which is quoted by Förstemann and reproduced in many German place-names. We may without hesitation derive Birstall from OE *burgsteall,* 'the place of the burh, or fortified homestead,' and we may explain Birley as 'the lea of the burh.' See Borough.

It should be noted that although Birstall gives its name to an ancient parish, it is not an ancient manor, being a member of that of Gomersal, a fact which agrees with the interpretation now put forward. The 'burh' of the manor of Gomersal was doubtless at Birstall, and the church which rose up in connection with the 'burh' was close at hand; hence the ancient church is at Birstall, not at Gomersal. Compare Kirkheaton and Kirkburton; and note that an early sculptured stone, probably of the 9th century, is preserved in the church at Birstall.

BIRTHWAITE, Barnsley.—Early records of the name are YR 1234 *Birketweyt,* WCR 1297 *Birchtwayt,* and in undated charters of Pontefract and Rievaulx *Birketwait* and *Birkewait.* The meaning is the 'birch-tree paddock,' from ON *birki-,* the birch-tree, and *thveit,* a paddock.

BLACKBURN,   BLACKSHAW,   BLACKWOOD, BLAXTON.—Woods and clearings darker or duller than the rest, and streams brown from the mosses and moors, were often described by the OE word *blac, blæc,* which meant dark as well as black. Another word almost the same in form but very different in meaning was OE *blāc,* bright, shining. The corresponding ON words *blakkr* and *bleikkr* are easy to distinguish, but the OE words are often indistinguishable.

BLACKBURN, 1321 *Blakeburn,* 1326 *Blakeborne,* a side stream joining the Calder near Elland, is simply 'dark brook,' from OE *burna,* a brook, and the weak form of OE *blac.*

BLACKBURN, DN 1161 *Blacaburna,* a small stream near Ecclesfield, has the same origin and meaning.

BLACKSHAW, Heptonstall, HW 1539 *Blakschey,* HW 1540 *Blakeshaye,* is the 'dark wood,' from OE *sceaga,* a copse or wood.

BLACKWOOD, Sowerby, is recorded in WCR 1308 as *Blacwode;* from OE *wudu,* a wood.

BLAXTON, Bawtry, PM 1294 *Blacstan*, PT 1379 *Bakestan*, is 'the dark stone,' from OE *stān*.

BLACKER, BLACKLEY, BLACUP, BLAKE LOW.—

The first element in Blacker and Blackley is sometimes pronounced 'black' and sometimes 'blake'; compare Ackroyd and Ackton. In Blacup the first element is always pronounced 'blake.'

BLACKER occurs in Darton, Hoyland Nether, Skelmanthorpe, and Crigglestone, DN *Blaker*, and is either 'the pale carr,' from ON *bleikkr*, pale, and *kjarr*, copsewood, brushwood, or 'the dark carr,' from ON *blakkr*.

BLACKLEY, Elland, BM *Blacklau*, SE *Blakeley*, has rejected its original termination—which came from *hlāw, hlǣw*, a mound, cairn, hill,—in favour of the more common -ley ; compare Ardsley.

BLACUP, Liversedge, 1783 *Blacup*, although it shows no sign of the original diphthong, is derived from ON *bleikkr*, pale, and *hōp*, a secluded valley.

BLAKE LOW, Kirklees, formerly *Blachelana* (that is *Blackelaua*), appears to be the same word as Blackley.

BLAMIRES, Northowram, is an almost exact counterpart of the ON *Blamyrr* which occurs in Havards Saga, and means 'the dark swampy place,' from ON *blār*, dark, and *mȳrr*, a moor or bog.

BOB.—Three places rejoice in the name Lingbob, one in Wadsworth, another in Wilsden, and the third near Mount Pellon. In addition Sowerby has the name Collon Bob, and Midgley the name Bob Hill. EDD explains the word as a knob, a lump.

BOGDEN, BOG GREEN, BOG HALL.—These are doubtless to be connected with the Gael. and Ir. *bog*, a marsh, a soft wet place.

BOGDEN, Lepton and Rishworth, means 'marsh-valley,' from OE *denu*, a valley. BOG GREEN occurs in Kirkburton, and BOG HALL in Kirkheaton.

BOLE EDGE, BOLE HILL.—The former occurs in Bradfield, the latter in Treton and Ecclesall. The first element is most probably the ON *bol*, a farm.

BOLSTERSTONE, BOLSTER MOOR.—There are local explanations of these names which call to mind the story of the patriarch at Bethel. The true explanation is, however, much more prosaic.

Of BOLSTER MOOR, Golcar, there are no early records, but of BOLSTERSTONE, Bradfield, we find YD 1375 *Bolstyrtone*, YD 1398 *Bolstyrston*, YD 1402 *Bolsterston*, YD 1425 *Bolstirston*. The first element comes from ON *bolstaðr*, a farm-house, a name which occurs frequently in the Landnama Book, and the terminal is -ton, not -stone. There are several places in Norway called Bolstad, and in the West Riding, in addition to Bolsterstone, we have two other names from the same source, namely, Bolster Moor, 'farmhouse moor,' and Bowstagill, near Settle, 'farmhouse valley.'

BOLTON.—This is a distinctively Northern name; Kimbolton in Hunts and Chilbolton in Hants are not true Boltons, the former being Cynebald's ton and the latter Ceolbald's. Of true Boltons I count eighteen examples, one in Scotland and seventeen in the six northern counties, three being in Lancashire and ten in Yorkshire. The West Riding has almost half of the whole number, namely, seven, and of these South-west Yorkshire had four, namely, Bolton by Bradford, Bolton on Dearne, Bolton Brow in Warley, and Bolton in Calverley. Early spellings are as follows :

BOLTON by Bradford	BOLTON on Dearne
DB 1086 *Bodeltone*	DB 1086 *Bodeltone, Bodetone*
KF 1303 *Bolton*	PR 1190 *Boulton*
NV 1316 *Boulton*	KI 1285 *Bolton*
CC 1328 *Boulton*	KF 1303 *Boulton*

It has been customary to refer the first element to OE *botl* or *bold*, a building, a dwelling-place, but this does not seem satisfactory, and an examination of collateral facts must be made.

1. Among modern place-names there are several with the termination 'bottle' or 'battle': (*a*) Harbottle, Lorbottle,

Shilbottle, Walbottle, in Northumberland; Newbottle in Durham (BCS 963 *Niubotle*); Dunbottle in Yorkshire; Newbottle in Northants; (*b*) Newbattle in Edinburgh; Battleburn in Yorkshire. These may fairly be linked to the OE *botl*.

2. There are, further, several modern place-names involving the form 'bold' or 'bald': (*a*) Bold in Peebles, Lancashire, and Salop; Newbold in Lancashire, Cheshire, Derbyshire, Staffs, Notts, Leicester, Northampton, and Warwick (DB 1086 *Newebold*); Parbold in Lancashire (LF 1202 *Perebold*); Wichbold in Worcester (BCS 692 *Uuicbold*); (*b*) Newbald in Yorkshire (DB *Niwebold, Niwebolt*). Obviously these go back to OE *bold*.

3. The DB spellings of the various Boltons assume quite regularly the form *Bodeltone*; compare DB *Bodelforde*, now Bolford. But the prefix 'Bodel' represents 'Bothel' which would naturally become 'Bol'; compare DB 1086 *Medeltone*, PF 1208 *Methelton*, KI 1285 *Melton*, now Melton.

It appears, indeed, that we must credit OE with three forms, *botl, bold, bothel*; and a reference to Torp shows that the three may all be referred to the Germanic form **bōthla, *bōdla*, a dwelling-place, while among cognate forms there are OFris. *bold* and *bodel*, and Du. *boedel, boel*.

In regard to the geographical distribution of the three forms we find Bottle almost wholly in Northumbria, chiefly in the northern part; Bothel almost wholly in Northumbria, but chiefly in the southern part; while Bold occurs chiefly in Mercia. It seems very probable, therefore, that the distinction between the three forms is tribal and of early origin.

BOOTH, BOOTHROYD, BOOTHTOWN.—Although to-day the word 'booth' has but one form, our South-west Yorkshire names formerly presented two, viz. *boude*, that is, *bouthe*, from ON *būð*, and *bothe* from ODan. *bōð*. This difference gradually disappeared and only the Danish form is now to be found.

The Icelandic Sagas have many references to booths and their uses. We find, for example, that at the meetings of the Icelandic Parliament, which lasted for two weeks, temporary dwellings were used—booths which remained empty the rest of the year. A passage in the Laxdala Saga tells us that when the sons of a certain Hoskuld got to the Thing 'they set up booths, and made themselves comfortable in a handsome manner.' And another passage in the same Saga gives an account of Hoskuld's landing in Iceland. Having unloaded his ship, he laid her up and built a shed over her; then 'he pitched his booths there, and the place is still called Boothsdale' (*Búðar·dalr*).

BOOTH occurs in Rishworth, PT 1379 *Bothe* ; in Austonley, WCR 1307 *Bothe*; and in Midgley, Birkenshaw, and Thurlstone.

BOOTHROYD occurs in Dewsbury, WCR 1275 and 1286 *Bouderode* ; in Rastrick, WCR 1274 *Botherode*, WCR 1298 *Bouderode* ; and in Thurstonland. The meaning is 'the clearing beside the booth ' ; see Royd.

BOOTHTOWN, Halifax, is referred to as *Bothes* in WCR 1274, *Bothes* in YF 1548, and *Bouthtowne* in HPR 1579. The termination is a recent addition, and therefore takes the form ' town ' not ' ton.'

BORD HILL, Saddleworth and Thurlstone.—OE *bord* meant a board, plank, shield ; and ON *bord* had similar meanings ; but a later signification was food, maintenance, ' board,' and Johnston explains the name Bordlands, South Scotland, as 'board or mensal land,' land held on the rental of a food-supply. On the other hand Middendorff has OE *bord*, a boundary, which appears to give the sense we now require.

-BOROUGH, -BURY.—These words come from the nominative and dative of the same OE word. The nominative *burh* or *burg* gave the ME *burgh, borw*, and the modern form 'borough.' The dative, *byrig*, gave such ME forms as *byrie, byry, biry*, and, under the influence of Norman scribes, DB *berie* ; compare Almondbury, DB *Almaneberie*, Dewsbury, DB *Deusberie*, Horbury, DB *Horberie*. The original meaning appears to have been a fortress or a fortified place, a homestead enclosed by a wall or mound : but later the word came to mean a walled town, a city. Signifying in the earliest days nothing more than a rampart of earth provided as the defence of some isolated farmstead, the word touched every stage of meaning until it attained the idea of a fortified town.

Along the valley of the Don there is a series of seven 'boroughs' which is thought to represent a line of early fortifications; the names are Templeborough, Masborough, Greasborough, Mexborough, Barnborough, Conisborough, and Sprotborough.

Barnsley is the centre of another series : Kexborough, Harborough, Measborough, Worsborough, and Stainborough.

Further north there is a cluster of three which lie on or near the Calder, namely, Horbury, Dewsbury, and Almondbury.

Still further north there are near Bradford two other examples, namely, Stanbury and Thornbury.

BOTANY BAY occurs as a local name in Lepton.

BOTTOMBOAT, BOTTOMLEY.—The first element is from OE *botm*, ME *botym, botum, bothem*, a foundation, bottom. Early spellings include the following :

PF	1202 *Stanliebothem*	WCR	1275 *Bothemlei*
WCR	1286 *Bothem*	WCR	1277 *Bothemley*

BOTTOMBOAT, Stanley, is unique in its terminal, which undoubtedly refers to a ferry-boat. Johnston has placed on record several Scotch names which involve the word, among them 'Boat of Forbes' and 'Boat of Inch'; and he explains them as referring to ancient ferries. Bottomboat is simply 'the boat at Bottom.' A ferry over the Calder is still in existence.

BOTTOMLEY, Barkisland, means 'the lea in the lowlying ground,' from OE *lēah*, a lea.

BOWER.—This name is found in Goody Bower, Wakefield, in Hall Bower, Almondbury, in Harry Bower, Kirkburton, and Bower Hill, Oxspring. It is derived from OE *būr*, ME *bour*, a dwelling, a store-room, a chamber.

BOWLING, Bradford.—See Ing.

BRADFIELD, BRADLEY, BRADSHAW.—In the ordinary course of development OE *brād* became *broad*; but before double consonants a shortening took place giving us instead the form *brad*. For the same reason OE *āc·tūn* has given us Ackton, whereas the two elements standing alone would have become *oak* and *town*. Hence we can tell at a glance that names like Bradfield and Morton are early, while Broadfield and Moortown are of later formation.

BRADFIELD, Sheffield, *Bradefeld* in KI 1215, YS 1297, NV 1316 and PT 1379, is the 'broad field,' from the weak form of OE *brād* and OE *feld*.

BRADLEY, Huddersfield, DB *Bradelei*, PF 1202 *Bradelai*, is the 'broad lea,' from OE *lēah*, a lea or meadow.

BRADLEY occurs also in Stainland.

BRADSHAW, which occurs near Slaithwaite, Halifax, and Holmfirth, is the 'broad copse,' from OE *sceaga*, ME *schagh*, a copse or wood.

BRADFORD, DB *Bradeford*, PC †1250 *Bradeford*, KI 1285 *Bradford*, PT 1379 *Bradforth*, is the 'broad ford,' from the weak form of OE *brād* and OE *ford*.

The place-names in the immediate neighbourhood of Bradford are to a large extent Anglian. No other part of Yorkshire has so many names like Bowling, Cowling, Cottingley, Cullingworth, Drighlington, Frizinghall, Girlington, Manningham, and Stanningley, names which contain a patronymic. Other Anglian examples are Allerton, Bolton, Clayton, Heaton, Horton, Thornton; Bierley, Calverley, Dudley Hill, Farsley, Rodley, Shipley; Pudsey, Wibsey; Birkenshaw, Buttershaw, Oakenshaw; Hunsworth, Shuttleworth; and, in addition, there are such names as Chellow, Denholme, Eccleshill, Lidget Green, Norwood Green, Owlcotes, Ryecroft, and Strangford.

As evidence of Scandinavian influence the following names may be mentioned: Leventhorpe, Priestthorpe, the two Gaisbys, Scholemoor, Slack and Toftshaw.

BRAITHWAITE, BRAITHWELL.—The first element in these names goes back to ON *breiðr*, broad; compare OE *brād*, broad.

BRAITHWAITE, Doncaster, HR 1276 *Braytweyt*, PM 1328 *Braithwaite*, is 'the broad clearing,' from ON *thveit*, a clearing.

BRAITHWAITE, Keighley, has doubtless the same origin and meaning.

BRAITHWELL, Doncaster, DB *Bradeuuelle*, CR 1232 *Braith-well*, HR 1276 *Braytewell*, YS 1297 *Braythewell*, NV 1316 *Braythewell*, SM 1610 *Brawell*, shows conflict between the first and all the remaining forms. DB *Bradeuuelle* is Anglian; it goes back to OE *brād*, broad, and *well*, a spring, and the modern name derived from it would have been Bradwell. The remaining forms are all Scandinavian, and I take it their meaning is the same.

BRAMLEY, BRAMPTON.—South-west Yorkshire has three Bramleys and three Bramptons:

BRAMLEY, Leeds, DB *Bramelei*, KC 1198 *Bramleia*, PC †1220 *Bramleia*, KI 1285 *Brameley*, NV 1316 *Bramlay*.

BRAMLEY, Rotherham, DB *Bramelei*, CR 1232 *Bramley*, NY 1316 *Bramlay*, IN 1324 *Bramley*.

BRAMLEY HALL, Handsworth.

BRAMPTON, Doncaster, DB *Brantone*, HR 1276 *Brampton*, YI 1280 *Brampton*, KC †1325 *Brampton*.

BRAMPTON, Morthen, DB *Brantone*, KI 1285 *Brampton*, NV 1316 *Brampton in Morthyng*.

BRAMPTON, Wath, DB *Brantone*, KI 1285 *Brampton juxta Wath*, IN 1307 *Brantonbirlagh*, CH 1483 *Brampton Birlagh*.

The Bramleys are probably derived from an early word *brame* which means a briar or bramble; compare OE *bræmel*, a bramble, and the dialect-word *brame* recorded in EDD. But it is possible they spring from a personal name Brami recorded by Nielson, whence the Scandinavian place-name Bramthorp.

The Bramptons, on the other hand, are perhaps derived from OE or ON *brant*, steep, high. This, however, could not apply to Brampton near Doncaster, and it should be noted that the Norman scribes often wrote *n* for *m*; hence *Brantone* may represent *Bramtone*; in which case the meaning would be 'bramble enclosure,' from OE *tūn*, an enclosure or farm.

BREARLEY, Sowerby Bridge.—See Brierley.

BRECK, BRINK.—The ON **brenka*, gave the Dan. *brink* and by consonantal assimilation the Icel. *brekka*, a slope, a hill. From the Danish word came our English word *brink*, a margin.

Instances of Brink as a place-name are quite numerous to the west of Halifax; it occurs, for example, in Cragg Vale, Langfield, Mytholmroyd, Midgley, and Sowerby. Further south it appears in a slightly altered form in Micklebring and perhaps Oxspring.

There are also several examples of the word Breck, but it is doubtful whether they represent Icel. *brekka*, a slope, or a dialect-word *breck* recorded in EDD and explained as a piece of un-enclosed arable land. The name New Break, Marsden, appears to represent the latter, and with this must be compared the phrase 'an essart called *Newebrekk*,' YD 1348, connected with Dalton, Rotherham, where we find to-day Brecks Hill.

BRETTON.—South-west Yorkshire has two places of this name, Monk Bretton near Barnsley and West Bretton near Wakefield. Early records are as follows:

WEST BRETTON.		MONK BRETTON.	
DB	1086 *Bretone, Brettone*	DB	1086 *Brettone, Bretone*
PF	1202 *Bretton*	YR	1233 *Britton*
YS	1297 *Bretton*	NV	1316 *Bretton*
WCR	1308 *Westbretton*	PT	1379 *Monckebretton*

In OE there was a personal name *Brytt*, a Briton, of which the gen. sing. was *Bryttes* and the gen. pl. *Brytta*. Though *Brettone* could not come from the former, it might be derived from the latter; that is, from *Bryttatūn*. More probably, however, it is formed on the pattern of *Bretland*, the land of the Britons. This would give the meaning 'the farm of the Britons,' from *tūn*, an enclosure or farm. Monk Bretton received its distinctive affix because of the monastic establishment—sometimes called Lund Priory—which formerly existed there.

BRIANSCHOLES, BRYAN LANE, BRYAN CLOSE.
—For Brianscholes, Northowram, WCR has 1337 *Brynscoles*, 1338 *Brynscoles*, and 1403 *Brynescholes*. Here the ending is from ON *skáli*, a hut, while the first element is probably from ON *brún*, the brow or projecting edge of a cliff or hill; compare Dan. *bryn*, the brow of a hill. Bryan in Bryan Lane, Fixby, and Bryan Close, Marsden, has probably the same origin.

BRIERLEY, BRIESTFIELD, BRIESTWISTLE.—
The first element in these words is from *brāra*, the gen. pl. of
OE *brār, brĕr*, a briar, bramble, witness the following early forms:

DB 1086 *Breselai, Breselie*	CH 1259 *Breretwisell*
YI 1255 *Brereley*	YD 1292 *Breretusil*
NV 1316 *Brerelay*	DN †1312 *Brertwisell*
PT 1379 *Brerelay*	KC 1348 *Breretwisell*

BRIERLEY, Barnsley, where the Domesday forms appear to
be inaccurate, is ' briar lea,' from OE *lēah*, a lea or meadow.

BRIESTWISTLE, Thornhill, derives its termination from OE
twisla, a fork, confluence, land between two uniting streams.
The change in the prefix—compare YD 1418 *Brestewesyll*—is
perhaps due to assimilation.

BRIESTFIELD, Thornhill, though less ancient than Briestwistle, is gradually taking its place. It seems to have been
formed under the impression that the proper division of its
predecessor was Briest-wistle, doubtless through the influence
of two well-known words.

Near Wath-on-Dearne there was formerly a place called
Breretwisel in YD 1253, and *Breretwysel* in YI 1323.

BRIGHOUSE, which lies between Huddersfield and
Halifax, is placed at the exact point where an ancient road—
possibly a Roman road—crossed the river Calder. Early
references are WCR 1275 *Brighuses* and *Briggehuses*, WCR
1307 *Briggehouses*, WCR 1308 *Brighouses*, WCR 1334 *Brighus*,
WCR 1392 *Brighous*. The name is doubtless derived from
ON *bryggja*, a bridge, and *hūs*, a house.

In South-west Yorkshire the only bridge recorded by DB is
that near Wakefield called *Hagebrigge*, Agbrigg. It is clear that
at such a point the bridge could have had only the most modest
dimensions. But, as the early place-names show, a bridge
existed at Brighouse as early as the year 1275, and we may
fairly assume that so great an engineering effort was caused by
traffic of considerable amount. Still earlier, in 1199, apparently
in connection with Elland and its bridge, we find the name
Brigrode in a charter of Fountains Abbey, and in the same
document the name *Ferybrigge* occurs, witnessing to bridges
over the Aire and Calder at the end of the 12th century.

BRIGHTHOLMLEE, Sheffield, DN 1337 *Brightomlegh*, 1342 *Brightonlegh*, may be 'Brihthelm's lea,' or to give the OE form of the name 'Beorhthelm's lea.' If so, the sign of the genitive has been lost as in such cases as Alverley and Alverthorpe.

BRIGHTSIDE, *BRIKSARD, Sheffield.—The former is found after the 15th century, the latter chiefly before that date; but it is certain that Brightside is not derived from Briksard, witness YF 1573 '*Brekesherth, Sheffield*, and *Bryghtsyde*.' Among other records we find the following:

YD	— *Brykesherith*	YF	1573 *Bryghtsyde*
YS	1297 *Brikeserd*	YF	1577 *Brightside*
HH	1383 *Brikserth*	YF	1595 *Brightsyde*
YF	1520 *Briksard*	CH	1638 *Brightside*

The termination in the earlier name is curiously fickle, and takes the forms *erth, arth, erd, ard, herth*, and *herith*. Cornard in Suffolk has forms in -*erth* followed by others in -*erd* and -*ard*, and Dr Skeat says 'the forms with *th* must be the more original.' If we regard *Brikeserth* as the correct ME form afterwards represented by *Briksard*, we shall be able to explain the name as 'Brik's plough-land' or Brik's land,' from OE *earð*, ploughland, or OE *eorðe*, ground, land. The forms in -*herth*, -*herith*, appear to show the influence of OE *heorð*, hearth, home.

BRIGHTSIDE probably bears the obvious meaning, from OE *beorht*, bright, and OE *sīde*, a side.

BRINCLIFFE, Sheffield.—CR 1251 has *Brenteclive*, HR 1276 *Branteclive*, YF 1574 *Brynclyff*, and a Will dated 1653 *Brendcliffe*. The first element has oscillated between *brente* and *brynk*, but the present day name is undoubtedly derived from the latter, and the meaning is 'slope cliff' from ON *brenka*; compare Dan. *brink*, Icel. *brekka*, a slope.

BRINK.—See Breck.

BRINSWORTH, Sheffield, must be compared with Brinsford, Staffordshire, which is situated four miles north of

Wolverhampton. Early spellings—those of the Yorkshire name first—are as follows:

DB	1086 *Brinesford*	994	*Brunsford*
PF	1202 *Brinesford*	994	*Brenesford*
KI	1285 *Brinford*	1227	*Bruneford*
YS	1297 *Brinisford*	1300	*Brunesford*
NV	1316 *Brineford*	1381	*Bruynesford*

Both names show a mutated form of the OE personal name Brun, which corresponds to the modern surname Brown, and both show oscillation between the strong and weak declension. These names have a special interest in that they must be considered whenever the site of the battle of Brunanburh, 937, is being discussed. It will be remembered that one of the names given to the battle is that of *Bruneford*.

BRISCOE LANE, Greetland.—The terminal in Briscoe is doubtless from ON *skōgr*, a wood, and the prefix is probably from Dan. *birk*, ON *birki-*, the birch-tree. In 1277 Briscoe near Guisbro' was *Birkescov*, the birch wood.

BROADSTONE HILL, Saddleworth, is mentioned in the 13th century deeds as *Brodeston* and *Bradeston*, spellings which fully justify the present name.

BROCKHOLES, Honley and Mixenden.—The former is recorded in WCR as *Brocheles* in 1275, and *Brokholes* in 1277 and 1284. The first element comes from *broc*, a badger, a word found in OE as *broc*, in Danish as *brok*, in Welsh and Cornish as *broch*, and in Irish and Gaelic as *broc*. The termination is from OE *holh*, a hole or hollow.

BRODSWORTH, Doncaster, DB *Brodesuurde*, *Brodesuuorde*, PC 1240 *Broddeswrd*, KI 1285 and PT 1379 *Brodesworth*, is plainly 'the farmstead of Brod,' from OE *weorð*, a farmstead or holding; compare the DB personal name Brode.

BROMLEY, BROOMFIELD, BROOMHEAD. — The OE *brōm*, ME *brom*, *broom*, means broom, brushwood; and Bromley, which occurs in Cumberworth and Wortley (Sheffield), means 'broom lea,' from OE *lēah*, a lea; compare BCS *Brōm-lēah*.

BROOMFIELD, Stocksbridge, and BROOMHEAD, near Bolster-stone, require no explanation.

BROOKFOOT, BROOKHOUSES, Brighouse and Laughton, have their prefix from OE *brōc*, a small stream. The latter is recorded in YS 1297 as *Brokhouses*.

BROTHERTON, Wadsworth, is 'the farmstead of Brother, from OE or ON *tūn*, a farmstead, and the personal name Brother recorded by Searle; compare ODan. Brothær, and note also WHS †1030 *Broðertun*, for Brotherton near Pontefract.

BROW.—This word occurs in such examples as Birkby Brow, Hopton Brow, and Berry Brow, and is derived from OE *brū*, a brow, the edge of a hill.

BRUNTCLIFFE, Morley, 1639 *Bruntcliffe*, may perhaps be derived from ON *brunnr*, a spring or well, and *klif*, a cliff.

BULLA TREE HILL, BULLY BUSH, BULLY TREES, occur respectively near Roche Abbey, in Thrybergh, and in Liversedge. EDD says Bully is a West Riding form of the word 'bullace,' and the association of the word with Tree and Bush gives support to such an explanation. The Hartshead Parish Registers have *Bulitrees* in 1783 for the Liversedge name.

BURFITTS LANE, Quarmby.—The termination -fitts is doubtless from ON *thveit*, a paddock, and the first element is probably ON *būr*, a storehouse or chamber. See Thwaite and Linthwaite.

BURGHWALLIS, Doncaster, DB *Burg*, NV 1316 *Burgh*, is frequently recorded in the 13th and 14th centuries as *Burgh-walais* and *Burghwaleis*. 'Stephen le Waleys'—that is, Stephen the foreigner—of *Burghwaleis* was living in 1294. See Borough.

-BURY.—See Borough.

BUSK, BUSKER.—These names call attention to an interesting difference between words of Scandinavian origin and words purely English, for 'busk' is the Scandinavian equivalent of our common word 'bush'; compare Dan. *busk*, Sw. *buske*, a bush, with OE *bysc, ME *bush*.

From a common ancestor, Teutonic *sk*, ON got *sk* and OE got *sc*; but at a time prior to the Viking settlements in England the OE *sc* was softened to *sh*, while the Scandinavian *sk* retained its original pronunciation. In consequence our language has several interesting doublets; among ordinary names, for example, we find 'scrub' and 'shrub,' and among place-name elements 'ask' and 'ash,' 'busk' and 'bush,' 'marsk' and 'marsh,' 'skelf' and 'shelf.'

It follows from this that we may with confidence claim those Yorkshire place-names which contain the sound *sk* as Scandinavian: the Scars and Scouts, the many examples of Schole or Scholes, all names ending in -scoe, and such further examples as Skeldergate, Skircoat, Skelmanthorpe and Skinner-thorpe. It would not, however, be equally safe to claim all place-names in *sh* as English, for there are occasional examples where a word of undoubted Scandinavian origin has not retained the sound *sk*.

BUSK occurs in Kirkburton and Hunshelf, in the latter case in the form Briery Busk.

BUSKER, Skelmanthorpe, is most probably the Scandinavian plural, and thus means simply 'bushes.'

BUTTERBUSK, BUTTERLEY, BUTTERNAB, BUTTERSHAW, BUTTERTHWAITE, BUTTERWORTH.—Names commencing with Butter- occur as far north as Perth and as far south as Devon, while a few are found in Ireland. The well-known Buttermere occurs in Wiltshire as well as in Cumberland; there are Butterwicks in Westmorland, Durham, Yorkshire, and Lincoln; and there are strange-looking examples like Buttergask (Perth), Butter-crambe (York), and Butterbump (Lincoln).

It is clear that the words come from various sources, some being English, others perhaps Celtic, and others certainly Scandinavian. There can be no doubt, for example, that Butterby (Durham) is Scandinavian, and so also are Butterbusk,

Butternab, Buttercrambe (DB *Bute'crame*), Butterthwaite, and Butterwick (DB *Butruid*), where the terminals are connected with Dan. *busk*, a bush, ON *nabbi*, a knoll, Norw. *krampe*, a thicket (Aasen), ON *thveit*, a clearing, and ON *viðr*, a wood. Buttergask, †1200 *Buthyrgasc*, is said to be derived from Gael. *bothar*, a road, and *gasc*, a hollow (Johnston). But the Butterleys (Derby, Hereford, York, and Devon) and the Buttertons (Stafford and Devon) seem clearly English.

In the case of the names with English terminations it will probably be right to derive Butter- from OE *butere*, butter; while in those with Scandinavian terminations there are two alternatives, (1) the Scandinavian personal name Buthar, DB Buter, and (2) the plural of the ON *būtr*, a log, tree-trunk, stump of a tree, Norw. *butt*. In each case the present form would be due to the influence of the common word 'butter'; derivations from *būtr* are, however, much more likely.

BUTTERBUSK, Warmsworth, PT 1379 *Buttirbuske*, derives its termination from the Norw. and Dan. *busk*, a bush, shrub.

BUTTERLEY, Fulstone, WCR 1274 *Buttreley* and 1307 *Butterley*, is probably 'butter lea,' from OE *butere*, and *lēah*; but compare Buttershaw.

BUTTERNAB, Lepton and Crossland, probably means 'the knoll covered with tree-stumps,' from ON *būtr* and *nabbr*.

BUTTERSHAW, Liversedge and Bradford, has an English termination, from OE *sceaga*, a copse, while its first element is most probably Scandinavian, from ON *būtr*.

BUTTERTHWAITE, Ecclesfield, YS 1297 *Butterwayt*, YD 1302 *Buttertwayt*, is often pronounced Butterfitt. It is either 'Buter's clearing' or 'the clearing among the tree-stumps.'

BUTTERWORTH, Norland, WCR 1297 *Buttrewrth*, is probably from OE *butere*, butter, and *weorth*, a holding or farmstead.

BY.—This is the best of all tests for Viking settlements. It is connected with both Norsemen and Danes, though of more frequent occurrence in districts settled by the latter. According to Flom it is to be found 600 or 700 times in Skane and Denmark, and 450 times in Norway, while according to

Jellinghaus Schleswig has about 80 examples. Among the Norwegian names given by Rygh we find Kirkeby, Lundby, Vestby, Dalby, Sörby, names which correspond exactly to well-known Yorkshire examples.

In South-west Yorkshire the earliest form of the word is -*bi*, which occurs quite regularly in DB ; compare also *Barnabi*, which occurs in 1023 in the list of Archbishop Ælfric's festermen. In East Anglia the name assumes other early forms. Alongside forms in -*by* like *Kerkeby, Malteby, Ormesby, Scrouteby* in Norfolk, and *Barneby* in Suffolk, the Domesday record gives forms in -*bei* and -*bey*, like *Clepesbei, Colebei, Essebei, Haringebei, Kerkebei, Ormesbei, Othebei, Stokesbei, Wilebei, Filebey, Maltebey, Ormesbey, Stokesbey*; and one example in -*be*, namely, *Clepesbe*. These words, -*bi*, -*be*, -*bei*, go back to a Teutonic stem *būvi, *bōvi, whence ON *bȳr, bær*, Norw. *by, bö*, Dan. *by*, Sw. *by* (Torp.).

The modern examples in South-west Yorkshire number thirty. Denby occurs four times ; Kirkby and Birkby thrice ; Fixby, Gaisby, and Scawsby, twice ; and the list is completed by the names Balby, Cadeby, Denaby, Firsby, Fockerby, Foulby, Haldenby, Hellaby, Maltby, Quarmby, Ringby, Rusby, Sowerby, Wragby. It is noteworthy that more than half are names of farms or hamlets, not of townships. It is also noteworthy that the word did not become a living element in the language ; there was apparently no creation of names in -by after the Norman Conquest.

CADEBY, Doncaster, DB *Catebi*, PF 1202 *Cathebi*, YI 1277 *Cateby*, NV 1316 *Cateby*, PT 1379 *Cateby*, has for its first element the ODan. personal name Kati, while the ending is from ON *bȳr*, a farm or village ; compare Dan. Kattorp formerly *Katathorp*, and see Shibden.

CALDER, COLNE.—The former rises on the Lancashire border, passes Todmorden and Wakefield, and joins the Aire at Castleford ; the latter, after passing Huddersfield, joins the Calder at Colnebridge. It should be noted that in Scotland there are six Calders, and in the North of England four.

Early spellings of our Yorkshire Calder include PF 1202

Kelder, LC 1296 *Keldre*, WCR 1308 *Calder*; and early spellings of the Colne include BM *Kalne* and *Kalnebotmes*. It seems probable that the two rivers contain the same element, Kal or Cal, and not improbable that early records of the Calder have been influenced by ON *keld*, a spring.

1. What is this element Kal?

Among German rivers Lohmeyer records the Kalle, Kahl Kallbach, Kellwasser, Kallenbach, Kallenborn, and Kalbe. He goes further and connects the root with ON *kalla*, OE *ceallian*, to call, and OHG *challon*, MHG *kallen*, to babble.

Turning to Holder we find enumerated as Celtic river-names the Callus, now la Chée, and two streams named Calla, now the Call and Callbach.

In connection with the Scottish Calders McClure quotes an early form *Caledofre*, and Johnston has *Caldovere*. It has been usual to refer the prefix to the Ir. *caill*, Gael. *coill*, a wood, which represent an older stem **kaldet-*, a wood. Note, however, that according to Hogan there are several Irish rivers which were formerly called *Callann*.

2. What are the endings -der and -ne?

Although the Yorkshire Calder appears in its early forms much crushed up, it is probably the same word as the Scottish Calders. In that case the terminal comes from the Celtic stem *dubro-* water, a root which has given Welsh *dwfr* and *dwr*, Corn. *dofer*, Ir. and Gael. *dobhar*. Corroboration is provided by records of the Lancashire Conder given in the Cockersand Chartulary: †1220 *Kondover*, †1250 *Kondoure*.

For the ending in Kalne compare the termination -*ana* -*ona* recorded by Holder in such names as *Isana, Lohana, Axona, Matrona*, early spellings of the Isen, Lahn, Aisne, and Marne.

CALLIS occurs in Callis Wood, Hebden Bridge, and Callis Lane, Penistone. A reference to the former appears in WCR 1375 where a certain 'Adam de Calys' is mentioned, and HW 1551 speaks of 'my playces called Calys.'

CALVERLEY, Leeds, DB *Caverleia, Cauerlei*, PF 1203 *Couerlee*, PC †1220 *Calverleia*, KI 1285 *Calverlay*, KC 1332 *Caluerlay*, PT 1379 *Caluerlay*, may have a personal name for the first element, though no suitable form is on record. More probably the meaning is 'the lea of the calves,' from OE *calf,* gen. pl. *calfra*; compare the examples *Cealfra·mere* and *Calfre· croft* recorded in BCS.

CAMPSALL, Doncaster, DB *Cansale*, PF 1208 *Camsal*, PC †1210 *Cameshale*, KC 1218 *Camsale*, NV 1316 *Camesall*, PT 1379 *Campsale.* The *p* is intrusive, as in the Bramptons, and the *n* in the DB spelling is probably an error of the Norman scribe. The first element is doubtless a personal name, but, though there is a modern surname Camm, I can find no such OE name. The second element comes from OE *healh*, a corner.

CANKLOW, Rotherham, PF 1202 *Kankelawe*, must be compared with the Frisian place-name Kankeber (Sundermann), and the Frisian personal name Kanke (Brons). Hence we may interpret Canklow as 'the burial-place of Canka or Kanki,' from OE *hlāw, hlǣw*, or from Prim. Norse *hlaiv*, a burial-mound, cairn, influenced by the OE word. Note, however, that a hamlet in Worcestershire is called Cank, which can scarcely be a personal name. Further search discovers a Norwegian word *kank*, a knot, clump (Aasen); compare ON *kökkr*, a clump.

CANTLEY, Doncaster, DB *Canteleia, Cantelie*, KC 1209 *Canteleia*, PF 1210 *Kantelai*, YR 1272 *Canteley*, is 'the lea of Canta,' a name given by Searle; compare BCS *Cantanleah.*

CARLECOTES, CARLETON.—From the OE *ceorl*, a peasant, we get the various Charltons found in the southern counties, as well as the common surname Charlesworth, which was originally a place-name. But Carlecotes and Carletons involve the corresponding Northern form, and are derived either from ON *karla*, gen. pl. of ON *karl*, a man, freeman, or from ON *Karla*, gen. sing. of the ON personal name Karli (Naumann).

CARLECOTES, Penistone, WCR 1277 and 1286 *Carlecotes*, signifies 'the cottages of the freemen,' from ON *kot*, a cottage.

CARLTON, Rothwell, DB *Carlentone*, CR 1251 *Carleton*, YI 1258 *Carleton*, is 'the homestead of the freemen,' from ON *tūn*, an enclosure or homestead.

CARLTON, Barnsley, DB *Carlentone*, YR 1233 *Carlton*, NV 1316 *Carleton*, PT 1379 *Carleton*, and
CARLETON, Pontefract, YI 1256 *Carleton*, have the same origin and meaning.

CARLINGHOW, Batley, is recorded in KF 1303 as *Kerlynghowe*, and in PT 1379 as *Kerlynghawe*. There is here no sign of the genitive; compare *kerlingar*, gen. sing., and *kerlinga*, gen. pl., of ON *kerling*, a woman, old woman. The meaning may possibly be 'the burial-mound of the women'; but more probably it is 'the burial-mound of Kerling,' from ON *haugr*, a howe or cairn. Compare Kellington and Thurstanland.

CARR, CARBROOK, CARCROFT, CARR HOUSE.— The word Carr is in frequent use as a field-name, especially to designate lowlying land beside a stream. It is derived from ON *kjarr*, copsewood, brushwood; compare Dan. *kær*, a bog, fen. There is a considerable number of compounds in which the word appears as a suffix, the form being either -kar or -ker; among them are Blacker, Bullcar, Cobcar, Deepcar, Durker, Elsecar, and Moscar. Other examples of the use of the word are Batley Carr, Birley Carr, and Carr in Saddleworth.

CARR HOUSE, Maltby, appears in a Fine of 1435 as *Carhouses*.

CARBROOK, Sheffield, HH 1383 *Kerbroke*, YF 1520 *Carbroke*, combines OE *brōc*, a brook, with ON *kjarr*.

CARCROFT, Doncaster, PC †1170 *Kerecroft*, PF 1204 *Kerecroft*, DN 1342 *Kercroft*, YD 1348 *Kercroft*, may have a personal name for its first element.

CARTWORTH, CORTWORTH.—The former is a township running into the heart of Holmfirth, and the latter a hamlet in Brampton Bierlow. Omitting the DB forms of Cartworth, we find such spellings of the two names as

WCR	1274 *Cartewrth*	YD	1486 *Corteworth*
WCR	1307 *Cartewrth*	YF	1515 *Cortworth*
PT	1379 *Cartworth*		

Seeing that the personal name Kort exists in Frisian (Brons) and in Norwegian (Rygh), Cortworth may possibly be 'the holding of Korti,' from OE *worth, weorth,* a homestead, farm, holding. Perhaps Cartworth has a similar explanation, though no suitable name is recorded by Searle, Rygh, Naumann, or Nielsen.

But the DB forms of Cartworth, *Cheteruuorde* and *Cheteruurde,* give pause; they remind us of early forms of Catterick, which appears in the Antonine Itinerary as *Cataractone,* in Bede as *Cetreht, Cetreht·tune, Cetreht·weorthige* (Miller), and in CR 1241 as *Cheteriz.* It seems possible, then, that the first element in Cartworth may be equivalent to *Catar-* in *Cataractone,* which according to Williams is an extension of the Prim. Celt. **cat-,* a battle. On the other hand the DB forms may be faulty, and represent *Cherteuuorde* and *Cherteuurde,* in which case the interpretation suggested above would hold good. But see Catcliffe.

CASTLE.—South-west Yorkshire has five names containing this word : Castleford, Castleshaw, Hardcastle, Horncastle, and Ladcastle. Borrowed from Lat. *castellum,* it took the form *castel* in both OE and ME, and signified a village or hamlet as well as a fortress. In the Third Gospel the different versions present at one point a very interesting comparison. Speaking of the two disciples going to Emmaus (xxiv. 13), the Authorized Version (1611) says they went 'to a village'; but Tyndale's Bible (1526) says 'to a toune'; Wycliffe's translation (1389) gives 'to a castel'; and the Anglo-Saxon Version (995) gives 'on thæt castel.' Possibly in some instances the name Castle Hill commemorates an ancient village rather than a castle.

CASTLEFORD stands at the point where the great Roman road called Erming Street crossed the Aire. It is the *Legeolium* or *Lagecium* of the Romans, and has therefore been a post of importance for well-nigh twenty centuries.

A place called *Ceasterforda* is mentioned in the AS Chron. under the date 948. The full passage, which relates to the struggle for supremacy between the Vikings of Northumbria and the English Kings, reads as follows : 'In this year Eadred king harried all the land of the Northumbrians because they had taken Yric[1] as their king. And then, during the pillage, was the great minster which Saint Wilferth built at Rypon consumed by fire. And when the king was on his way homeward, the army of the Danes from within York attacked the king's army from behind at Ceasterforda, and made great slaughter. Then was the king so enraged that he would have marched his forces in again and the land with all destroyed. When the Witan of the Northumbrians understood this, then forsook they Hyryc, and made with king Eadred reparation for the deed.'

The name *Ceasterforda* has sometimes been explained as referring to Chesterfield; but Oman and McClure agree in identifying it with Castleford, and early records of the name leave no room for doubt, witness the following:

PC	†1220 *Castelforda*	WCR 1274 *Castelford*	
CR	1230 *Castreford*	WCR 1285 *Castilforth*	
PC	1235 *Castleforda*	WCR 1297 *Castelford*	
PM	1258 *Kasterforde*	WCR 1307 *Castilford*	

Thus, the element 'caster,' which comes from Lat. *castrum*, has been displaced by 'castle,' which is derived from Lat. *castellum* through OFr. *castel*[2]; and Castleford may therefore be explained as 'fortress-ford.'

LEGEOLIUM appears to have for its first element a Celtic river-name of the form *Lege*. The Flemish river Lys, a tributary of the Scheldt, is recorded by Holder under such early forms as *Lege*, *Legia*, and *Leie*, *Leia*; and among possible examples in our own country we may mention the Hertfordshire Lea, formerly *Lyge* (Skeat), and the Argyllshire stream-names Dubh-lighe and Fionn-lighe, the black and white rivers (Gillies). According to Holder the Celtic word is probably cognate with OHG *lahhan*, a cloth or sheet.

But there is a further point of much interest. Just as the Roman name Isurium was lost, and displaced by the Anglian name Aldborough, so the Roman name Legeolium disappeared,

---

[1] Eric Blood-Axe, an elder son of Harold Fairhair.

[2] The word *castel* reached us from the North of France, the corresponding form in use in the more central parts of France being OFr. *chastel*, from which comes the Fr. *château*.

and its place was taken first by Casterford and afterwards by Castleford. In each case the Roman fortress still remained, a mysterious relic of the past, and the new names, 'Old fortress' and 'Fortress-ford,' bore witness to the fact. But in neither case does the Anglian name contain any hint of its predecessor, *Isurium* or *Legeolium*, and we are left to infer that for a season each place lay desolate.

In the case of Doncaster, as at Aldborough and Castleford, the fortress remained down to Anglian times, witness the second element in the name; but the story of Doncaster is in another respect quite different from those of Aldborough and Castleford, for a remnant of the old population seems to have lived on, and so the Romano-British name *Danum* was preserved as the first element in the Anglian *Danecastre*, and is still maintained in the modern Doncaster.

CASTLESHAW, Saddleworth, stands on the Roman road which led from *Mancunium* (Manchester) to *Cambodunum* (Slack), and it possesses the remains of a Roman camp. The present name, spelt *Castylshaw* in 1544, means 'the copse beside the fortress,' from OE *castel*, a fortress or village, and OE *sceaga*, a copse or wood, ME *schagh, schawe*.

CATCLIFFE, CATBEESTON, CAT CLOUGH, CAT HILL, CAT MOSS, CATSHAW, CAT STONES, CHAT HILL, CHATTS WOOD.—We find Cat Clough in Hepworth and Stocksbridge; Cat Hill in Hoylandswaine; Cat Moss in Rishworth; Catshaw in Liversedge and Thurlstone; Cat Stones near Bingley; Chat Hill in Thornton; and Chatts Wood in Hunsworth. The available spellings are not very early, and prove of little assistance; they include

CATCLIFFE, Rotherham, PM 1255 *Catteclif*, HH 1366 *Catcliff*;
CATBEESTON, Beeston, *Cattebeston, Catebeston, Cadebeston*;
CATSHAW, Liversedge, *Catcheye*.

Further, WRM 1391 gives the name *Catekeldre*, where Keldre is plainly the river Calder, and PF 1209 has *Cadtheweit* for some place in the vicinity of Morley and Beeston.

Catcliffe is perhaps 'the hill of Kati,' from ON *klif*, and the

ON personal name Kati; but this and other names may perhaps be connected with the wild-cat, OE *cat, catt*, ME *cat*. Neither of these explanations would prove satisfactory, however, in the case of the first element in WRM 1391 *Catekeldre*, where *Keldre* is obviously the river Calder. Is any other explanation possible?

There is an important Brythonic word cognate with Engl. 'heath' which occurs in certain Gaulish and British place-names in the form *cēto-*; compare Cetobriga, Letocetum, and Utocetum. In Welsh this word appears as *coed*, a wood, which corresponds to Corn. *cuit*, Bret. *coit, coat*; and Stokes gives the primitive form *keiton, a wood, forest, heath. Connected with these there appears to be quite a considerable body of English and Scotch place-names. Thus, in Scotland, it seems probable that the following names involve the word:

Keith	1169 *Keth*
Kincaid	1238 *Kincaith*, 1250 *Kyncathe*
Pencaitland	1145 *Pencet-*, 1150 *Pencat-*
Dalkeith	1140 *Dalkied*, 1145 *Dalketh*

And in England the following:

Chatcull, Staffs.	†1200 *Chatkull*
Culcheth, Lancs.	1201 *Culchet*, 1311 *Culcheth*
Penketh, Lancs.	1292 *Penketh*, 1296 *Penket*
Penge, Kent	1067 *Penceat*
Lichfield, Staffs.	†200 *Letocetum*, later *Liccedfeld*
Kesteven, Lincs.	1086 *Chetsteven*, 1170 *Chetsteuene*

Particularly interesting are the early records of Chatteris in the county of Cambridge:

Ramsey Chartulary	*Ceatrice, Chateric, Chaterik*
Domesday Book	*Cetriz, Cietriz*
Inquisitions	*Cetriz, Chetriz, Cateriz*

And equally interesting are references in the Charter Rolls: 1248 *Forest of Chett*, 1270 *In bosco de Cett*, 1290 *Chetwod*, where we find the duplication of meaning so common when a name from one tongue is adopted by another.

If we seek to interpret the names given above as Celtic words we find that Penketh and Penge may be explained as 'head of the wood,' from Welsh *penn*, head; that Kincaid has the same meaning, from Gael. *ceann*, head; that Chatkull and Culcheth

G.   7

appear to mean 'back of the wood,' compare Welsh *cil*, Ir. and Gael. *cul*; while Dalkeith may be explained as 'wood-place,' from Gael. *dal*.

It seems clear that the great variety of forms—*cat, cet, chat, chet, caith, keith*—with which we have just been met may fairly be linked with Prim. Celt. **keiton*, Gaul. and Brit. *cēto-*, and not improbable that some of the names enumerated at the head of this note may also be derived therefrom.

It remains only to add that the material used above is drawn chiefly from McClure's *British Names in their Historical Setting*, and the books by Skeat, Duignan, Wyld, and Johnston on the place-names of Cambridge, Stafford, Lancashire, and Scotland.

CATHERINE.—This name is found in Catherine Slack which occurs thrice near Halifax, viz. in Cragg Vale, near Queensbury, and near Brighouse; it also occurs in Catherine House, Midgley (Halifax). Similar names are found elsewhere, and among Yorkshire examples we find Catherine House in Bransdale and Catherine Closes in Gowthorpe. I am unable to give early forms, and can only suggest comparison with such Scottish names as Loch Katrine, the town Catrine in Ayr, a mansion called Catter in Dumbarton, Catterline in Kincardine, Catterlen and Blencathara in Cumberland, and Catterick in North Yorkshire, the ancient *Cataractonium*. See Catcliffe.

CATTERSTON, CATTERSTORTH.—The first element in these words is probably the Danish personal name *Katær recorded by Nielsen.

CATTERSTON, Almondbury, RE 1634 *Catterston*, is 'Katær's farm,' from ON *tūn*, an enclosure, farm.

CATTERSTORTH, Stannington, HS 1637 *Catterstorth*, is 'Katær's wood,' from ON *storth*, a young plantation or wood.

CAWTHORNE, Barnsley.—Early forms, followed by those of Cawthorne near Pickering and Cawton in Ryedale, are as follows:

DB	1086	*Caltorne*	DB	1086	*Caltorne*	DB	1086	*Calvetone*
PC	†1160	*Calthorna*	PF	1202	*Kaldthorn*	KI	1285	*Calveton*
CR	1230	*Calthorn*	KI	1285	*Calthorne*	NV	1316	*Calveton*
PT	1379	*Calthorne*	KF	1303	*Calthorn*	RC	1332	*Calvetona*

Cawton is plainly 'calf enclosure,' and the North Riding Caw-thorne appears to be 'cold thorn.' Our own Cawthorne is obviously not derived from OE *cealf*, a calf. Neither can it be connected with OE *calu*, callow, bare, for its early forms show no sign of the second syllable in *calu*; compare Callow Hill, Staffordshire, formerly *Caluhull* and *Kalewhull*. Far more likely is a derivation like that of the northern Cawthorne from OE *ceald*, ME *cald*, *cold*, the final consonant having disappeared at an early date before the succeeding dental ; compare Owston, Ulley, and Methley.

CHAPELTHORPE, CHAPELTOWN, WHITE-CHAPEL.—The word chapel has a curious origin. It comes from OFr. *chapele*, LL *capella*, a little cloak or cope (*capa* or *cappa*). NED tells us that the word is derived from the *capella* or cloak of St Martin which was preserved as a sacred relic by the Frankish Kings. This *capella* was borne before the kings in battle and was used to give sanctity to oaths. Later the name was transferred to the sanctuary in which the cloak was kept ; afterwards to any sanctuary containing holy relics ; and last of all to any oratory or lesser church.

CHAPELTHORPE, Wakefield, 1285 *Schapelthorpe*, DN 1447 *Chapelthorp*, the 'chapel village,' from ON *thorp*, received its name from a chapel-of-ease to the parish church of Sandal.

CHAPELTOWN, DN †1277 *Capella*, HH 1366 *Capell*, YF 1554 *Chappell*, is so-called from a chapel-of-ease to the parish church of Ecclesfield.

WHITECHAPEL, Cleckheaton, was called *Heton Chapel* in Saxton's Survey 1575. It was a chapel-of-ease in the ancient parish of Birstall.

CHARLESTOWN is the name given to a portion of the borough of Halifax high above North Bridge, and also to a district lying near the Calder west of Hebden Bridge. The ending -town suggests that the name is of late origin ; see Carlton.

CHAT HILL, CHATTS WOOD.—See Catbeeston.

CHEERBARROWS, Cleckheaton.—An early form which seems to correspond to this name is HR 1276 *Chirebarwe*; this relates, however, to a place near Barnsley. Other records involving an early form of Cheer or Chare are *Penchare* (1381) for Pencher in Durham, *Smallchare* (1600) in the Wath-on-Dearne Register, and *Offechere* in the Baslow Court Rolls. In EDD 'Chare' is explained as (1) a narrow lane or alley, (2) marshy land ; compare Norw. *kjerr*, a fen (Aasen).

CHELLOW, Bradford, DB *Celeslau*, SC 1252 *Chelleslawe*, YI 1288 *Cheleslawe*, PT 1379 *Chellowe* and *Chellow*. The first element comes from a well-known OE personal name Cēol, and the ending from OE *hlǣw*, *hlāw*, ME *lawe* or *lowe*, which means a mound, cairn, hill. In their edition of the Crawford Charters Professor Napier and Mr Stevenson tell us that *hlǣw* is almost invariably joined to a personal name, 'no doubt recording the person buried there.' We may fairly explain Chellow as 'the burial-mound of Cēol.'

CHEVET, WENT.—These words, in common with other Yorkshire names like Dent and Elmet, possess a termination of much interest. Among early records of Chevet, Went, and Dent, we find the following :

DB	1086 *Cevet, Ceuet*	DN	—	*Wenet*	CR	1251 *Deneth*
YI	1243 *Chevet*	DN	—	*Wenetes hill*	YD	—	*Denet*
WCR	1275 *Chyvet*	KC	—	*Wenet*	YS	1297 *Dent*
NV	1316 *Chevet*	WCR	1307	*Wentebrigge*	PT	1379 *Dent*

Elmet appears as *Elmet* in Bede, *Elmed* in BCS, *Elmete* in PF 1212; and Barnside near Penistone is recorded in WCR 1274 as *Barnedeside*. Further, there are several Lancashire names presenting a similar appearance, namely :

Thornley	1262 *Thornedelegh*,	1289 *Thornedeley*
Cuerdley	1331 *Keuerdelegh*,	1411 *Kiuerdeley*
Dinckley	1247 *Dunkythele*,	1369 *Dynkedelay*

It seems probable that an earlier form for *Barnedeside* was *Barnede*, for *Thornedelegh Thornede*, and that *Keuerdelegh* and *Dunkythele* sprang from earlier forms like *Keuerde* and *Dunkythe*. Thus we appear to be in the presence of a termination which takes the form *-ythe, -ethe, -ede, -et, -t*.

Gallée[1] and Jellinghaus[2] have long lists of names with a similar termination ; and a few typical examples may well be given.

Oelde	890	*Ulithi*	1277	*Olede*
Drumpt	850	*Thrumiti*	1200	*Drumthe*
Eschede	1046	*Ascete*	1212	*Eschethe*
Braamt	1241	*Brameth*	1250	*Bremet*
Eekt	1307	*Ekit*	1320	*Eket*

In the last three examples the stem is obviously a common tree-name : Eschede is from *esch*, an ash; Braamt from *brame*, a briar or bramble ; Eekt from *eke*, an oak. The termination, according to Gallée, comes from *thja* (Indo-Germ. *tio*), which apparently had a collective meaning.

But there is a similar termination of Celtic origin, witness the ancient names Reged and Guened. Among early river-names there are many instances :

Churnet, Staffs.	*Chirnete* in 1284
Teme, Worcs.	*Temede, Tamede,* in early charters
Kennet, Berks.	**Cunetio, Cyneta*

And still more interesting are certain place-names where an English termination has been added to the river-name :

Tenbury, Worcs.	*Tamedeberie, Tametdeberie,* in 1086
Kintbury, Berks.	*Cheneteberie* in 1086
Ribbesford, Worcs.	*Ribbedford* in 1023

Probably all these river-names are Celtic; but in any case the Berkshire Kennet is of that origin.

CHEVET, Wakefield, is probably Celtic ; compare Chevin, which must, I think, be connected with Welsh *cefn*, Gaulish *kebenna*, a ridge.

WENT, a stream which passes Wentbridge, has been linked with the site of the battle where Oswy defeated Penda in 655. In Bede the stream connected with this battle is recorded as the *Winwed,* and in the AS Chronicle the place is called *Winwidfelda.* As to the forms *Winwed, Winwid,* Mr Quiggin says they are undoubtedly Brythonic, and the second syllable must be compared with the last element in the stream-names *cinguid, annouid,* found

---

[1] NGN III, 362.　　　[2] *Westfälischen Ortsnamen,* pp. 26–29.

in the Book of Llandaff, and also with the modern Welsh stream-
name Ebwydd (generally written Ebbw). But in Geoffrey of
Monmouth the river *Winwed* appears as *Winned*, a form closely
approximating to *Wenet*, yet possibly not the same. I am
disposed to look upon Elmet as equivalent to Elmt in North
Brabant, 1179 *Elmeth*, from the Teut. stem *elma*, an elm
(Förstemann), and to explain Went similarly from the Teut.
stem **venjo*, pasture-land, OE *wyn(n)* pasture-land ; compare
Wendhagen, 1234 *Wenethage*, 1259 *Winethage* (Jellinghaus).

CHEVIN, CHEW.—These words are probably connected
with Welsh *cefn*, Gaulish *kebenna*, a ridge.

CHEVINEDGE, Southowram, is a ridge of land stretching
towards the Calder, and the name is obviously related to the
Celtic words given above.

CHEW HEAD, 1486 *Blackchew Head*, is a summit on the
borders of Saddleworth ; compare the Somerset names Chewton,
IL 1241 *Cheuton*, Chew Stoke, IL 1350 *Cheuestoke*, Chew, IL
1350 *Chyw*.

CHICKENLEY, Dewsbury, WCR 1277 and 1298 *Chykenley*,
DN 1461 *Chekingley*, is 'Chicken's meadow.' The name John
Chickin occurs in WCR 1309 in connection with Horbury.

CHIDSWELL, Batley.—This was originally 'Chid's hill,'
as is shown by spellings given in WCR : *Chydeshyll* 1275 and
Chideshill 1298. At a later date the name ceased to give any
sign of its connection with a hill; in YF 1550 it was spelt
Chydsell, and in 1577 *Chitsele*. The popular imagination forth-
with interpreted the word as 'Chid's well,' and the spelling
assumed its present form. The personal name comes from an
OE name Cidd recorded by Searle.

CHISLEY, Hebden Bridge.—WCR records the name in
1296 as *Chesewaldeley*, 1307 *Chesewelley*, 1308 *Cheswalleye*, 1309
Chesewalleye. The first element is probably from Norw. *kjesa*,
the dwarf-birch (Aasen); compare Cheer, as in Cheerbarrows,
from Norw. *kjerr*, a fen (Aasen). For the termination in
Chesewell, *Chesewall*, see Wall, Well, and note that *d* is intrusive
in *Chesewaldeley*.

CHOPPARDS, Holmfirth, appears to be derived simply from a personal name. There is a French surname Chopard, and WCR 1274 has *Robert Chobard*.

CHURWELL, Morley, DN 1226 *Cherlewall*, WCR 1296 *Chorelwell*, CC 1499 *Chorlwell*, CH 1616 *Churwell*, is 'the peasants' well,' from *ceorla*, gen. pl. of OE *ceorl*, a peasant, and *well*, a well. See Carleton.

CINDERHILLS, Sheffield, YS 1297 *Scynderhill*, YD 1306 *Sinderhilles*, derives its prefix from OE *sinder*, ME *sinder*.

CLAPGATE, CLAPPER HILL, CLAPPERS.—The first occurs in Sowerby and Rothwell, the second in Midgley (Halifax), the third in Thurgoland. Madsen gives Dan. *klöpp*, and explains it as a low, flat rock; compare ON *klöpp*, pl. *klappir*, which meant a pier-like rock projecting into the sea, or stepping-stones over a stream. Near Windermere there is a place called Clappersgate.

CLAYTON occurs thrice in South-west Yorkshire and is derived from OE *clæg*, clay, and *tūn*, an enclosure or homestead. In DB CLAYTON near Bradford was *Claitone*, and CLAYTON WEST near Wakefield was *Claitone* and *Clactone*. CLAYTON near Hooton Pagnell was *Clayton* in an inquisition of 1264.

CLECKHEATON, Bradford.—In the earliest records we find the first syllable omitted ; compare

DB 1086 *Hetun, Hetone*   YI 1254 *Hetun*

At a later period we find the name oscillating between two forms, the affix Clack sometimes preceding, sometimes succeeding, the original name :

KI 1285 *Claketon*	KF 1303 *Heton Clak*
KC 1348 *Clakheton*	NV 1316 *Heton Cleck*
YF 1514 *Clakheton*	YD 1355 *Hetonclak*

Heaton is derived from OE *hēah, hēh*, high, and *tūn*, a farmstead; but Cleck- can scarcely be derived from OE *clæg*, clay, as is sometimes suggested. There are, however, two other alternatives.

First, Cleck- may come from a personal name, and Neilsen records the ODan. name Klakki, while Searle has the name Clac. Heton Clak would then correspond to such a name as Chipping Ongar, where the affix comes from the OE personal name Ongær ; but no connection between Cleckheaton and any person of this name has been discovered. The second and more probable alternative is that the affix is connected with the Danish word *klak*, a marshy place (Blandinger IV, 243).

*CLEGGCLIFFE, CLEGGFORD BRIDGE, Halifax and Dewsbury.—The former, spelt *Clegclyve* in WCR 1275 and *Clegcliff* in 1345, is mentioned in a deed of 1553 quoted by Watson in the phrase 'le Bekyn super altitudine montis de Gletclif.' The first element in both names is probably derived from Dan. *kleg*, clay.

CLEWS MOOR, Queensbury.

-CLIFFE, CLIFTON.—The word 'cliff' comes from OE *clif* or ON *klif*, and often means a steep hill. It occurs as a termination in the following names: Attercliffe, Bilcliffe, Birchencliffe, Brincliffe, Bruntcliffe, Catcliffe, Cleggcliffe, Cowcliffe, Dovecliffe, Eddercliffe, Endcliffe, Hartcliffe, Hinchcliffe, Lightcliffe, Raincliffe, Rawcliffe, Shirecliffe, Stenocliffe, Sutcliffe, Thorncliffe, Topcliffe, Whitcliffe, Yarncliffe.

CLIFTON, Brighouse, DB *Cliftone*, PM 1307 *Clyfton*, is 'the cliff farmstead,' from OE or ON *tūn*, an enclosure, homestead.

CLIFTON, Conisborough, DB *Cliftune, Clifton*, NV 1316 *Clyfton*, has the same origin and meaning.

CLOUGH.—This characteristic word comes from OE *clōh*, a ravine with steep sides, usually forming the bed of a stream or river. In Kirkheaton BM records such names as *Gate-brigge Cloh* and *West-hau-cloh*, and near Rochdale the Whalley Coucher Book records *Blakeclogh* and *Midilclogh*. The word is of frequent occurrence in South-west Yorkshire, examples being Cat Clough, Thurlstone ; Magdalen Clough, Meltham ; Pennant Clough, Stansfield ; Stainery Clough, on Broomhead Moors ; Strines Clough, near Holmfirth.

CLUNTERGATE, CLUNTERS.—In NED a dialect-word *clunter* is explained as a big lump. This word is connected with MDu. *klonter*, EFris. *klunter* = *klunt*, a lump; and Halliwell gives a verb *clunter* which means to turn lumpy. But in Cluntergate, Horbury, the termination—from ON *gata*, a path or road—suggests the possibility of a Scandinavian origin, and Larsen gives us the Danish and Norwegian *klunt*, pl. *klunter*, a log or block; thus Cluntergate probably means 'the road paved with logs.' The name Clunters is found in Sowerby and Stansfield.

COATES, COTE, COTTONSTONES.—There is a village in Derbyshire called Cotton which is recorded in DB as *Cotun*, and in the North Riding one called Coatham formerly spelt *Cotum* and *Cottum*. Both may be interpreted 'the cottages,' from the dative plural of OE *cot*, *cott*, a dwelling, house, cot, or from ON *kot*, a cot, hut. The short vowel was lengthened in an open syllable and so the forms Cote, Cotes, Coates were obtained. The word occurs as a terminal in Carlecotes, Kebcote, Owlcotes, Silcoates, and Skircoat.

COATES occurs in Oxspring, COTE HILL in Warley, and COTTONSTONES in Sowerby.

COLDEN, COLEY, COLLIN, COLLON, COWLEY.—All, save the last, occur in the neighbourhood of Halifax. In Sowerby there is Collon Bob; in Wadsworth Collon Flat and Collon Hall; in Soyland Collin Hill; in Greetland Collin Lane; near Hebden Bridge the Colden Valley.

COLLON or COLLIN is explained by EDD as 'stalks of furze-bushes which remain after burning,' but no etymology is given. Apparently the word comes from ON *kol*, charcoal, with the addition of the suffixed article.

COLDEN, HW 1521 *Colden*, HW 1539 *Coldon*, HW 1514 *Coldenstokkbridge*, is most probably 'charcoal valley,' from OE *col*, charcoal, and *denu*, a valley.

COLEY, Northowram, which is frequently recorded in the 13th and 14th centuries as *Coldeley* and *Coldelay*, is derived from OE *ceald*, cold, and *lēah*, a lea or meadow.

COWLEY, Ecclesfield, YF 1554 *Colley*, YF 1572 *Colley*, appears to be 'coal lea,' that is 'charcoal lea.'

COLNE, COLNEBRIDGE, Huddersfield.—See Calder.

COMBES, COWMES.—South of the Aire we find four examples of this name, together with a possible fifth example embedded in the name Alcomden.   The words are derived from the Prim. Celt. **kumb-*, a valley or dingle; compare W *cwm*, Bret. *cum, cwm*, Corn. *cum*, Ir. *cum*.

CAULMS occurs in Dewsbury.

COMBS, TPR 1682 *Cowmhill*, 1694 *Cowms hill*, is the name of a hollow near Thornhill Church.

COWMES, Bradfield, is recorded in YS 1297 as *Cumbes*, in PT 1379 as *Caume*, and in HS 1637 as *Cowmes*.

COWMES, Kirkheaton, may perhaps be indicated by Burton (in his notes on Fountains) as *Newcombgill*.

CONISBOROUGH, Doncaster, with its well-known castle towering above the valley, stands immediately opposite the confluence of the Dearne and the Don.

It is referred to by Geoffrey of Monmouth, who has the description 'oppidum Kaerconan quod nunc Cunungeburg appellatur.'   It is also referred to by Pierre de Langtoft, who tells us that King Ambrosius 'took the city of Conaun with all the treasure that belonged to Sir Hengist,' and that at the beginning of a certain summer King Egbert with all his household went 'to the burgh Conane':

> "Egbrith après le yver, en entraunt le sée,
> Est al burge Conane alez of sa meyne."

This may perhaps refer to the year 829, for in 827 Egbert had led his army as far north as Dore, while in 828 he had subdued the Britons of North Wales.

Further references to Conisborough show that, though in 1066 it was part of the 'terra' of King Harold (DB), but a short time before, in 1004, it was in the hands of a subject, namely, Wulfric Spot (KCD 1298).

In the British name *Kaerconan* the second element, *conan*, is possibly to be connected with the river. Among the stream-names of Britain there is a very large number where the first element is *con* or *can*. Scotland has the Conon in Ross, the Cannich in Inverness, the Conglass in Banff, the Conrie and the Cannie in Aberdeen, the Cona and Cannel in Argyll, the Connat and Conait in Perth, the Cander in Lanark, and the Connal in Ayr. In Wales there is the Conway; and England has the Can in Essex, and the Conder in Lancashire.

Fortunately, a certain number of early spellings are forthcoming, thus in 1220 the last-named was written *Kondover*; the Conon in Ross occurs in the 16th century name *Strachonane*, that is, Strathconon; and a stream-name in Glamorgan appears in GC 1253 and 1256 in the phrase 'per rivulum *Canan*,' while GC 1203 has the name *Polcanan*.

It seems clear that these names are connected with Welsh *cawn*, reeds, from Prim. Celt. **kāno-*; compare OIr. *connall* (Stokes). For the termination in Conan see Allan.

Having referred hitherto only to stream-names it will be helpful to go a step further. Among other names involving the stem *can, con,* the most interesting are Candover in Hants., KCD 903 *Candefer*, and Condover in Shropshire, where the meaning is simply 'reed-water.' Canford and Conford in Hants., and Canford in Dorset, tell their own tale. Perthshire has a loch called Con of which the greatest depth is only nine feet; Wigtown has a loch called Connell, and Hereford has a place called Cananbridge. There are, indeed, scores of names which appear to be derived from this source.

Early forms of Conisborough may well be compared with those of Coniston in North Lancashire :

DB	1086	*Coningesborc, Cuningesburg*	†1163	*Coningeston*
AR	†1216	*Cunesburc*	1196	*Koningeston*
CR	1232	*Cuningesburgo*	†1272	*Conyngeston*
KI	1285	*Cunynggesburgh*	1337	*Kunyngeston*
NV	1316	*Conyngesburgh*	1401	*Cunigestun*
PT	1379	*Conesburgh*	1404	*Cuningeston*

Obviously the first element in both names is from ON *konungr*, a king, rather than OE *cyning*; obviously also the terminal in

Coningesborc is from ON *borg*, a fortified place, rather than OE *burh*—though the latter has given all the remaining forms. We may therefore put forward the interpretation 'king's fortress' without hesitation, Coniston being 'king's farmstead or enclosure.' On the other hand it seems not improbable that the full history of the name has three stages, (1) the Celtic *Conan* or *Conaun*, (2) the Anglian *Conanburh, catching up the Celtic name as in the case of Doncaster, (3) the wholly Scandinavian *Konungsborg and the partly Scandinavian *Coningesburg*.

COOPER BRIDGE, Huddersfield.—A bridge 'over Keldre between *le Couford* and the grange of Bradeley' is mentioned in WCR 1336. Again, in HW 1483 we find a sum of 6s. 8d. is left for the repair of '*Cowford brigge*.' It is plain that where Cooper Bridge now stands there was in earlier days a way across the Calder called 'the cow-ford.'

COPLEY, Halifax, is spelt *Coppeley* in WCR 1275 and 1297, and *Coplay, Copelay* in PT 1379. As the place lies in a valley it seems impossible to connect it with OE *coppa*, a summit, peak. On the other hand the interpretation '*Coppa's lea*' may well be correct; compare KCD *Coppanleah*.

COPTHIRST, Holmfirth, WCR 1307 and 1308 *Coppedhirst*, is 'the pollard wood,' from OE *copped*, polled, lopped, and *hyrst*, a copse, wood.

CORNHOLME, Todmorden.—See Holme.

CORTWORTH.—See Cartworth.

COTTINGLEY, Bingley, DB *Cotingelai, Cotingelei*, CR 1283 *Cotingeleye*, PT 1379 *Cottynglay*, is 'the lea of the family of Cota'; compare DB *Cotingeham*, now Cottingham.

COTTONSTONES, Sowerby.—See Coates.

COWCLIFFE, Huddersfield, WH *Cawcliffe*, RE 1716 *Cawcliff*, is probably to be explained as 'the bare cliff,' from OE *calu*, callow, bare.

COWICK, Snaith, SC *Cuwic, Cowyk* and *Cowyck,* DN 1250 *Cowicke,* YI 1251 *Kuwyke,* YI 1280 *Couwicke,* is 'the cow enclosure,' from OE *cū,* cow, and *wīc,* an enclosure, house, village.

COWLERSLEY, Linthwaite, appears in WCR 1277 as *Colleresley* and WCR 1308 as *Collereslay,* while certain 15th century deeds give *Collerslay* and *Collersley.* DB has the personal names Colle and Collo; from these we may assume the form Coller, and explain the place-name as 'Coller's meadow,' OE *lēah,* a meadow. With OE *Coller compare the present-day surnames Coller and Collar.

COWLEY.—See Colden.

COWMES.—See Combes.

COXLEY, Horbury.—The Rievaulx Chartulary gives the name *Cockesclo,* which appears to be 'Cock's clough.' Perhaps Coxley, 'Cock's meadow,' existed concurrently. The personal name Coc is recorded in DB.

CRABTREE occurs as a place-name near Sheffield.

CRACKENEDGE, Dewsbury, DC 1579 *Crackenedge,* DC 1588 *Crakenedge,* must be compared with Crackenthorpe, Westmorland, which is plainly Scandinavian. Perhaps Cracken comes from the Norw. *krøkjen,* crooked, bent (Aasen); but, whether English or Scandinavian, it undoubtedly goes back to the Germanic *kraken,* something crooked (Torp).

CRAGG is of Celtic origin, being connected with Welsh *craig,* Irish and Gaelic *creag.* It occurs in Hardcastle Crags and Cragg Vale near Hebden Bridge ; in Cragg Lane, Thornton ; in Wharncliffe Craggs; as well as in Harden, Shipley, and elsewhere. The word may, of course, be simply a modern borrowing.

CRAWSHAW, is derived from OE *crāwe,* a crow, and *sceaga,* a copse or wood. There are three places of the name, one near Sheffield, PT 1379 *Crauschagh* ; a second in Saddleworth, DN 1388 *Crawshagh* ; and a third in Emley, PF 1208 *Croweshagh,* SE 1715 *Crawshaw.*

CRESWICK, Ecclesfield, YD 1322 *Creswyk*, IN 1342 *Cresewyk*, HH 1349 *Creswick*, is plainly 'cress village,' from OE *cærse*, *cresse*, cress, and *wīc*, an enclosure, habitation, village; compare Cressbrook, Creskeld, Creswell.

CRIDLING, Pontefract.—See Ing.

CRIGGLESTONE, Wakefield, is very probably a name of similar type to Doncaster. Early spellings are as follows:

DB	1086 *Crigeston, Crigestone*	WCR	1275 *Grigelston*
PF	1199 *Crigleston*	HR	1276 *Crickeliston*
PF	1202 *Crikeleston*	NV	1316 *Crigheleston*
WCR	1274 *Crigeliston*	PT	1379 *Grigelston*

I can find no OE personal name corresponding to Crigle-, and there can be little doubt that the word is of Celtic origin; compare Crugyll in Anglesey, Cruggleton in Wigton, and *Crugleton* in Shropshire (recorded in DM). All these names are doubtless connected with the Welsh word *crug*, a heap, barrow, stack; compare Corn. *cruc*, Bret. *crug*, *krugell* (Stokes). Thus the -*s*- in Crigglestone would be intrusive, and Crigle- would represent an earlier Crugel- of which the sense would be very appropriate, namely 'little hill.' See the note on Crimes, Crimbles.

CRIMES, CRIMBLES, CRIMICAR, CRIMSHAW, CRIMSWORTH, CROMWELL BOTTOM, CRUMACK, CRUTTONSTALL, KRUMLIN.—Crimes occurs in Hepworth and Slaithwaite; Crimble in Thornhill; Crimbles in or near Pudsey, Lofthouse, Kirkheaton, Slaithwaite, Upperthong, Norton and Stocksbridge; Crimicar in Hallam; Crimshaw in Bolton (Bradford); Crimsworth near Heptonstall; Cromwell Bottom near Elland; Crumack Lane in Haworth; Cruttonstall near Hebden Bridge; and Krumlin in Barkisland.

Of the four first-named no early records have come to hand; but Crimbles near Cockerham in Lancashire was *Crimeles* in DB, *Crumles* in LF 1206, *Crumeles* in LF 1209, and *Crimbles* in LF 1241.

Cruttonstall is probably to be connected with DB *Cru'betonestun*; it is spelt *Crumtonstall* in WCR 1308 and *Cruntonstall*

in WCR 1342.   Of Crimsworth and Cromwell Bottom we have
the following early spellings :

WCR	1275 *Crumliswrthe*	YD	1277 *Crumbewellebotham*
HW	1551 *Crymmysworthe*	WCR	1326 *Crumwelbothume*
WH	1775 *Crimlishworth*	WCR	1332 *Cromwelbotham*

Apart from Krumlin, which seems definitely Celtic, and
Crumack, which is probably Celtic, there is nothing to show
whether we have to do with derivatives of OE *crumb*, crooked,
or of Prim. Celt. **krumbo-s*, bent, crooked.   It should be noted,
however, that according to Spurrell there is a Welsh word *crim*,
a ridge, and another Welsh word *crimell*, a sharp ridge.   In the
forms Crimes and Crimbles the earlier *u* has given place to a
later *y* by mutation, and the same process accounts for the vowel
in the first element of Crimicar and Crimshaw.

CRIMICAR, formerly *Crimeker*, has for termination the ON
kjarr, brushwood, copsewood.

CRIMSWORTH is 'the farmstead at Crimbles,' from OE *weorth*,
a holding, farmstead.

CROMWELL BOTTOM is probably of similar origin, Cromwell
being 'the well at Crum'; compare Crumton.

CRUMACK may perhaps be 'the curved or sloping place,' the
termination being, it would seem, the common Celtic collective
suffix -ach ; compare the Welsh word *crwmach*, which according
to Evans means convexity, or a convex.

CRUMTON in Cruttonstall is most probably 'the farmstead
at Crum.'

KRUMLIN, Barkisland, appears to be the 'crooked pool,' or
'winding stream'; compare Ir. *linn*, Gael. *linne*, W *llyn*, Bret.
lenn, a pool.   To-day the name is given to a hillside district, but
formerly it was doubtless applied to the stream below.   The
name Crumlin occurs both in Wales and Ireland; compare CR
Cremlin, *Cremlyn*, now Crymlyn, Anglesea.

CRODINGLEY, Netherthong.—Searle has on record the
personal names Croda and Crodo, from which a patronymic
Croding could be formed, and hence the place-name Crodingley,
that is, 'the lea of the Crodings,' OE *lēah*, a lea.

CROFT, CROFTON, CUSWORTH.—Early records of Crofton and Cusworth include the following:

DB 1086 *Scroftune, Scrotone*	DB 1086 *Cuzeuuorde, Scuseuurde*
CR 1215 *Crofton*	PF 1208 *Cucewordh*
YS 1297 *Crofton*	YS 1297 *Cuscewrth*
PT 1379 *Crofton*	PT 1379 *Cusseworth*

In both names the Domesday record shows an intrusive *s*, due to the Norman scribe. Other examples of the same kind are DB *Sclive* for Cliff, DB *Stablei* for Tabley, DB *Stimblebi* for Thimbleby (Zachrisson).

CROFT occurs as a terminal in Barcroft, Carcroft, Havercroft, Ryecroft, Scholecroft, Thurcroft. It comes from OE *croft*, a small enclosed field.

CROFTON, Wakefield, is simply 'croft farm,' from OE *tūn*, an enclosure or farm.

CUSWORTH, Doncaster, shows early forms which must be compared with those of Kexborough. It means 'Cuthsa's holding,' from OE *weorth*, and a personal name *Cuthsa. See Wilsden.

CROMWELL.—See Crimes.

CROOKES, CROOKHILL.—ON *krōkr*, a corner, nook, has given us many northern place-names, including Crooke and Crooklands, as well as our South-west Yorkshire examples.

CROOKES, Sheffield, YS 1297 *Crokis*, PT 1379 *Crokes, Crekes*, YF 1532 *Crokys*, requires no elucidation.

CROOKHILL, Edlington, AR *Crocwell*, PT 1379 *Crokewell*, YF 1575 *Crokwell*, should be compared with Chidswell, where popular etymology has produced a result diametrically opposite. For the termination see Wall, Well.

CROSS, CROSLAND, CROSSLEY, CROSS STONE, OSGOLDCROSS, STAINCROSS.—The word 'cross' is one of the most interesting of our place-name elements. The native word was *rōd*, found to-day in Holyrood and roodscreen; but the Normans brought into England a derivative of Lat. *crucem*, namely *croiz* or *crois*, while before the Norman Conquest another derivative of *crucem* had been made use of, the word *cruche*. Examples of the latter still exist in the name Cruche

Stoke, Norfolk, and in the description Crutched Friars. At an earlier date than either *crois* or *cruche* a third derivative of *crucem* had come into use, namely, *cros*, derived from the Old Irish word of the same form, and brought to England by Norsemen, who settled in considerable numbers in Cumberland, Westmorland, Lancashire, and the West Riding. See p. 29.

Early records of the three names in 'cros' which occur in the Domesday Survey are as follows:

DB	1086 *Crosland, Croisland*	DB	1086 *Osgotcros*	DB	1086 *Staincros*
PC	1212 *Croslanda*	PF	1167 *Osgodescros*	PF	1166 *Steincros*
WCR	1286 *Croslande*	DN	1251 *Osgodcrosse*	PF	1170 *Steincros*
NV	1316 *Crosseland*	HR	1276 *Osgotecrosse*	LC	1296 *Staincross*

CROSLAND, Huddersfield, is 'the land or estate where there is a cross,' from ON *land*, land, estate.

OSGOLDCROSS, the name of the wapentake in which Castleford and Pontefract are situate, is 'Osgod's cross,' Osgod or Osgot being the DB form of the ON personal name Asgautr.

STAINCROSS, the name of the wapentake in which are Barnsley and Penistone, is simply 'stone cross' from ON *steinn*, a stone.

CROSSLEY occurs four times in South-west Yorkshire—in Hipperholme, WCR 1326 *Crosslegh*; near Bradford, PC †1246 *Crosley*; in Ecclesfield, YD 1290 *Crosselay*, YI 1298 *Crosseley*; and in Mirfield. The meaning is 'cross lea,' from OE *lēah*, a lea or meadow.

CROSS STONE, Todmorden, can scarcely be connected with DB *Cru'betonestun*, which should rather be linked with Cruttonstall. WH 1682 has *Crostone*, and HW 1537 speaks of 'the chapel builded at the *Crosse Stone* in the parish of Heptonstall.' Both the Crossleys and Cross Stone must be claimed as English.

CROW BROOK, CROW EDGE, CROW HILL, CROW NEST, CROW POINT, CROW ROYD, CROW WOOD.—

Minor names involving the word Crow are very common. We find Crow Brook and Crow Edge in Thurlstone; Crow Hill in Sowerby, Fulstone, and Hepworth; Crow Point in Queensbury; Crow Royd in Thornhill; Crow Wood in Stainton; and

Crow Nest in Erringden, Lightcliffe, Dewsbury, Beeston, and Worsborough.

CROW NEST appears in YD 1307 as *Crovnest*, where the interpretation is the obvious one from OE *crāwe*, a crow, and *nest*, a nest.

CROW HILL, Sowerby, like Crookhill, Edlington, shows a change from 'well' to 'hill,' witness the following early forms: HPR 1562 *Crowelschais*, WH *Crowell shaws*. A similar early form occurs in connection with Fixby, namely, WH *Crowallsike*. Doubtless *Crowell* and *Crowall* are to be interpreted as 'crow well'; compare Ouzelwell and Spinkwell.

CRUMACK, CRUTTONSTALL.—See Crimes.

CUDWORTH, CULLINGWORTH.—The termination comes from OE *worth, weorth, wyrth*, which may be explained as a property or holding, and was applied to a homestead or farm. Early spellings of the two names are

YR	1233	*Cudewrth*	DB	1086 *Colingauuorde*
NV	1316	*Cutheworth*	PF	1235 *Cullingwurth*
YD	1318	*Cuttheworth*	CH	1236 *Cullingwurthe*
PT	1379	*Cotheworth*	PT	1379 *Collyngworth*

CUDWORTH, Barnsley, is 'the homestead of Cutha,' from the recorded personal name Cutha which appears also in Cutsyke.

CULLINGWORTH, Bingley, is 'the homestead of the Cullings,' that is, of the sons of Culla; Searle gives the name Culling. In *Colingauuorde* the Norman scribes wrote *o* for *u*, and *d* for *th*; compare Cudworth, Cumberworth, and Kimberworth.

CUMBERWORTH, KIMBERWORTH.—Early records of these names, which occur respectively near Huddersfield and Rotherham, are as follows:

DB	1086 *Cu'breuuorde, Cu'breuurde*	DB	1086 *Chibereworde*
CR	1226 *Cumberwrthe*	KI	1285 *Kimberworth*
YS	1297 *Cumberworth*	NV	1316 *Kimberworth*
NV	1316 *Cumbreworde*	PT	1379 *Kymbirword*

Dr Skeat explains the Hertfordshire name Cumberlow as the 'barrow of Cumbra,' and he says the original sense of the personal name, like that of the Welsh Cymro, was 'Welshman.'

CUMBERWORTH may therefore be explained as ' the holding of Cumbra,' the Welshman, from OE *worth*, a holding or farmstead.

KIMBERWORTH has a similar meaning, but is derived from Cymbra, a secondary form of the name Cumbra.

The story suggested by these names is obvious.   In the early days of the Anglian settlement the two places were each in the possession of a Briton.   Two other places near at hand, West Bretton and Monk Bretton, have a similar signification.

CUSWORTH.—See Crofton.

CUTSYKE, Whitwood, CH †1235 *Cutthesik*, appears to be ' Cutha's stream,' from OE *sīc*, a runnel ; see Cudworth.

DAISY GREEN, DAISY HILL, DAISY LEE.—The first occurs in Linthwaite ; the second in Heaton, Dewsbury, and Morley ; the third near Langsett, Lindley, and Holmfirth. It is possible that the names mean just what they appear to mean, but it is not in every case probable ; Daisy Lee, Langsett, for example, is very exposed and stands 900 feet above the level of the sea.

DALE, DALTON.—The word ' dale' may be derived either from OE *dæl* or ON *dalr*, a valley.   Professor Skeat says it is ' as much Scandinavian as Anglo-Saxon ' ; but it is certain that many of the north country -dales are due to Viking settlers. In South-west Yorkshire the word is not common ; it occurs in the two Daltons, in Barnsdale, Brocadale, Cockersdale, Magdale, and Wooldale.

DALTON, Kirkheaton, DB *Dalton, Daltone*, NV 1316 *Dalton*, is ' the farm in the dale,' from OE or ON *tūn*, an enclosure, farm.

DALTON, Rotherham, DB *Dalton, Daltone*, KI 1285 *Dalton*, has the same meaning.

DAMFLASK, Sheffield.—See Flash.

DARFIELD, DARLEY, DARTON.—Of Darley in Worsborough there are no early records, but of Darfield and Darton, which are near Barnsley, we find the following :

8—2

DB 1086 *Dereuuelle, Dereuueld*   DB 1086 *Dertone, Dertune*
PC 1155 *Derfeld*   YR 1234 *Derton*
YR 1228 *Derfeud*   NV 1316 *Derton*
YS 1297 *Derfeld*   PT 1379 *Derton*

It is quite impossible that either Darfield or Darton should obtain its first syllable from the river on which they stand, the Dearne. The most likely etymology, indeed, would derive it from OE *dēor*, ME *der, dere*, an animal, a wild beast. But it is important to notice that while Der- is the first element in the Domesday name for Darton, in Darfield it is Dere-. Hence, though we may explain Darton as *Dēortun*, ' deer enclosure,' we must explain Darfield as *Dēorafeld*, ' deer field,' or *Dēoranfeld*, ' the field of Deora'; compare BCS *Dēoran-treow*. Darley is probably from *Dēorlēah*, ' deer lea.'

The DB record of Darfield shows a form *Dereuuelle* which would mean ' deer well' or ' Deora's well,' from OE *wella*, a spring.

DARNALL, Sheffield, YS 1297 *Darnale*, YI †1301 *Dernhale*, HH 1366 *Darnale*, YF 1560 *Dernall*, comes from OE *derne*, ME *dern*, secret, and OE *halh* or *healh*, a corner or meadow. According to Professor Skeat a Cambridgeshire Dernford, 1372 *Dernford*, derives its prefix from the same source; compare BCS *Derneforde*.

DARRINGTON, Pontefract, DB *Darnintone, Darnitone*, PC †1090 *Dardintona*, PC 1159 *Dardingtona*, PF 1205 *Darthingtone*, CR 1230 *Dardinton*, LC 1296 *Darthingtone*, NV 1316 *Darthyngton*. The first element of the word is in the form of a patronymic. Searle gives the name Deoring, which however would not account for the DB and other early spellings. But the name Deornoth, with the termination -ing added, would fully satisfy all conditions ; so we may explain Darrington as ' the homestead of the sons of Deornoth.' See Dirtcar.

DEAN, DEN, DENHOLME, DENROYD, DEN-SHAW.—When alone OE *denu*, a valley, became first *dene* or *deyne* and afterwards *dean* ; but in compounds it became *den*, as in Denshaw and Hebden. In South-west Yorkshire the names ending in -den include Agden (2), Alcomden, Arunden, Bagden, Bogden (2), Colden, Dwariden, Erringden, Ewden, Harden, Hebden, Hewenden, Howden, Lewden, Marsden, Mixenden, Moselden, Ogden, Ovenden, Prickleden, Ramsden, Ribbleden, Ripponden, Scammonden, Shibden, Skirden, Snailsden, Stiperden, Stubden, Sugden (2), Todmorden, Twizleden, Wessenden, Wilsden, Wickleden.

DENHOLME, Bradford, KC *Denum, Dennum*, YF 1564 *Denholme*, comes from the dative plural of OE *denu*, a valley, or *denn*, a den, cave, swine-pasture.

DENROYD, Denby, is probably 'the clearing in the valley.'

DENSHAW, Saddleworth, referred to in 1544 as *Denshaw*, and in 1727 as *Deanshaw*, is 'valley-copse,' from OE *sceaga*.

DEARNE.—This stream rises near Cumberworth, and after passing Barnsley joins the Don at Conisborough. Early records of the name are PC 1155 *Dirna*, CR 1230 *Dirna*, 1316 *Dirne*, 1413 *Dyrne*, YF 1495 *Dern*. A Wiltshire stream-name is given in BCS as *Dyre·broc*; in Oxfordshire there is the Dorn, and in Banff the Durn. Probably all these names come from a Celtic stem *dur* recorded by Holder and Förstemann. Among river-names derived therefrom Holder gives an early Irish example *Dūr*, and the German Thur, formerly *Dura*.

DEEPCAR occurs in Woodsetts, Wilsden, and Stocksbridge. Early records give *Depeker* for the second and *Depecarr* for the third. The word is derived from ON *djūpr*, deep, and *kjarr*, copsewood, brushwood.

DEER HILL, DEERPLAY, DEERSHAW, DEERSTONES.—The obvious explanation may perhaps be the true one, but it is not altogether convincing. The Gazetteer shows in Scotland two villages with the simple name Deer, while in Devon and Aberdeen the same word is applied to streams.

DEER HILL is in Marsden, and DEERSTONES in Sowerby.

DEERPLAY, Sowerby, HW 1560 *Dereplay*, WH *Derpley*, is paralleled by Dirplay, near Bacup, recorded in the De Lacy Compoti as *Derplaghe* in 1294. I suggest that the termination in *Derpley* is not *-pley* but *-ley*, the stem *Derp-* being a stream-name corresponding to such LG examples as Alpe, Marpe, Wilpe (Jellinghaus). Thus the -p- in Deerplay would represent *apa*, water, while Deer would represent the stem *dūr* mentioned in the note on Dearne.

DEERSHAW, Fulstone, corresponds to BCS *Deor·hyrst*, deer-coppice, for OE *sceaga*, like OE *hyrst*, signifies a copse.

DEFFER, DEFFERS.—In WRM mention is made of 'a piece of ground, lying near Kirkthorpe on the other side of the river, called Deffers'; and the same authority gives the name in 1342 and 1391 as *Defford*. Defford in Warwickshire was *Depeford* in DB and *Deopford* in a pre-Conquest charter; hence we may explain Deffers as 'the deep ford.'

Perhaps the early name *Cuindever* connected with Hoyland-swaine should be linked with Deffer Wood, Cawthorne. These recall the Hampshire names Candover and Mitcheldever, spelt *Candefer* and *Myceldefer* in early charters. The termination *-defer* appears to be connected with the W *dyfyr*, a form of *dwfr*.

DEIGHTON, DEIGHTONBY. Huddersfield and Barnsley, present the following early forms:

WCR	1284 *Dychton*	DB 1086 *Dictenebi*
YS	1297 *Dicton*	CH 1486 *Dicthenbi*
PT	1379 *Dyghton*	

The first name means 'dike farmstead or enclosure,' from OE *dīc* or ON *dīk*, a dike, and OE or ON *tūn*, an enclosure or farm; compare Beighton and Boughton, formerly *Bēctūn* and *Bōctūn*. In the second name the first element is a weak personal name formed from ON *teinn*, a twig or stake; compare Benteinn (Naumann).

DELPH, Saddleworth.—In the district around Halifax 'delf' is the usual name for a quarry. The word is derived from late OE *dælf*, ME *delf*, a trench, ditch, quarry.

DENABY, DENBY. — South-west Yorkshire has four examples of the name Denby, but of two near Keighley there are no early records. Of the rest, with Denaby, we find the following ancient spellings:

DENABY, Conisborough	DENBY, Penistone	DENBY, Whitley
DB 1086 *Denegebi, Degenebi*	DB 1086 *Denebi*	DB 1086 *Denebi*
CR 1277 *Dennyngeby*	YS 1297 *Deneby*	KF 1303 *Deneby*
KI 1285 *Denigby*	NV 1316 *Deneby*	NV 1316 *Deneby*
NV 1316 *Denyngby*	PT 1379 *Denby*	CH 1323 *Deneby*

OE *Dene*, the Danes, had two genitives plural, a shorter form Dena, and a longer Deniga. Denby may well, therefore, be 'village of the Danes,' and Denaby may have the same meaning, though 'village of the sons of Dene, the Dane,' is perhaps more probable. Falkman gives the personal name Dening.

DENHOLME, DENROYD, DENSHAW.—See Dean.

DERWENT.—Rising in Featherbed Moss, this stream is for some miles the county boundary; afterwards it flows past Chatsworth and Matlock to Derby, and joins the Trent near the borders of Leicestershire. There are four English streams of the name, and of two we have, directly or indirectly, early records.

The Antonine Itinerary names a Roman station *Derventione* which is usually connected with the Derwent of the East Riding, a river mentioned by Simeon of Durham as the *Dirwenta* and *Dyrwente*.

Ptolemy speaks of a *Derventione* which has been in the same way connected with the Cumbrian Derwent; and Bede mentions the place under the form *Derventio*.

Both these names are connected by Stokes and Holder with a primitive Celtic **derv*, meaning an oak-tree, a word which appears in Welsh as *derw*. The ending is well-known as the river-name suffix *-entia* or *-antia*. Examples in which this suffix appears are the Argenza, Paginza, and Elsenz in Germany; the Durance and Charente in France; the Trent and Carant in England. Early spellings of the last are BCS 778 *Carent*, BCS 780 *Cærent*; compare the Gaulish *Caranto-magus* (Stokes).

DEWSBURY.—The Domesday forms, *Deusberia* and *Deusberie*, have given rise to derivations from Lat. *deus*, while later forms, YR 1230 *Dewesbire*, DC 1246 *Dewesbury*, YR 1252 *Dewebyre*, WCR 1277 *Dewysbiry*, NV 1316 *Deuuesbury*, DC 1349 *Dewesbury*, PT 1379 *Dewsbyry*, lead quite naturally to the interpretation 'Dewe's stronghold.'

Let us first be clear about the termination -berie or -bury, forms which occur also for Almondbury and Horbury, DB *Almaneberie* and *Horberie*.   Professor Skeat has shown that Norman scribes frequently put *e* for OE *y*; hence -berie should be read -byrie.   This brings all the terminals into harmony, and links them directly with OE *byrig*, the dative of *burh*, a fortified post, a stronghold.

The first element can scarcely be the Lat. *deus*, though the Domesday scribe may have imagined that it was; indeed it is almost certainly a personal name, and though Searle and Naumann give no such form, Brons comes to the rescue and announces a Frisian name Dewe.   It is noteworthy that the surname Dews is quite common in the neighbourhood.

Ancient sculptured stones in the parish church take us back to the 9th century, and tradition asserts that Paulinus baptized on this spot.

A sentence in the description of Dewsbury given in 1828 by Clarke in his Gazetteer shows what the attractions of its site were a century ago.   'The appearance of the town from the Wakefield Road,' he says, 'bursting at once unexpectedly upon the sight, is as beautiful as interesting.'

DIGGLE, Saddleworth.—13th century charters give *Dighull* and *Diggell*, and a deed of 1468 has *Dighil*.   The termination, like that in Adel, Idle, Nostell, appears to be -el, but I am unable to explain further.

DINNINGTON, Rotherham, DB *Dunnintone, Dunnitone, Domnitone*, CR 1200 *Dunyngton*, YR 1271 *Dynington*, NV 1316 *Donyngton*, PT 1379 *Dynnyngton*, oscillates in its first element between Dunning and Dynning.   Both forms are represented in English place-names; there are Dinningtons in Northumberland and Somerset, and Dunningtons in Warwick and the East Riding; and, further, Holland has the place-name Dunninge. The sense is probably 'the farmstead of the sons of Dunne or Dynne,' for Searle gives both names; but see Ing.

DIRTCAR, DURKER.—The village Dirtcar or Dirker near Wakefield is mentioned in WCR 1284 as *Drytkar*, WCR

1285 as *Drytker*, and WCR 1297 as *Dritker*, while in a Fine of
1514 it is *Dirtcarre*.   The word is Scandinavian, from ON *drit*,
dirt, and *kjarr*, copsewood, brushwood.   There are many other
names apparently of the same origin, including Durker Wood,
Meltham ; Dirker Bank, Marsden ; and Dirk Carr, Northowram.
It should be noted that when three or more consonants come
together the one enclosed generally disappears.   In this way
Northland has become Norland, and *Northmanton* Normanton ;
and in the same way *Dirtcar* or *Dirtker* would quite naturally
become Dircar or Dirker.

**DOBB, DOB CARR, DOBCROSS, DOBBING, DOB-
ROYD.**—We find Dob or Dobb in Sowerby, Cartworth, and
Keighley ; Dobroyd in Calverley, Hepworth, Denby Dale, and
Todmorden ; Dob Carr in Bradfield ; Dobbing in Ecclesall.
The only early records are HS 1637 *Dobinge* for the last-named,
WCR 1308 *Dobberode* for Dobroyd, Hepworth, and CC 1482
Dobrode for Dobroyd, Calverley.

According to NED the word 'dob' is an obsolete form of
'dub,' which means a muddy or stagnant pool, a deep pool in a
river.   The origin of the two words is uncertain.

It should be noted that in every case the yokefellow of Dob
is possibly Scandinavian : Carr from ON *kjarr*, copsewood ;
Cross from OIr. *cros* ; Ing from ON *eng*, a meadow ; Royd from
ON *ruð*, a clearing.

DODWORTH, Barnsley, DB *Dodeswrde, Dodesuuorde*, PC
†1090 *Dodewrd*, PC 1122 *Dodewrdam*, NV 1316 *Dodeworth*, PT
1379 *Dodworth*.   The present name is the descendant of the
early *Dodewrd* which means ' the homestead of Doda,' from
OE *weorth*, a holding, farmstead ; but the DB names are derived
from a strong form of the personal name and may be interpreted
as ' the homestead of Dod,' the genitive of Dod being Dodes,
while the genitive of Doda is Dodan, later Dode.   Compare the
Dutch place-name Dodewaard recorded in NGN III, 77 under
the date 1107 as *Dodewerda*.

DOGLEY, DOGLOITCH WOOD, Kirkburton and Soothill.—Skeat quotes two OE place-names involving the name of

the animal, *Doggi·thorn* and BCS 941 *Doggene·ford*, ford of the dogs ; and probably the two names above are derived from the same source, OE *docga*, ME *dogge*. The termination -loitch is a dialectal form of ME *lache*, a pond, pool, swamp ; compare *Blakelache*, *Grenelache*, *Wyggelache*, all found in the Whalley Coucher Book.

DOLE, DOLES.—In Woolley we find the name Common Doles ; in Dalton (Huddersfield) Red Doles ; in Clifton (Brig-house) Doles Lane ; in Saddleworth Dolefield ; in Snydale Doles Close ; in Throapham Doles Wood ; and in Braithwell Fordoles. Watson speaks of a *fordoll* in Fixby ; a Cresswell deed of 1318 has *fordoles* ; a Pickburn deed of 1208 has *haluedol*; Aughton deeds speak of *mapeldoles* and *moredoles* ; and in other cases we find the names *bierdoll*, *shrovedole*, and *waterdole*. From EDD we learn that a dole, OE *dāl*, is a portion of a common or undivided field. A deed dated 1238 in the Pontefract Chartulary speaks of ' duas seliones que vocantur fordolis.' The word doles is in fact an interesting survival bearing witness to the ancient method of land tenure and cultivation called the ' common field ' system.

DON, DONCASTER.—In the Antonine Itinerary the Roman station was called *Danum* ; KCD 1004 has *Donecestre*; DB 1086 *Donecastre*, PF 1202 *Danecastre*, WCR 1298 *Danecastre*, KF 1303 *Donecastre*. The ending -caster comes from Lat. *castra*, a camp, a word which appears in different parts of England under the forms chester, cester, caster ; and the first element is the river-name. Thus the meaning of Doncaster is ' the camp beside the Don.'

The river-name is undoubtedly Celtic, but its origin is not certain. Perhaps it represents the Prim. Celt. *danos, a beater, fighter, which Stokes suggests as the origin of *dan* in *Rodanos*, the Rhone ; compare the Ir. *dana*, bold, strong. Among names recorded in the Gazetteer there are the river Dane in Cheshire, Lough Dan in Wicklow, Dean Burn in Linlithgow, and Dean Water in Forfar.

Pierre de Langtoft has an interesting early reference to Doncaster. King Egbert of Wessex, he tells us, came to

Conisborough—probably in the year 829—and from Conisborough he advanced to the Tweed where he gave battle to the
Danes, but with unsatisfactory results.   Later, the Northmen
appeared at Adlingfleet with thirty-five ships ; and Egbert,
assisted among others by Haldan de Danekastre, once more
gave battle and gained a great victory.   After the battle Egbert
entered Doncaster in triumph.

Near Doncaster the river is bordered by a series of villages,
hamlets or farms whose names are plainly of Scandinavian
origin.   Among them are the following : Almholme, Armthorpe,
Balby, Barnby, Bessacarr, Braithwaite, Braithwell, Bramwith,
Cadeby, Denaby, Edenthorpe, Eskholme, Goldthorpe, Hexthorpe,
Kilholme, Langthwaite, Micklebring, Scawsby, Scawthorpe,
Shaftholme, Thornholme, and Wilby.

DOVE, DOVECLIFFE.—Dove, RC *Duva, Duvé,* is the
name of a stream which flows through Worsborough Dale.
There is another stream of the same name in the Cleveland
Hills, and a third which separates the counties of Stafford and
Derby.   The origin of these names may perhaps be the Celtic
stem *dubos,* very dark, black ; compare OW *dub,* W *du,* Ir. and
Gael. *dubh,* Gaulish *Dubis,* the last of which has given the French
river-name Doubs.   Hogan places on record several rivers in
Ireland formerly called *Dub* or *Dubh* ; one of these is now called
the Duff.

DOVER.—At Holmfirth, beside a stream now called the
Ribble, we find Dover Wood and Dover Mills.   Perhaps these
names are modern borrowings ; in any case the ultimate origin
is the Prim. Celt. *dubron,* water, whence Ir. and Gael. *dobhar,*
Corn. *dofer,* Bret. *dour,* OW *dubr,* W *dwfr, dwr.*   In the
Antonine Itinerary we find Dover (Kent) described as *Dubris,*
'at the waters.'

DRANSFIELD HILL, Hopton, WCR 1275 *Dranefeld,*
WCR 1307 *Dronesfeld,* and *Dranefeld* and *Dransfeld* in early
deeds, gets its prefix from OE *drān,* a drone, used perhaps as a
personal name.

DRIGHLINGTON, Bradford, provided a puzzle for the Domesday scribes, and they wrote -*s*- for the guttural, just as they did in the case of *Lastone*, now Laughton, *Lestone*, now Leighton, and *Distone*, now Deighton.   They also wrote *e* for *y*. Early forms are as follows :

DB	1086	*Dreslintone*	NV	1316 *Brightelington* (*br*=*dr*)
DB	1086	*Dreslingtone*	PT	1379 *Drithlyngton*
PF	1202	*Drichtlington*	CC	1444 *Dryghtlyngton*
KI	1285	*Drithlington*	CH	1478 *Drightlyngton*
KF	1303	*Drighlington*	IN	1551 *Drighlington*

The first element in the modern name, and in the forms given by DB, KI, KF, PT, IN, represents a patronymic Drygeling formed from the recorded personal name Dryga by the addition of -el and -ing ; compare Cridling.   This patronymic actually occurs in the Domesday name of Little Driffield, *Drigelinghe*, and Dr Wyld, dealing with the name Droylsden, postulates a personal name Drygel.   The first element in the forms given by PF, NV, CC, CH, shows the influence of OE *dryht*, a host, troop, company ; this seems, however, to be unoriginal.   It should be noted that to-day 'drigh' rhymes with 'rigg'; compare Hagg, Haigh, and Magdale.

DRUB, Gomersal.—No early forms of this name are forthcoming and the meaning is not clear.

DUDFLEET LANE, DUDLEY HILL, DUDWELL LANE, occur respectively in Horbury, Bradford, and Halifax. For the first of these WRM has 1653 *Dudfleete*, 1728 *Dudfleet*. The terminations come from OE *flēot*, a stream, *lēah*, a lea, and *wella*, a spring ; but no definite explanation of the first element can yet be given.   We find, however, in EDD a noun 'dud' explained as a teat, and a verb 'duddle' which means to boil, bubble up, simmer.   It is possible, therefore, that 'dud' is an ancient word meaning a bubbling spring.

Compare Dudbridge, a hamlet in Gloucestershire, and the Dudleys and Dudwells which occur in various parts of Great Britain.   Note also the river-name Duddon which according to Wyld has such early forms as *Duden, Doden, Dodyne, Dodine—*

forms which appear to involve the Celtic suffix -*ina* found in many ancient river-names, among them *Sabrina*, the Severn.

DUDMANSTONE, Almondbury, RE 1634 *Dudmanston*, RE 1716 *Dudmanstone*, is 'Dudeman's farm,' from OE *tūn*, an enclosure or farm, and the personal name recorded by Searle. Popular etymology has been busy with this name, and is responsible for RE 1780 *Deadman Stone*, which is doubly inaccurate.

DUNBOTTLE, Mirfield.—See Bolton.

DUNFORD BRIDGE, Penistone, is recorded in DN 1282 simply as *Dunneford*, and the bridge is probably therefore of later date. We may interpret Dunford as 'the ford of Dunna,' the personal name being well known.

DUNGWORTH, Bradfield, written *Dongworth* in 1311 and *Dungwith* in an undated deed, is from the OE *dung*, and *weorth*, a farmstead.

DUNKIRK, Denby, Golcar, Northowram, Sowerby, and Whitley Lower.—There is no evidence to connect the name with the French port, but I think that must be its source. Dunkirk was besieged in 1793 by the Duke of York, who, however, was defeated at Hondschoote, and compelled to withdraw. Is it possible that this can be the event which gave rise to our Yorkshire names?

DUNNINGLEY, DUNSLEY.—The modern terminations are from OE *lēah*, a lea, a meadow.

But DUNNINGLEY, Morley, is recorded in WCR as 1285 *Donynglowe*, 1297 *Donigelawe*, 1298 *Doniglawe*, where the termination comes from OE *hlǣw*, a mound, cairn, hill. The prefix corresponds to the name Dunning, recorded by Searle.

DUNSLEY, Holmfirth, WCR 1308 *Dunesleye*, is 'Dun's lea.'

DUTCH RIVER.—This is the name of a portion of the river Don near Goole. It is so-called because it was made navigable in the time of Charles I by Cornelius Vermuyden and his Dutch settlers (Clarke).

DWARIDEN, Bradfield.—This name is found in a charter of 1311 as *Dwarriden,* in 1335 as *Dweryden,* in 1398 as *Dwaryden.* Doubtless OE *dweorga·denu,* valley of dwarfs, gives the true derivation, as suggested by Mr Henry Bradley.

EARLSHEATON, Dewsbury, like Kirkheaton and Cleckheaton, takes the accent on the second syllable. This is due to the fact that the first syllable is a comparatively late addition. Early spellings are

DB	1086 *Etone, Ettone*	NV	1316 *Heton*
WCR	1286 *Heton Comitis*	WCR	1483 *Erlesheton*
WCR	1308 *Erlesheeton*		

Obviously the Domesday forms are at fault in omitting the aspirate. They should be read as *Hetone* and *Hettone,* and their meaning would then be 'high farm,' from OE *hēah, hēh,* high, or 'heath farm,' from OE *hǣth,* a heath, and *tūn,* an enclosure, or farm. See Heaton.

At the Ossett Court in 1297 (WCR) it was reported that ' Richard del Dene of Heton dug stone to burn, and sold it, to the detriment of the Earl.' The 'stone to burn' was of course coal ; Heton appears to have been Earlsheaton ; and the Earl was the lord of the manor, the then Earl of Warren and Surrey. It was from this family that the name Earlsheaton received its distinctive affix.

EARNSHAW, Bradfield and Stansfield, seems to mean ' eagle copse,' from OE *earn,* an eagle, and *sceaga,* a copse ; compare BCS *Earnalēah,* Earnley, Sussex.

EASTFIELD, EASTTHORPE, EASTWOOD.—We are reminded by these words of a well-marked difference between English and Scandinavian. From a common ancestor, Teutonic *au,* ON got *au* and OE got *ēa* ; and, in consequence, while the ON word for east is *austr,* the OE is *ēast.* Because of this variation our modern place-names possess such doublets as Austhorpe and Eastthorpe, Austwick and Eastwick. See Austerfield and Austerlands.

EASTFIELD, Thurgoland, is mentioned in PC as *Estfeld,* and in PT 1379 under the strange form *Hestofeld.*

EASTTHORPE, Mirfield, may perhaps be connected with the *Esthaghe* referred to in DN 1346 where we find *Mirfield, Hopton,* and *Esthaghe* named together. The meaning of this word is 'the east enclosure or homestead,' OE *ēast·haga.*

EASTWOOD, Todmorden, YD 1336 *Estwode,* YD 1364 *Estewod,* PT 1379 *Estwode,* and EASTWOOD, Rotherham, YS 1297 *Estwod,* PT 1379 *Estwode,* come from OE *wudu,* a wood.

EASTOFT, near the Lincolnshire border, CR 1251 *Estofte* HR 1276 *Eshetoft,* DN 1304 *Esketoft,* WCR 1308 *Essetoft* DN 1338 *Esketoft,* shows extraordinary variations. The termination is evidently Scandinavian, from ON *toft,* a green knoll, a grassy mound, a homestead; and the modern prefix seems to come from the ODan. personal name Esi, but early forms of the name were probably influenced by OE *æsc* and Dan. *æske,* an ash-tree.

EAU.—Near Doncaster there is a stream called the Old Eau River. NED gives a dialect-word *ea,* which means a river or running water. This word is applied in the fen country to canals for drainage, in which sense, says NED, it is usually spelt *eau* as if from French *eau,* water.

ECCLES, ECCLESALL, ECCLESFIELD, ECCLES- HILL, EXLEY.—In these names we are brought face to face with a familiar crux. The name Eccles, whether in composition with other place-name elements or in its simple form, is to be found over a wide area. It occurs as far north as the Firth of Forth, and as far south as the borders of the English Channel; and there are even instances across the North Sea. The simple name is found in Kent, Norfolk (2), Lancashire, Yorkshire (Eccles Parlour, Sowerby), the Lowlands of Scotland (2), and in North-east France; and in composition there are the following English examples :

> Ecclesborough in Berkshire
> Ecclesbourne in Sussex, Hampshire, and Derby
> Ecclesbrook in Worcester
> Eccleston in Cheshire (2) and Lancashire (2)
> Eccleswall in Hereford

Beyond the Cheviots there are Ecclescraig in Kincardine and Ecclesmachan in Linlithgow; and, lastly, there are three names where the form Eccle occurs, Eccle Field in Yorkshire (Woolley), Ecclerigg in Westmorland, and Ecclefechan in Dumfries. Among the early forms recorded in BCS and KCD we find the following:

| Æceles·beorh | Ecles·broc | Eccles·ford |
| Æcles·mor | Ecles·burne | Eccles·hale |

while others of post-Conquest date include

Eccles, N.E. France,	1339 *Eccles*	
Eccles, Norfolk,	1086 *Eccles, Heccles*	
Eccles, Berwick,	1297 *Hecles*	
Eccles, Lancashire,	1235 *Ecclesie de Eccles*	
Exley, Halifax,	1274 *Ecclesley*,	1286 *Ekelesley*
Ecclesall, Sheffield,	1267 *Ecclissale*,	1297 *Ekilsale*
Eccleshill, Bradford,	1086 *Egleshil*,	1216 *Ekelishill*
	1316 *Eccleshill*,	1379 *Eccleshill*
Ecclesfield, Sheffield,	1086 *Eclesfeld*,	1161 *Eglesfeld*
	1190 *Ecclesfeld*,	1287 *Ekelesfeld*

It has been customary to derive the name, whether standing alone or in composition, from Lat. *ecclesia*, a church. Dr Moorman is disposed to accept this explanation in the names with which he deals, Eccleshill and Ecclesfield; but Dr Wyld refrains from expressing any opinion in regard to the Lancashire Eccleston, and unfortunately Dr Skeat does not deal with the Berkshire name Ecclesborough. It is certain, however, that the Celtic races made considerable use of *Ecclesia* in their early place-names. In Cornwall the word became *Eglos*, as in Egloshayle and Egloskerry; in Ireland it is represented by *aglish*, as in Aglishcormick and Aglishdrinagh; in Wales it became *eglwys*, as in Eglwysbach and Eglwysfair; in Scotland it is represented by *eaglais*, reduced to *les* in Lesmahagow; in Brittany by *ilis*, as in Bodilis and Lannilis. Moreover, Hogan gives several examples of the early form *eclas*, as in *Eclas Peatair* for S. Peter's, Rome.

The frequent occurrence of the simple name provides perhaps the best argument for accepting *ecclesia* as the source, for a comparison of such compound names as Ecclesborough, Ecclesfield, Eccleston, with similar compounds in -borough, -field and -ton would lead at once to the conclusion that the first element in each case is a personal name in the genitive case; and although no such name is on record there are many ancient names which lack only the -*l* ending, Æcce, Æcci, Ecca, Ecci, Ecco, for example. It is of course, not impossible that the simple name Eccles should be merely a genitive. But it is most improbable that so exceptional a use should occur so frequently, and therefore when the name is used alone some other sense must be found— probably indeed also when it is used in composition. Why not *ecclesia*, after all? Is there, indeed, any alternative? The answer to this question is not yet clear, but there are hints which may lead to a solution. Note, therefore, the following names involving the element *ec-* or *eck-* :

Eck, a loch in Argyll,	1595 *Heke*
Eckford, in Roxburgh,	1200 *Eckeford*
Eccup, near Leeds,	1086 *Echope*

and note also the following examples where the element *ac-* or *ack-* is involved :

Hackforth, near Bedale,	1086 *Acheford, Acheforde*
Acle, in Norfolk,	1086 *Acle*
Acklam, near Malton,	1086 *Aclum, Achelu'*
Acklam, near Middlesborough,	1086 *Aclum, Aclun*

Here then we have seven names, one denoting a lake, two denoting fords, and the remaining four attached in each case to places where there are streams. It seems not improbable, indeed, that the element *ac- ec-* is a term applied to water, and Förstemann actually records a stem AK as occurring in German rivernames, giving as an example the Agger, formerly *Ackara*.

A glance at the compounds in Eccles- shows that the terminal in four cases is -bourne or -brook, and in one it is -ford. One more point. In his notes on the possessions of Fountains Abbey in Kirkheaton (*Heton*), Burton speaks of '*a sichet called Eccelds*' —a little stream called Eccelds—and I take *Eccelds* to be *Eccels* with an intrusive *d*, due perhaps to the influence of ON *keld*, a fountain.

EDDERCLIFFE, EDDERTHORPE, Liversedge and Darfield.—The latter is spelt *Edricthorp* in YD 1253 and YS 1297, and *Edirthorp* in DN 1377. We may explain it as

Edric's village,' an Anglian personal name—from OE Eadric—being joined to the Scandinavian *thorp*. Possibly the first element in Eddercliffe is the same, but it may be OE *eodor*, a hedge, fold, enclosure.

EDENTHORPE, Doncaster.—I have not found any early record of this name, which may perhaps be interpreted 'the village of Eden.' A charter dated 1240 has a witness named Thomas Hedne, and Brons gives the Frisian name Eden, while LN has the ON name Edna.

-EDGE.—NED explains the word 'edge' as 'the crest of a sharply-pointed ridge, the brink or verge of a bank or precipice,' the meaning in Scotland being given as 'a ridge or watershed.' It comes from OE *ecg*; compare ON *egg*, MHG *egge*, *ecke*, MLG *egge*, OS *eggia*. As to the significance of OE *ecg* Wyld says that in place-names it appears to mean 'edge, point, cliff, declivity,' also probably 'ridge.'

The examples in South-west Yorkshire include Crackenedge, Hullenedge, Netheredge, Liversedge, Stanedge, as well as Bole Edge, Chevin Edge, Crow Edge, Elland Edge, Hove Edge, Quick Edge, Thornhill Edge, Winter Edge.

EDDISH.—HS 1637 has *Eadish feild* in Ecclesfield, SE 1715 has *Edish Close* in Emley, while a field called *Eddish Hawkswell* appears in the Kirkheaton Tithe Award dated 1845. The OE word from which these are derived is *edisc*, a pasture or park.

EDLINGTON, Conisborough.—See Adlingfleet.

EGBOROUGH, a thinly-populated parish near Snaith, appears in early records under the following forms:

DB 1086 *Egeburg, Acheburg, Eburg*		PF 1202 *Egburgh*	
KC 1194 *Eggeburg*		CR 1249 *Eggeburg*	
KC 1199 *Eggeburg*		NV 1316 *Eggeburgh*	

The termination comes from OE *burh*, a fortified place, a fortress, and the first element is doubtless a personal name, probably ON Eggi. The Domesday forms *Acheburg* and *Eburg* appear to be faulty.

EGERTON, Huddersfield, IN 1311 *Eggerton*, PT 1379 *Hegerton*, DN 1461 *Egerton*, is pronounced Edgerton, and probably means 'Ecgheard's farmstead,' from OE *tūn*, an enclosure, farmstead. The sign of the genitive has been lost as in Alverley and Alverthorpe.

EGYPT.—This name, whether due to ancient or modern events, occurs in Gomersal and Thornton.

ELLAND, Halifax, is not to be connected with OE *elland*, a foreign country, seeing that early records of the name show quite regularly only one *l*:

DB 1086 *Elant, Elont*	KI	1285 *Eland*
PR 1167 *Eiland*	NV	1316 *Eland*
FC 1199 *Eland*	WCR	1422 *Eland*
PF 1202 *Elande*	WCR	1613 *Ealand*

The first element is the same as that found in the Lancashire Emmott, 1295 *Emot*, and in one of the Bedfordshire Eatons, DB *Etone*, HR *Etone*; it represents OE *ēa*, ME *ee, e*, a stream, river. The second element is from OE *land*, which meant an estate or territory as well as land. It will be noticed that in both Elland and Emmott the initial vowel is now short. The spelling *Eiland* in PR 1167 corresponds with that in a second Bedfordshire Eaton, DB *Eitone*, HR *Eyton*. Its origin is OE *ēg*, ME *ey*, an island. Obviously Elland means 'the estate beside the water' rather than 'island estate.'

ELMHIRST, Worsborough, PT 1379 *Elmerst*, YD 1415 *Elmehyrst*, comes from OE *elm·hyrst*, a small wood of elms.

ELMSALL, South Kirkby, DB *Ermeshale*, YI 1264 *Elmesale*, YR 1268 *Suth Elmeshale*, NV 1316 *Elmesall*, PT 1379 *Elmeshale*. The Domesday scribes sometimes wrote *r* for *l*; hence we may re-write the DB form as *Elmeshale*, and so obtain a consistent series of forms meaning 'Elm's corner,' from OE *healh*, a nook, corner, meadow. Seeing that Naumann records the Danish name Almi, from ON *ālmr*, an elm, we have support in postulating an OE name Elm; compare also the East Riding name Emswell, DB *Elmesuuelle*.

9—2

ELSECAR, Barnsley, has a Scandinavian terminal, from ON *kjarr*, copsewood, brushwood, and its prefix may well be a Scandinavian personal name; Falkman gives Elsa and Nielsen has Elso.

EMLEY, Wakefield, has usually been explained as derived from the elm, and attempts have been made to change the name to Elmley.   Early spellings, however, show there is no warrant for such a course, witness DB *Ameleie, Amelai,* YR 1238 *Emmele,* YI 1266 *Emmelay,* HR 1276 *Emmele,* NV 1316 *Emeley,* PT 1379 *Emlay.*   The explanation is 'Æmma's lea,' from OE *lēah,* a lea or meadow, and the ancient personal name Æmma.   See Adlingfleet.

EMMET BRIDGE, Bradfield.—A very common Norwegian place-name is Aamot, which means a confluence, a meeting of the waters; and Emmet is the corresponding English word. This is sometimes spelt Emmot or Emmott, as in Emmott near Nelson, LC 1295 *Emot* (Wyld), and in the East Riding name Emmotland.   The etymology is from OE *ēa,* ME *e, ee,* water, a stream, river, and OE *gemēt,* or OE *mōt,* a meeting.   The two vowels, though formerly long, are now short; see Elland.

ENDCLIFFE, Sheffield, provides a curious instance of popular etymology, for the name was *Elcliffe* in 1333 and *Elclyff* in 1577 (Gatty).   The word probably comes from Dan. *el, elle,* an alder (Falkman), and ON *klif.*

-ER.—Along the western border no termination shows itself more frequently than this.   From the names of woods and hills, and cloughs and lanes, we might gather a hundred examples. Unfortunately, owing to the absence of early records, attempted explanations cannot claim to be more than suggestions; yet it is clear that the words are derived from more than one source.

1. Some of the words appear to be Scandinavian plurals, among them the following:

Asker Wood, Birstall,	ON *askr,* an ash-tree
Busker Lane, Skelmanthorpe,	ON *buskr,* a bush
Clapper Hill, Midgley,	ON *kloppr,* a rock
Stocker Gate, Shipley,	ON *stokkr,* a trunk

2. Others may well come from ON *erg*, a shieling, summer pasture, among them a certain number where the termination is -ar. Compare the list of names given on page 30, which includes Salter, Winder, Potter, Docker, Torver, Cleator, Feizor, Medlar, Golcar.

3. A certain number of names show palatalization in the final consonant of the first element. Among such names we find the following:

Badger Hill, Rastrick	Rotcher, Marsden
Gaukrodger, Sowerby	Rotcher, Holmfirth
Ledger Lane, Lofthouse	Rodger Leys, Mixenden
Ratcher, Stansfield	Scatcher, Liversedge

4. Still a fourth group is doubtless connected with the Teutonic termination -er dealt with by Jellinghaus; among the examples which he gives we find Atter, Diever, Erder, Eller, Kilver, Schieder, Wewer.

Further examples are the following, no attempt being made to classify them:

Bagger Wood, Stainborough	Pepper Lane, Bramley (Leeds)
Bloomer Gate, Midgley	Pepper Hill, Shelf
Capper Clough, Saddleworth	Pinnar Lane, Southowram
Cocker Edge, Thurlstone	Ramper Road, Laughton
Cooper Lane, Shelf	Roker Lane, Pudsey
Corker Lane, Bradfield	Roper Farm, Queensbury
Currer Laithe, Keighley	Sagar Lane, Stansfield
Deffer Hill, Denby	Screamer Wood, Bradley
Draper Lane, Wadsworth	Seckar Wood, Woolley
Drummer Lane, Golcar	Soaper Lane, Shelf
Farrar Height, Soyland	Silver Wood, Ravenfield
Fryer Park, Whitley	Slipper Lane, Mirfield
Hamper Lane, Hoylandswaine	Stotter Cliff, Penistone
Haychatter, Bradfield	Swiner Clough, Holme
Heater, Cumberworth	Tinker Hill, Bradfield
Hepper Wood, Whitley	Toller Lane, Wilsden
Hunter Hill, Ovenden	Trimmer Lane, Stansfield
Ibber Flat, Keighley	Trister Hill, Cawthorne
Knowler Hill, Liversedge	Waller Clough, Slaithwaite
Liner Wood, Whiston	Weather Hill, Lindley
Nicker Wood, Todwick	Wicker, Sheffield
Nopper Head, South Crosland	Wither Wood, Cumberworth
Oliver Wood, Hopton	

ERRINGDEN, Hebden Bridge, shows changes of an unusual character. WCR 1277 has *Ayrykedene*, and WCR 1308 *Ayrikedene*; in 1336 we find *Heyrikdene*, 1348 *Hairweden*, 1447 *Ayringden*, and 1560 *Airingden*. The first three of these appear to involve the ON personal name Eirikr, that is, Eric; the fourth reminds one of certain forms of the Gael. *airigh*, a shieling, found in such names as Golcar; the fifth and sixth show the influence of such names as Manningham and Trimingham. The meaning is 'Eric's valley,' from OE *denu*.

EWDEN, EWES, EWOOD.—Many other place-names have the same prefix, Ewhurst and Ewshott for example. They go back to OE *ēow*, ME *ew*, a yew-tree.

EWDEN, Bradfield, CR 1290 *Udene*, YD 1307 *Udene*, may perhaps be 'yew-tree valley,' from OE *denu*, a valley.

EWES, Firbeck and Worrall, the former *Ewes* in 1543, means simply 'the yew-trees.'

EWOOD, Hebden Bridge, WH 1536 *Ewwod*, HW 1548 *Ewewood*, signifies 'yew-tree wood.'

EXLEY.—See Eccles.

-EY.—This termination comes from two different sources: (1) OE *ēa*, water, a stream or river, (2) OE *ēg*, *īeg*, an island. Discussing the Hertfordshire name Ayot Dr Skeat says, 'It is an interesting fact in philology, that this AS *īeg* arose from a fem. Teutonic type **ahwia*, the exact equivalent of the Lat. *aquea*, a fem. adjectival form; just as the AS *ēa*, a stream, arose from a Teutonic type *ahwa* (Goth. *ahwa*), the exact equivalent of the Lat. *aqua*. Thus the original sense of *īeg* was merely "watery," which perhaps helps to explain why it seems to have been applied to a peninsula, or a place with watery surroundings, just as freely as to a piece of land completely isolated.' Further, in his explanation of Colney and Odsey in the same county, Dr Skeat says AS *ēg*, *īeg*, 'meant not only an island in the modern sense, but any elevated piece of land wholly or partially surrounded by marshy country or flooded depressions.'

It is probably from the second of these sources, of which the Anglian form was *ēg*, that we obtain the terminal in such

South-west Yorkshire names as Arksey, Fenay, Pudsey, Wibsey, and Pugneys.

FALHOUSE, Whitley Lower, has records as follows: YS 1297 *Falles*, KF 1303 *Falles*, DN 1335 *Falehes*, YF 1534 *Fallowes*, TPR 1582 *Fallowes*, TPR 1671 *Falhouse*. There is great variation between the different forms of the word, but the interpretation is almost certainly 'the fallows,' from ME *falwes*; compare OE *fealh*, ploughed land. See Faugh, and compare the name with Barrow and Hallows.

FALLINGWORTH, a farmstead in Norland, is given in PT 1379 as *ffaldingworth*, and in a deed †1399 as *Faldyngworth*. The meaning is 'the farm of Falding, or the Faldings,' from OE *weorþ*; compare the name Westfaldingi which was used of the men of Vestfold in South Norway.

FALTHWAITE, Stainborough.—Locally pronounced Faulfitt. Early forms of the word are PF 1235 *Falgthwayt*, CH 1333 *Falghthweit*, DN 1386 *Falthwayt*, PT 1379 *ffaltwaith*. The name combines the OE *fealh*, ME *falghe*, ploughed land, and ON *thveit*, a paddock or clearing.

FARNLEY, FARNLEY TYAS.—Here are some of the early spellings of these words, spellings which show how essential it is to obtain records as early as possible.

FARNLEY, Leeds	FARNLEY TYAS, Huddersfield
DB 1086 *Fernelei*	DB 1086 *Ferlei, Fereleia*
PC †1220 *Farnelei*	PF 1236 *Farlegh*
KI 1285 *Farneley*	NV 1316 *Farneley*
NV 1316 *Farneley*	DN 1361 *Ferneley Tyes*

In the former case the explanation is 'lea of the ferns,' from OE *fearn*, a fern, and *lēah*, a lea or meadow; but in the case of Farnley Tyas the recorded spellings are in obvious conflict, and two other meanings are possible: (1) 'the far lea,' from OE *feorr*; (2) 'the lea of the boars,' from OE *fearr*.

PF 1236 connects 'Farlegh' with 'Baldwin le Teys,' whence the name Tyas, and a charter about the same date shows that

Roger de Notton granted all his lands at Farnley and Notton to 'Baldwinus Teutonicus.'

FARSLEY, Bradford, DB *Fersellei*, PF 1203 *Ferselee*, PC †1220 *Ferseleia*, KI 1285 *Ferselay*, NV 1316 *Ferslai*, is probably 'the gorsey meadow,' from *fyrsa*, gen. pl. of OE *fyrs*, furze, gorse, and *lēah*, a meadow. Note that the Norman scribes wrote *e* for OE *y*, and that BCS 938 has *Fyrsleage*.

FARTOWN, Huddersfield, is a name of comparatively late formation, witness the termination -town. The meaning 'distant farm,' from OE *feorr*, far, is probable; but the meaning 'sheep farm,' from ON *fær*, a sheep, is also possible. Note that *far*, sheep, and *far-pastures*, sheep-pastures, are found in the dialect of the county.

FAUGH.—In EDD the word 'faugh' is explained as fallow ground, and is derived from OE *fealh*. See Hale, and compare 'faugh' and 'fallow' from OE *fealh* with 'haugh' and 'hallow' from OE *healh*. The name occurs in Huddersfield, at Midgley (Halifax), and elsewhere; and Falhouse is a corruption of another form of the word.

FEATHERSTONE, FEATHERBED MOSS, FEA-THER TEAM.—There are no early records of the second and third, which are situate in Saddleworth and Rishworth; but of Featherstone, Pontefract, we find DB *Ferestane*, PC 1155 *Federstana*, PF 1166 *Fetherstan*, PC 1192 *Fethirstana*, CR 1215 *Fetherstan*, YI 1299 *Fethirstan*, NV 1316 *Fetherstan*. These prove that the termination is from OE *stān*, a stone, not OE *tūn*, a farmstead.

Three other points are clear. In the first place the first element in Featherbed and Feather Team can scarcely be a personal name. In the second place the first element in Feather-stone—Feather, not Feathers—need not be a personal name, though OE has the name Fæder, and ODan. Fathir. And thirdly, it is most unlikely that OE *feðer*, a feather, should be involved.

On the other hand there are many Scottish place-names where the early forms have some such prefix as 'fethir' or 'fother'—compare Fetternear, 1157 *Fethirneir*; Fetteresso, 1251 *Fethiresach*; Forteviot, 970 *Fothuirtabaicht*; Fordoun, 1100 *Fothardun*; Fettercairn, 970 *Fotherkern* (Johnston). McClure suggests that this Fother or Fethar means 'woodland,' and that Furness, formerly Futher·ness, preserves the same word in a contracted form. As to our West Riding names it is not easy to speak with assurance.

FELKIRK, FOULBY, both near Wakefield, are particularly interesting names presenting as they do the same variation of vowel in their early forms:

CR	1215 *Felkirke*	YD	1318 *Folby*
CR	1226 *Folkirke*	PT	1379 *ffelby*
YR	1252 *Felechirche*	WRM	1391 *Felby* (surname)
YD	1318 *Folkirk*	YD	1398 *Folby*
DN	1555 *Felkirke*	YF	1553 *Folbye*

Comparing these with WC *Felebrige* and WH *Felinge*, it seems certain that all the four terminations are Scandinavian, and that we must therefore look for a Scandinavian source for the first element. There seems no other possible word than ON *fjöl* (stem *fjal-*) which is explained as a thin board or deal, and from which comes ON *fjala-bru*, a bridge of planks. Thus we may interpret Felkirk as 'the plank church,' Foulby as 'the farmhouse of planks,' Felebrige as 'the bridge of planks,' and Felinge as 'the field where planks are stored.'

In regard to the variation of vowel shown in the early forms it should be noted that Prim. Norse *e* was under certain circumstances 'broken' into *ia, io*, and that in East Scandinavian these diphthongs were liable to the so-called 'progressive i- mutation,' through which *ia* became *iæ*, and *io* became *iø*[1]. According to Torp ON *fjol* goes back to a Prim. Germ. **felo*, and we may therefore take the early forms *Folkirk* and *Folby* as evidence that the 'breaking' had sometimes taken place at the time when the Danes made their settlements in the West Riding.

---

[1] Björkman, *Scandinavian Loan-words*, II, p. 292.

FENAY, FENWICK, which may well be considered together, show the following early form:

WCR	1274	*Fyney*	PF	1206	*Fenwich*
WCR	1295	*Feney*	PF	1208	*Fenwic*
WCR	1308	*Fynee*	CR	1251	*Fenwyke*
DN	1347	*Finey*	IN	1296	*Fenwyk*
DN	1393	*Fenay*	YF	1496	*Fenwyk*

FENWICK, Snaith, is obviously ' fen village,' from OE *fenn*, a fen, marsh, moor, and *wīc*, an enclosure, habitation, village.

FENAY, Almondbury, stands on a ridge almost surrounded by deep valleys. Its terminal comes from OE *ēg*, ME *ey*, an island. But it is not easy to decide what is the origin of its first element. Possibly it comes from OE *fina*, a woodpecker, or OE *fin*, a plant-name.

FERRYBRIDGE, Pontefract, is a place of historic interest. In 1461 it was the scene of an important skirmish which preceded the battle of Towton. The name is recorded as follows: DB *Fereia, Ferie*, PC 1192 *Feri*, FC 1199 *Ferybrigge*, HR 1276 *Ferye*, WCR 1326 *ffery*, DN 1343 *Ferribrig*, and it is derived from ON *ferja*, a ferry. Under the date 1316 the Pontefract Chartulary has a reference to this place which speaks of a portion of ground as 'abuttant super Limpit,' an interesting reference to what has long been a well-established industry.

-FIELD.—This termination comes from OE *feld*, ME *feld, feud*. In its original sense it denoted a plain, land naturally open, unenclosed country, as opposed to woodland or land cleared of forest; to-day, however, it is used to signify an enclosure. In OE the word was sometimes used as a prefix: *feldcirice* was a country church, *feldbēo* a locust, *feldhūs* a tent, and *feldminte* wild mint.

Examples: Austerfield, Bradfield, Briestfield, Broomfield, Darfield, Dransfield, Eastfield, Ecclesfield, Hemingfield, Huddersfield, Langfield, Mirfield, Oldfield, Ravenfield, Scholefield, Sheffield, Stansfield, Wakefield, Warmfield, Westfield.

FINKLE EDGE, FINKLE STREET, FINK HILL.— Fink Hill occurs in Barkisland, and Finkle Edge on the moors between Holmfirth and Penistone; but Finkle Street is to be found in or near Brighouse, Sowerby, Pontefract, and the southern Wortley.

NED gives a word 'finkle' meaning fennel, which comes to us from Lat. *fœniculum* through ME *fenecel*. It is possible that this word accounts for some of the names under discussion; but it is certain that the Dan. word *vinkel*, an angle or corner, is beside the mark, for the corresponding English form would be 'winkle.'

FINTHORPE is in Almondbury.

FIRBECK, Tickhill, HR 1276 *Frithebek*, PT 1379 *ffirthbek*, YD 1403 *Frythbeke*, comes from ON *bekkr*, a brook, and OE *frith*, *fyrhthe*, a wood, wooded country; see Holmfirth.

FIRSBY.—See Friezland.

FISHLAKE, Thorne, is usually presented to us in an Anglian form, e.g. DB 1086 *Fiscelac*, HR 1276 *Fishelak*, CH 1398 *Fyshelake*, where the meaning is simply 'fish stream,' from *fisca*, gen. pl. of OE *fisc*, a fish, and OE *lacu*, a stream, a channel. In YR 1269 we find, however, a Scandinavian form, *Fiskelake*, doubtless of the same meaning; and DB 1086 has a form *Fisccale* which seems corrupt, but may involve ON *skāli*, a shieling, shed.

FITZWILLIAM, Hemsworth, is 'a village about half-a-mile from Kinsley, formed for the accommodation of the miners working at Hemsworth colliery' (Kelly).

FIXBY, Huddersfield, DB *Fechesbi*, WCR 1274 *Fekesby*, DN 1293 *Fekisby*, NV 1316 *Fekesby*, YF 1570 *Fekesbye*, is 'the homestead of Fek'; compare the name Fech recorded in DB, and the Frisian name Feke recorded by Brons. There is on record, however, no Scandinavian name of the form. A second FIXBY occurs in Whitley Lower.

FLANSHAW, Wakefield, WCR 1274 *Flanshowe*, WCR 1277 *Flansowe*, 1369 *Flansowe*, 1391 *Flanshagh*, shows a change in its ending. In the early forms the terminal comes from ON *haugr*, a burial-mound, cairn; but in the later it is from OE *sceaga*, a copse, a wood. The prefix is plainly a personal name, and ON has Fleinn while ODan. has Flen, but neither of these would give Flan. On the other hand the name Flann is of frequent occurrence in ancient Irish history, borne among others by kings and abbots. Perhaps some Irish prince came over from Dublin with the Norsemen, and meeting his end was buried at the spot ever after called by his name—Flann's how, the cairn of Flann.

FLASH LANE, FLASK, DAMFLASK, occur respectively in or near Mirfield, Widdop, and Sheffield. In NED 'flash' and 'flask' are explained as 'a pool, a marshy place.' While 'flash' is said to be of onomatopœic origin, influenced by Fr. *flache*, which is of the same meaning, the *sk* in 'flask' seems to indicate a Scandinavian origin. Madsen records Dan. *flaske* as occurring in place-names, with the meaning 'meadows' or 'small bays encompassed with meadows.' The prefix in Damflask means a bank for restraining water, and comes from ON *dammr*, a dam, Dan. *dam*, Sw. *damm*; compare the Swedish place-names Damhus and Pildammen.

FLATT, FLATTS.—There are many place-names of which this is the second element—Crown Flatts, Dewsbury; High Flatts, Denby; Collon Flatt, Wadsworth; Cross Flatts, Bingley and Southowram. According to NED the word comes from ON *flatr*; compare Sw. *flat*, Dan. *flad*. It means a piece of level ground, a stretch of country without hill.

FLEET.—The name Fleet occurs in Skelmanthorpe near the river Dearne; and elsewhere the word is used as a termination as in the case of Adlingfleet, Ousefleet, and Trumfleet. It is from OE *flēot* or ON *fliōt*, a channel, running stream, river.

FLOCKTON, Wakefield, DB *Flochetone*, PF 1201 *Floketon*, YI 1287 *Flocton*, NV 1316 *Floketon*, may be 'Floki's farm,' from ON *tūn*, an enclosure, farmstead, and the ON personal name Floki; or it may be 'the farm of the flocks,' from OE *flocc*, a flock, and *tūn*, an enclosure, farmstead.

FLUSH, FLUSHDYKE, FLUSHHOUSE, are found respectively in Heckmondwike, Ossett, and Austonley. In his Scottish Dictionary Jamieson explains the word 'flush' as a morass, a piece of moist ground, a place where water frequently lies, and EDD gives a similar explanation.

FLY FLATT, near Midgley (Halifax), must be connected with ON *flōi*, a marshy moor, Norw. *fly*, a moor, a marshy tableland, Sw. *fly*, a marsh, moss. According to Madsen the name Flye occurs also in Denmark with the meaning 'marsh' or 'morass.' The word goes back to the Teutonic stem **fluhjō*, a marshy tableland (Torp).

FOCKERBY, on the Lincolnshire border, YI 1256 *Folkehuardeby*, 1304 *Folquardby*, 1358 *Folkquardby*, PT 1379 *Fowewardby*, is 'Folkward's homestead.' The first element is a personal name, and ON has Folkvarðr (Naumann), while ODan. has Folkwarth (Nielsen); in addition Naumann records the weak form Varði. The second element is from ON *bȳr*, a farmstead, and the first probably represents the genitive of *Folkvarði.

-FORD is from OE *ford*, a road or passage through a stream. The word occasionally becomes -forth or -worth as in Stainforth and Brinsworth. Among the Domesday place-names in South-west Yorkshire while there is only one bridge and one ferry, there are many fords. To the wayfarer the ford was a matter of the deepest concern, and the all-important questions must continually have been asked : 'How deep is it?' 'What sort of bottom has it got?' 'Who lives beside it?' 'How is it marked out?' Accordingly we find such names as Defford and Shalford, the deep ford and the shallow ford; Cuford, not too deep for a cow; Horseford, which may be crossed on horseback; and Wainford where a waggon could be got from bank to bank. Further, we find names like Rufford, where the river bed was rough and uneven; Sandford with its sandy bottom; and Stainforth where perhaps it was paved with stones. Other typical names are Wudel's ford, Werm's ford, and Creve's ford, Salford near the willows, Milford near the mill, Stapleford marked by a post, and Castleford defended by a castle.

Examples in South-west Yorkshire include Addingford, Battyeford, Bradford, Brinsworth, Castleford, Cleggford, Cooper, Deffer, Dunford, Keresforth, Salford, Stainforth, Strangford, and Woodlesford.

FORDOLES is the name of a farm near Rotherham ; see the note on Dole.

FOULBY.—See Felkirk.

FOULSNAPE.—See Snape.

FRIARMERE is one of the four 'meres' into which the township of Saddleworth is divided. Prior to 1468, the earliest date at which the name Friarmere occurs, we find *Hildebrighthope* 1293, *Ildbrictop* 1297, and in the 14th century *Hildebrighthope*, *Hilbdebrighthope*, and *Hillbrighthorpe*. This means 'Hildebeorht's secluded valley,' from OE *hop*; but Friarmere, which tells plainly of the connection with Roche Abbey, is from OF *frere*, Lat. *fratrem*, and OE *mǣre*, a boundary, border.

FRICKLEY, Doncaster, DB *Friceleia, Frichehale,* YI 1246 *Frikeley,* KI 1285 *Frikelay, Frykelay,* NV 1316 *Frikley,* appears to have as its first element either the OE *fricca*, a herald, or OE *freca*, a warrior—more probably, indeed, the former—used, as Moorman suggests, as a personal name. No name of the form required is recorded either by LV, Searle, or Naumann, but Brons has the Frisian names Frike and Frikke.

FRIEZLAND, FRIZINGHALL, FRYSTON, FIRSBY. —In describing the conquest of Britain in the 5th and 6th centuries, Bede says 'those who came over were of the three most powerful nations of Germany—Saxons, Angles, and Jutes' (1, 15), and the AS Chronicle mentions only the same tribes. There can be no doubt, however, that among the early invaders there were Frisians. A passage in Procopius, which dates from a time nearly two centuries earlier than Bede's work, reads as follows : 'The island of Brittia contains three very populous nations, each of which has a king over it. The names borne by these nations are Angiloi and Phrissones and Brittones, the

last having the same name as the island[1].'   At a later period
we find Frisians assisting the Danish invaders; Henry of
Huntingdon, after speaking of the impiety of the Anglo-
Saxons, declares that 'The Almighty therefore let loose upon
them the most barbarous of nations, the Danes and Goths,
Norwegians and Swedes, Vandals and Frisians.'   Under these
circumstances we may expect to find traces of the Frisians in
our place-names, and the Frystons of Yorkshire and Lincoln,
the Frisbys of Leicester, and Friseham in Devon have been
quoted as examples.

The OE name for the Frisians was Frisan or Fresan, and
the personal name Frisa or Fresa originally denoted a member
of that nation.   Other forms of the name given by Schönfeld
are Frisii and Frisiones as well as OFris. Frisa and Fresa.
Dumfries, which appears in Nennius as Caer Pheris, is explained
by Skene as 'fort of the Frisians.'

On the Aire near Ferrybridge there are three Frystons,
Monk Fryston on the northern bank, Water Fryston and Ferry
Fryston on the southern.   Among early records we find the
following:

BCS	963 *Fryssetune*[2]		PM	1247 *Fristone*
WHS	†1030 *Fristun*		CR	1255 *Friston*
DB	1086 *Frystone*, *Fristone*		NV	1316 *Frystone*
PC	1155 *Fristona*		PT	1379 *ffryston*

FRYSTON appears therefore to be 'the homestead of Frisa,
the Frisian,' from OE *tūn*, an enclosure or farmstead, an explanation accepted by Middendorff.

FRIEZLAND, spelt *Frezeland* in the Parish Registers of
Saddleworth, may well be the 'land of Fresa, the Frisian.'

FRIZINGHALL, Bradford, YI 1287 *Fresinghale*, DN †1287
Fresinghal, DN 1424 *Frisinghall*, YF 1567 *Frysynghall*, is 'the
corner of the sons of Fresa,' from OE *healh*, a corner or meadow.
Compare BCS 951 *Frisingmæde*, referred by Middendorff to
Fresa, Frisa; and note also Fressain, Pas de Calais, formerly
according to Mannier written *Fresingahem*.

---

[1] Quoted from Chadwick, *The Origin of the English Nation*, p. 55.
[2] BCS III, 345; the spelling *Frythetune* in BCS III, 695 has no support.

FIRSBY, Doncaster, WCR 1275 *Friseby*, YF 1504 *Frysby*, is 'the homestead of Frisi,' ON *býr*. Nielsen gives the ODan. personal name Frisi and quotes two place-names derived therefrom, namely, Frislev and Fristrup.

FULSTONE, Holmfirth, occasionally spelt Foulston, appears in DB as *Fugelestun*, and later as follows:

WCR 1274 *Fugeliston*	WCR 1307 *Fouleston*
WCR 1298 *Fugeleston*	WCR 1309 *Fouleston*
WCR 1306 *Fugheleston*	WRM 1577 *Foulston*

As in Silkstone and Penistone the terminal comes from OE *tūn*, an enclosure or farmstead—not from OE *stān*, a stone. The meaning is 'Fugel's farm,' and the personal name comes from OE *fugol*, a fowl.

GANNERTHORPE WOOD is in Wyke.

-GATE.—This suffix may be derived from OE *geat*, a gate, door, or from ON *gata*, a road, way, path. Thus whenever the word has the latter meaning it may safely be declared of Scandinavian origin, as in the case of Clapgate, Cluntergate, Howgate, Kirkgate, Skeldergate, and Slantgate. The Lidgates on the other hand are of Anglian origin.

GAWBER, Barnsley, PM 1304 *Galghbergh*, PT 1379 *Galbergh*, YF 1526 *Galbarre*, is 'gallows hill,' from OE *gealga* or ON *galgi*, gallows, and OE *beorg*, a hill, mound, or ON *berg*, a rock. In the olden days every 'Thing' had its gallows-hill, and perhaps Gawber was the place of execution for the wapentake of Staincross.

GAWTHORPE.—South-west Yorkshire has two places of this name, the prefix being the ON personal name *Gauki.

GAWTHORPE, Ossett, HR 1276 *Goucthorpe*, WCR 1298 *Goukethorpe*, means simply 'Gauki's hamlet,' from ON *thorp*.

GAWTHORPE, Lepton, YS 1297 *Goutthorp*, DN 1324 *Gawkethorp*, PT 1379 *Gaukethorp*, shows a variation in the prefix which represents a change from the personal name Gauti to Gauki.

GIB, GIBB.—This word occurs in Gibb Hill, Ovenden, Gib Slack, Wadsworth, Gibriding, Austonley, and stands alone as Gibb in Honley. NED records a word *gib* which means a male cat or a male ferret, and another word of the same form which signifies a hump.

GILCAR, GILDERSOME, GILROYD, GILTH-WAITE, UGHILL.—The ON *gil*, a valley or ravine, occurs in all these words, as well as in Gill Hey and Gill Lane near Holmfirth, and Gill Sike near Wakefield. These names are interesting because Gill comes to us from the Norsemen, rather than from the Danes.

GILCAR occurs in Emley, SE 1715 *Gilcar*, in Elland, and in Sheffield, and means 'the carr in the valley,' from ON *kjarr*, copsewood, brushwood.

GILDERSOME has been explained as 'the home of the Gelders,' and the explanation has been supported by the statement that 'Dutch settlers introduced the manufacture of cloth in 1571.' The story revealed by the name itself is, however, of quite a different character. The early spellings include

YI	1249 *Gilhusum*	DN	1461 *Gildosome*
WCR	1294 *Gildusme*	YF	1504 *Gyldersom*
DN	1435 *Guildesham*	YF	1563 *Gyldersum*

It is plain that so far as the place-name is concerned the legend of the Gelders must be abandoned, the sense being simply 'the houses in the gill, or valley,' from ON *hūs*, a house. The history of the word shows two other interesting points, the intrusion of *d* as a supporting consonant, and the development of *-er* from the indefinite vowel of the second syllable.

GILROYD occurs in Barkisland, Linthwaite, Morley, Dodworth, and Bradfield, and its meaning is 'the clearing in the valley'; see Royd.

GILTHWAITE occurs in Skelmanthorpe and near Rotherham, early records of the latter being YD 1342 *Gilthwayt*, YD 1409 *Gylthewaite*. The sense is 'the clearing in the valley,' from ON *thveit*, a clearing.

UGHILL, Bradfield, DB *Ughil*, YS 1297 *Wggil*, 1337 *Ughill*, YF 1536 *Ughill*, is, I imagine, another place-name involving the ON *gil*; compare Raygill near Skipton, DB *Raghil*.

GIRLINTON, Bradford, PT 1379 *Gryllyngton*, is 'the farmstead of Gyrling or the Gyrlings.' In DB we find the personal name Gerling where *e* represents OE *y*, and Skeat gives the ME forms of 'girl' as *gyrle, girle,' gerle*, all referred to OE *gȳr-el*.

GLEADLESS, GLEDHILL, GLEDHOLT.—In addition to the meaning joyous or glad, OE *glæd*, ME *glad, gled*, had the signification bright or shining; that, indeed, was the original sense; compare ON *glaðr*, bright. In *The Flower and the Leaf*, a poem of the 15th century, we find the leaves of trees referred to as 'Som very rede, and some a glad light grene.' A Bedfordshire place-name, Nares Gladly, spelt *Gledelai* in DB, is derived from this source by Professor Skeat.

GLEADLESS, Sheffield, HH †1277 *Gladeleys*, HH †1300 *Gledeleys*, YF 1549 *Gledeles*, YF 1561 *Gledles*, SM 1610 *Gledles*, means 'the bright leas,' OE *lēah*, a lea or meadow. The modern ending is quite extraordinary, yet a similar case occurs near Scarborough, where the ancient *Kirkelac*, church lea, is represented by the modern Kirkless.

GLEDHILL, Halifax, WCR 1275 *Gledehull*, WCR 1277 *Gledehyll*, is 'the bright hill." The name Gledhill occurs also in Almondbury, Hartshead, and Warmfield.

GLEDHOLT, Huddersfield, LC 1296 *Gledeholt*, WCR 1298 *Gledeholt*, WCR 1308 *Gledeholte*, is 'the bright wood,' OE *holt*, a grove or wood.

GODLEY, Greetland, Rishworth, and Northowram.—In 1307 and 1308 WCR gives the second as *Godelay*, and in PT 1379 *Godlay* occurs. The meaning is 'the lea of Goda.'

GOLCAR, Huddersfield.—'Letters, like soldiers, are very apt to desert and drop off in a long march,' said Horne Tooke; and it would be very difficult to find a better illustration of the fact than that presented by Golcar. Here are some of the early spellings :

DB	1086 *Gudlagesarc, Gudlagesargo*	WCR	1308 *Gouthelaghcharthes*
WCR	1272 *Gouthelaghcharthes*	WCR	1309 *Goulaghcarthes*
YI	1286 *Goutlackarres*	DC	1349 *Gouldelakekerres*
WCR	1306 *Gouthlacharwes, Guthlacharwes*	DC	1351 *Gouldelakkerres*

DC 1356 *Goullakarres*   HW 1481 *Goulkery*
DN 1398 *Guldecar, Guldeker*   YF 1535 *Golcar*
YD 1438 *Gowlkar*   SE 1715 *Gowker*

This is perhaps the most interesting of all our South-west Yorkshire place-names.   Since the 14th century the terminal has taken the form 'ker' or 'car' as though from ON *kjarr*, copsewood, Dan. *kær*, a bog, fen; but obviously the original name was neither 'Guthlac's carr' nor 'Guthlac's scar.'   The Domesday forms appear to represent a Nom. *Guthlages·arg* and a Dat. *Guthlages·arge*, in which the first element is the ON personal name Guthlaug, while the second represents ON *erg*, a shieling.   A passage in the Orkney Saga, describing an event in the year 1158, makes use of this word and equates it with the ON *setr*; and a second passage speaks of 'some deserted huts which are called *Asgrim's erg*.'   The word is derived from ÓIr. *airge*, a place where cows are kept, with which is connected the Gaelic word *airigh* or *airidh*, a shieling or hill pasture (Macbain).

Golcar is by no means an isolated example of the use of this word.   In a great crescent stretching from Whitehaven in the north to Liverpool in the south, ancient names in which it occurs as the second element number quite a score[1].   All these names have a most interesting story.   During the 9th century bands of Vikings rounded the north of Scotland and took possession of parts of the Western Islands and Ireland, as well as Argyll, Galloway, and the Isle of Man.   They made the conquered Celts their thralls; and when their descendants crossed the Irish Sea and landed on the shores of Morecambe Bay, or at the mouth of the Ribble, they brought with them not only their thralls, but also words of Celtic origin such as *erg* and *cros*.   While some of these Vikings were content to settle in Cumberland or Westmorland or Lancashire, others made their way eastward into Yorkshire; and to-day the story of their wanderings is written in indelible characters on the map of our country.   Such place-names as Golcar and Crosland, Staincross and Osgoldcross, bear witness to the fact that among our ancestors some were Norsemen who sailed down the west coast

---

[1] See pp. 29—30.

of Scotland, settled most probably for a time in Ireland, and thus before they crossed to the north-western shores of England were 'in familiar contact with the Celtic race[1].'

GOLDTHORPE, GOWDALL.—Goldthorpe is mentioned in DB as *Goldetorp*, *Guldetorp*, *Godetorp*, and later spellings of the two names are as follows:

KI	1285 *Goldthorpe*	PC	1220 *Goldale*
CH	1307 *Goldethorpe*	YI	1280 *Goldale*
YF	1528 *Goldethorpe*	DN	1353 *Goldhale*

GOLDTHORPE may fairly be explained as 'Golda's village,' from ON *thorp*. Golda and *Guldi appear to be respectively the OE and ON forms of the personal name (Naumann).

GOWDALL is 'Golda's corner,' from OE *healh*, a meadow or corner, and the personal name Golda recorded by Searle.

GOMERSAL, Bradford, DB *Gomershale*, *Gomeshale*, DC 1246 *Gumereshale*, HR 1276 *Gumereshale*, KI 1285 *Gomersalle*, NV 1316 *Gomersall*, is 'Gummær's corner,' from the recorded personal name Gummær and OE *healh*, a corner or meadow.

GOOLE is situated on a bend of the Ouse near its confluence with the Don. Early records are YF 1553 *Gowle*, YF 1558 *Goole*, YF 1564 *Gowle*. It is impossible to say definitely what is the origin of the word. Perhaps it may be the AF *gole*, which from the 16th to the 19th centuries took the forms *goule*, *gool*, *goole*, and which among other significations meant a small stream, a ditch, or sluice.

GRAIN.—In moorland districts this word occurs with some frequency as the designation of small streams. It is found in its simple form in Wadsworth, Mixenden, and Holme, while Widdop has Grainings, and Rishworth Oxygrains. The word comes from ON *grein*, a branch or arm.

GRANGE.—Instances of this name occur at Denby Grange near Wakefield, and Grange Hey in Saddleworth. The word is

---

[1] *West Riding Place-names*, pp. 216—8.

derived from AF *graunge*, which meant not only a building in
which grain was stored or cattle housed, but also a farm with its
outbuildings. It was applied more particularly to the outlying
farms of religious houses; and it is interesting, therefore, to find
a historic connection between the two granges just mentioned
and a great abbey, lands in Saddleworth being formerly held by
Roche, while lands in Denby belonged to Byland.

In his *Dictionnaire d'Architecture* Viollet-le-Duc gives many
interesting details relating to the French Abbey granges.
He tells us that the great abbeys took care to surround their
granges with walls fortified by watch-towers and pierced by
strong gateways. The granges were occupied by monks who
were 'sent down' as a penance for some fault, as well as by lay
brothers and peasants. Near the gateways were outbuildings in
which when night came on the wayfaring man was able to find
shelter. Little by little groups of cottages gathered round, and
thus the nucleus of a village was formed. In time of war the
villagers would shut themselves up, he says, within the encircling
walls of the grange and there defend themselves; but occasionally, when incited by some lord at feud with the monks, the
peasants would turn round and give it up to pillage, or even—
what was little to their profit—deliver it to the flames. Not a
few places in France which bear the name La Grange have had
an origin and history just such as this.

GREASBOROUGH, Rotherham.—An examination of the
early forms reveals two types, *Greseburg* and *Gresebroc*:

DB 1086 *Greseburg*	HR 1276 *Gresebroc*	YD 1482 *Gresbroke*
DB 1086 *Gresseburg*	KI 1285 *Grissebroc*	VE 1535 *Gressebroke*
DB 1086 *Gersebroc*	YD 1390 *Gresbroc*	WPR 1678 *Greasebrough*

The common word 'grass' appears in OE as *gærs*, *græs*, and in
ME as *gras*, *gres*, *gresse*. If this word were the first element we
should expect such forms as *Gresburg* and *Gresbroc*. Hence
I suggest instead a weak personal name formed from the common
noun; compare Graso given by Förstemann, and the English
place-names Grassington, 1259 *Gersington*, and Gressingham,
1202 *Gersingeham*. The terminals come from OE *burg*, *burh*, a
fortified place, and OE *brōc*, a brook.

GREENFIELD, GREENHEAD, GREENLAND, GREENSIDE, GREENWOOD.—The first is situate in Saddleworth; the second in Huddersfield; the third in Queensbury, Cowick, and Attercliffe; the fourth in Pudsey, Kirkheaton, Thurstonland, and Ecclesfield; the fifth in Heptonstall, WCR 1275 *Grenwode*, WCR 1297 *Grenewode*. Compare OE *grēne*, ME *grene*, green.

GREETLAND, Halifax, DB *Greland*, YS 1297 *Gretland*, WCR 1308 *Gretelande*, comes from OE *grēot* or ON *griōt*, gravel, stone, and OE or ON *land*, land, estate; compare the Norwegian place-name Grjotlid, 'the gravel slope.' The word Greetland is found somewhat frequently as a field-name.

GRENOSIDE, Ecclesfield, is spelt *Granhowside* in a document quoted by Dodsworth, while HH †1260 has *Gravenhou*, and HS 1637 has *Grenowside*. These are sufficient to prove that the second element is from ON *haugr*, a burial-mound. The first element is probably connected with ON *gröf*, a pit; compare the names Graven and Gravensfiord which occur in Norway.

GREYSTONES, Ecclesall, YF 1564 *Greyestones*, is derived from OE *grǣg*, gray, and *stān*, a stone.

GRICE, Shelley.—Early 14th century charters connected with Shelley speak of *Richard de Gris*. The name is derived from OF *greys*, *greis*, which meant a flight of steps, and came from Lat. *gradus*. In Yorkshire, according to EDD, this word assumed such forms as *grise* and *grice*, and a secondary meaning of the word was an ascent or slight slope.

GRIFF.—Near Keighley is Griff Wood; in Birstall Griff Well; in Ardsley (Wakefield) Griff House. YD 1348 gives *le Gryff* as the name of a field in the last-named township. The word comes from ON *gryfja*, a hole or pit. Compare Mulgrave and Stonegrave, DB *Grif* and *Steinegrif*.

GRIMESTHORPE, GRIMETHORPE, Sheffield and Brierley.—In DB we find near Sheffield *Grimeshou*, 'the burial-mound of Grim,' from ON *haugr*, a cairn or mound. At a later date the name Grimesthorpe comes into view, YS 1297 *Grimestorp*, DN 1369 *Grimesthorp*. These warrant the interpretation 'Grim's village,' from ON Grīmr, while Grimethorpe is from the personal name Grīma; compare the Danish place-names Grimstrup and Grimetune. Grīmr was a common Norse personal name; it was also a name of Odin, who was so-called because he went about in disguise, ON *grīma* being a hood.

GUNTHWAITE, Penistone.—This word has cast off first one burden and then another, until at last only two syllables remain of the four or five with which the journey began. Early forms are WCR 1284 *Gunyldthwayt*, PT 1379 *Gunhullewayth*, 1389 *Gunletwayt*, 1490 *Gunthwait*, YF 1585 *Gunthwaite als. Gumblethwaite*. The meaning is 'Gunnhilda's clearing,' from the ON personal name Gunnhilda and ON *thveit*, a clearing or paddock; compare the Norwegian place-names Gunnildrud and Gunnerud.

HADDINGLEY.—This name occurs in Sandal and Shelley. Nielsen gives the ON name Haddingr and ODan. Hadding, and we may therefore assume a corresponding OE name. But, further, Searle has the name Hædde; hence Haddingley may be explained as 'the lea of the sons of Hædde.' Compare Haddington in Lincolnshire and East Lothian.

HADES, Marsden and Holmfirth.—Things are not always what they seem. A dialect-word *hades* which means 'a place between or behind hills and out of sight' is recorded by EDD; and the same authority gives another word *hade* as meaning 'a headland or strip of land at the side of an arable field upon which the plough turns.' The latter word is also recorded in NED and is explained as 'a strip of land left unploughed between two ploughed portions of a field.' But NED has also a verb *hade* which means to incline, to slope, and this seems to be connected with the Norw. dialect-word *hadd*, pl. *haddir*, explained by Aasen as a slope or incline. In 1534 Fitzherbert has the following expression: 'Horses may be teddered vpon leys, balkes, and hades.'

HAGG occurs as the name of a hamlet in Honley and else-where.   In EDD a dialect-word Hag explained as a cutting in a wood is derived from ON *höggva*, Sw. *hagga*, to hew ; and a word Hagg meaning a copse or wooded enclosure is said to be a form of OE *haga*, an enclosure.

HAIGH—unlike Haigh in Lancashire which rhymes with 'play'—is pronounced so as to rhyme with 'plague.'   It occurs (1) in Elland FC 1199 *Hagh*, WCR 1314 *Hagh*; (2) near Barnsley, PT 1379 *Hagh*, YF 1569 *Haghe*; (3) in South Elmsall, 1828 *Hague*.   For the form of the word compare *legh, leigh*, a meadow, and *hegh, heigh*, high, and for the meaning compare OE *haga*, an enclosure or small farm, ON *hagi*, a hedged field or pasture.   See EDD under Hag, sb. 2.

HAINWORTH.—See Haworth.

HALDENBY, Goole, CR †1108 *Haldaneby*, PF 1198 *Haldanebi*, CR 1257 *Haldaneby*, HR 1276 *Haldaneby*, NV 1316 *Haldanby*, is a Scandinavian name meaning 'the farm of Halfdane or Halfdene,' from ON *bȳr*.   Possibly the reference is to one of the two Danish Kings of York called Halfdene in various early records.

HALE, HALGH, HALLAS, HALLOWS, HAUGH.— At first sight words like Snydale and Wheldale seem to come from ON *dalr*, a dale, while Ecclesall and Hensall appear to be connected with OE *heall*, a hall.   In both cases appearances are deceptive.   Many places interpreted as the 'hall' of this or that person have a meaning quite different.   If we gather together all possible examples in South-west Yorkshire we shall find that the majority show early forms in *-hale* or *-ale*, e.g.

Backhold	1277 *Bachale*	Sandall Par.	1285 *Sandhale*
Beal	1086 *Beghale*	Skellow	1200 *Scalehale*
Darnall	†1301 *Dernhale*	Snydale	1202 *Snithale*
Ecclesall	1267 *Ecclissale*	Stancil	1086 *Steinshale*
Gomersal	1086 *Gomershale*	Thornhills	1333 *Thornyhales*
Gowdall	1353 *Goldhale*	Tyersal	1203 *Tireshale*
Hensall	1086 *Edeshale*	Wheldale	1252 *Weldale*
Sandall Mag.	1202 *Sandale*		

Compare with these the following examples where -ll appears :

Skellow	1086	*Scanhalla*
Newhall, Wath	1086	*Niwehalla*
Woodhall, Darfield	1297	*Wudehall*

It will doubtless be right to explain the last two forms as derived from OE *heall*, a hall, Skellow being plainly exceptional ; but in the first set of examples the terminal appears to come from OE *halh* or *healh*. The dative singular of this word, *hale*, would account for all the terminations in this list. It should be noted, however, that in Skellow, Snydale and Wheldale, the prefixes are probably Scandinavian, and the words are either hybrids or -hale represents the ON *hali*, a tail, Dan. *hale*, a tongue of land ; compare the Norw. place-name Refsal, that is, Refshali (Rygh), and the Danish place-names Ulvshale and Revshale (Madsen). We must examine other early forms :

1. Whitehaughs, in Fixby, was formerly *Wytehalge* (WH).
2. Westnal, Bradfield, was *Westmundhalgh* in 1329.
3. A hill in Erringden is called *Greenhalgh* by Watson.
4. Upper and Nether Haugh appear to be indicated by BM '*haleges* de Rumareis.'
5. Hallas, Bingley, is recorded in BPR 1625 as *Hallowes*.
6. A 16th century field-name in Liversedge was spelt *Linhallowes*.
7. Thornhill had many field-names of the form ; compare SE 1634 *Hallowe* and *Hallowes*.

Altogether there are four clearly defined forms, 'hale,' 'halgh,' 'haugh,' 'hallowes.' NED deals with 'hale' and 'haugh,' describing the former as derived from OE *halh, healh*, a corner, nook, secret place, and the latter as connected with the same word, and as denoting a piece of flat alluvial land beside a river forming part of the floor of a river valley. EDD explains 'hale' as flat alluvial land beside a river, or a triangular corner of land ; Stratmann gives the interpretation meadow or pasture land ; and Skeat says that one of the special applications of 'hale' was a nook of land at the bend of a river, or a piece of flat alluvial land. But none of these authorities recognise the form 'hallow' which stands to *halh* as 'hollow' to *holh*, and 'barrow' to *beorh*.

The name HALES occurs in the township of Rawcliffe, and HALLAS in Bingley, Kirkburton, and Hoylandswaine.

HALIFAX.—This is a difficult name, and unfortunately it does not appear in the Domesday record, the earliest mention being found in a Charter of William de Warrene granting the church at *Halifax* to the Priory of Lewes. Later we find the form *Halifax* with great frequency, in YR 1268, for example, in WCR 1274, HR 1276, NV 1316, and PT 1379. Occasionally there are other forms, such as DC 1586 *Hallyfaxe*.

In his history of Halifax Watson gives two interpretations current in his day. (1) The first goes back to Camden. A maiden, we are told, was slain by the monkish lover she had repulsed, and her head was hung upon a yew-tree. Afterwards, 'the little veins, which, like hairs, were spread between the bark and the tree' were believed to be the very hairs of the maiden. The place became a great resort of pilgrims; and, though formerly called Horton, was now called *Halig·fax*, that is, holy hair. (2) The second appears to be due to the author of a book published in 1708. It begins by saying that Halifax was in early days 'an hermitage of very great antiquity' and it goes on to explain the name as Holy-face, the church being dedicated to St John Baptist, and his face being 'as they pretend' kept there.

No early record of either story is known, and Camden's is discredited by the fact that nowhere can any trace be found of a change from the name Horton to that of Halifax. Moreover, it is very improbable that a town should be called either Holy-hair or Holy-face; yet on the other hand an early spelling given by Watson, *Halifaxleie*, tends to remove this improbability; see Warley.

But the further suggestion has been made that the terminal has the same meaning as in Carfax. This, however, is quite impossible, early forms showing no points of contact with the ending in Fr. *carrefour*, OFr. *carrefors*, Lat. *quadrifurcus*, four-forked.

It will be well to compare the name with parallel forms, and we may fitly begin with the curious North Riding name *Belly-faxe*. This occurs in RC 1538 in the Ministers' Accounts, where the full expression is '*cum certis pasturis vocatis Bellyfaxe.*' Apart from Carfax, already discussed, no other place-name with the ending *-fax* or *-faxe* has come within my knowledge. What is the meaning of this ending?

According to Aasen there is a Norwegian word *faks*, which is connected with ON *fax*, a mane, and has the meaning 'heiregræs,' that is, 'brome-grass'; and, according to Rygh, dialectal Swedish has the same word in the form *faxe*, while in South Germany the corresponding word is written *fachs* and means 'poor mountain grass.' Among place-names involving the word there are the Danish name Faxe, 1370 *Faxæ* (Madsen) and the Norwegian Faxfalle (Rygh). But, further, there is a similar use of OE *feax*, hair. Middendorff quotes from an OE charter the expression '*on west healfe ealdan hege to feaxum*,' where *feaxum* is the dat. pl. of *feax* and means tufts of grass and shrubs. It appears therefore, that, whether Anglian or Scandinavian, there was an early Yorkshire word *fax* which might mean 'a place covered with rough grass.' This, I take it, is the meaning in the name Halifax.

Parallels to the first element, Hali-, are more numerous, and attention must be called to Hallikeld in Yorkshire and Halliwell in Lancashire, of which the following are early forms:

DB	1086 *Halichelde*	1246	*Haliwell*
YI	1286 *Halikeld*	1292	*Haliwall*
YI	1290 *Halykeld*	1332	*Haliwalle*

Both these names have duplicates. The Whitby Chartulary records a *Halikeld* in Liverton near Guisborough; the Guisborough Chartulary records a *Haliwelle* in the county of Durham; and Middendorff quoting from OE charters gives *hālgan·welle*, as well as *hālgan·forde* and *hālgan·āc*. Halikeld and Halliwell appear to have the same meaning, namely, 'holy well,' from OE *hālig*, ME *hali*, holy, sacred, and ON *keld*, OE *well*, a well, spring, fountain. The first element in Halifax agrees entirely with a similar origin, and a careful examination of other suggestions leads to their rejection. Perhaps further light may be thrown on the name by historical research.

The neighbourhood of Halifax shows a greater proportion of names probably Celtic than any other part of South-west Yorkshire. Among them we may put the simple names Calder, Hebble, Krumlin, and Spink. In most cases, however, the Celtic word now precedes one of Anglian or Scandinavian origin, as for example in Allan Gate, Allan Wood, Chevin Edge, Cragg Vale, Hanna Wood, Howcans Wood, Mankin Holes, and Pennant Clough.

Of Scandinavian names there is a far greater proportion. Four names end in -by, Birkby, Ringby, Scawsby, Sowerby; there are three thorpes, namely, Gannerthorpe Wood in Wyke and two Thorpes in Sowerby; there are many Booths, Flatts, Holmes, Lumbs, Nabs, Rakes, Scars, Scouts, Slacks, Storrs, and Whams; and in addition to all these there is still the following long list: Baitings, Blamires, Boothroyd, Brianscholes, Brighouse, Briscoe Lane, Clapgate, Clapper Hill, Clipster, Erringden, Fly Flatt, Gaukrodger, Heliwell, Howgate, Keelam, London, Rotten Row, Scaitcliffe, Scholes, Skeldergate, Skircoat, Slithero, Staups, Strines, Swithens, and Woolrow.

HALLAM, Sheffield, DB *Hallun*, DN 1161 *Halumsira*, PF 1202 *Hallum*, HR 1276 *Halumshire*, IN 1342 *Hallom*, YD 1359 *Hallum*, PT 1379 *Hallomshire*, YF 1564 *Hallomshyre*. These early forms are in conflict with the present termination; they have no connection with OE *hām* or *hamm*, being in fact representations of an early dative plural in *-um*. Names of this character occur quite frequently in the Icelandic sagas and in Anglo-Saxon charters. There are several examples in the West Riding; Hillam was formerly *Hillum*, the hills; Byram was *Birum*, the cowhouses; Malham was *Malgun*; Owram was *Ufrun*; Throapham was *Trapun*; and Denholme *Denum*. Hallam is either the dative plural of ON *hallr*, a slope or hill, or of OE *heall*, a hall, mansion.

Three districts within the West Riding were formerly dignified with the title shire: Sourbyshire, the district around Sowerby; Borgscire, around Aldborough; and Hallomshire, around Hallam. It will be observed that in each case the first element may be of Scandinavian origin.

HALLAS, HALLOWS.—See Hale.

HALSTEAD, Thurgoland, Thurstonland, and Woolley.— WCR 1308 refers to a place called *Hallestede*; this is the ME form of the modern Halstead, which means 'hall place,' from OE *heall*, hall, mansion, and *stede*, place, site, position.

HAMPOLE or **HAMPHALL STUBBS**, Doncaster, has been made famous by its connection with Richard Rolle, the 'Hermit of Hampole,' who wrote *The Pricke of Conscience*. The name is given in DB as *Hanepol*, and the same spelling occurs in YR 1230, YR 1253, and HR 1276, as well as many other early documents. As soon as the *n* came into contact with the *p* it became the labial *m* to agree with the labial *p*. Hampole may be either Anglian or Scandinavian; its meaning is 'Hana's pool' or 'Hani's pool,' from OE *pol* or ON *pollr*, and the OE personal name Hana or the ON Hani.

HANDSWORTH, Sheffield, DB *Handesuurde, Handesuuord, Handeswrde*, HR 1276 *Le Boure de Handesworth*, KI 1285 *Handesworth*, YD 1316 *Handisworth Wodehousis*, 1389 *Handesworth Wodhous*. Though Searle has no such name as Hand, there is a modern surname of the form, and we may therefore explain Handsworth as 'Hand's holding,' from OE *weorth*.

HANGING HEATON, Dewsbury, DB *Etun*, IN 1266 *Hingande Heton*, HR 1276 *Hengende Heton*, DN 1293 *Hangand Heton*. The distinctive prefix *Hingande* or *Hengende* reminds us of the extraordinary change which has taken place in the termination of the present participle—to-day -ing, but formerly -*ende* or -*ande*. There is a field in Batley called Hanging Field, and fields called Hanging Royd occur in Heptonstall and Carlton. The prefix evidently refers to a steep hill-side 'hanging' above the lower ground. Heaton is the 'high farm,' from OE *hēah*, high, and *tūn*, an enclosure, farmstead.

HANNA MOOR, HANNA WOOD.—See Anna.

HARBOROUGH, HARDEN, HARLEY, HARROP.— The first element, Har-, might spring from OE *hara*, a hare, OE *hār*, a boundary, OE *here*, an army, or the personal name Hæra. As the recorded spellings of the names give little assistance it is quite impossible to interpret them with confidence.

HARBOROUGH HILL, Barnsley, probably derives its first element from OE *here*, a predatory band, troop, army. In the Saxon Chronicle this word was commonly used of the Danish army; hence Harborough Hill may have been one of the encampments of the Viking invaders. The terminal comes from OE *burg*, *burh*, a fortified place.

HARDEN occurs near Bingley, Meltham, and Penistone, and early spellings of the first are RC 1234 *Hardene*, 1236 *Heredene*, RC 1332 *Hardene*, RC 1538 *Harden*. The termination is from OE *denu*, a valley.

HARLEY, Wentworth and Todmorden, YS 1297 *Harlay*, YI 1303 *Hareley*, for the former, and PT 1379 *Harley* for the latter, may perhaps be 'hare lea,' from OE *lēah*, a lea.

HARROP, which occurs in Wilsden and Saddleworth, is recorded by WCR in the latter case as *Harrop* in 1274, *Haroppe* in 1308, *Harehoppe* in 1309. Its termination is OE *hop*, a hollow between hills, a secluded valley.

HARDCASTLE, HARDWICK.—Connected with the ancient manorial system we find two companion terms, Berwick and Herdwick, that is, 'barley place' and 'herd place,' the latter from OE *heord*, a herd or flock.

HARDCASTLE, Hebden Bridge, obtains its terminal from OE *castel*, a village, and its first element is probably from OE *hierde*, ME *herde*, a shepherd, cowherd.

HARDWICK, Pontefract, DB *Harduic*, *Arduuic*, CR 1226 *Herdwick*, YI 1258 *Herdwyke*, and

HARDWICK, Morthen, 1305 *Herdwyck*, PT 1379 *Hardewyk*, both mean 'herd enclosure,' from OE *wīc*, an enclosure, dwelling.

HARLINGTON, Mexborough, CR 1280 *Herlington*, YF 1345 *Herlyngton*, PT 1379 *Herlington*, YF 1495 *Harlyngton*, must be compared with the Frisian name Harlingen, 1228 *Herlinge*, 1323 *Harlinge*, 1355 *Herlinghe* (NGN), and with the three Harlings in Norfolk. In DB the latter appear as *Herlinga*, the genitive of *Herlingas*, which is itself a plural form meaning the 'sons of Herl.' We may explain Harlington as 'the farm of the Harlings,' from OE *tūn*, an enclosure or farm. There are other examples of the name in Bedford and Middlesex.

HARTCLIFFE, HARTHILL, HARTLEY, HARTS-HEAD.—The name Hartshead occurs near Dewsbury, Horbury, and Sheffield, but in early records we find mention only of the first, DB *Hortesheue, Horteseue*, PF 1206 *Hertesheved*, YI 1258 *Herteshevede*, WCR 1286 *Hertesheved*. Unlike the names Hartcliffe, Harthill, and Hartley, which refer to the animal, OE *heorot*, ME *hert*, Hartshead should be explained as ' Heort's headland.'

HARTCLIFFE, Penistone, CH †1280 *Hertclyve*, PT 1378 *Hertclif*, is ' the hart's cliff.'

HARTHILL, Worksop, DB *Herthil*, PF 1191 *Herthille*, NV 1316 *Herthill*, is ' the hart's hill.'

HARTLEY, Ecclesfield, is given as *Hertelay* in YS 1297, while Hartley, Todmorden, is written *Herteley* in WCR 1297 and *Hertlay* in WCR 1308. The meaning is ' the lea of the harts,' from the gen. pl. of OE *heorot* and OE *lēah*.

HASSOCKS, Honley and Marsden, is derived from OE *hassuc*, which means sedge, coarse grass.

HATFIELD, Doncaster, is the site of the great battle in which Edwin, the Christian King of Northumbria, was overthrown by the heathen Penda, King of Mercia. The AS Chron. says of the year 633 ' In this year was Eadwine King slain by Cadwalla and Penda at *Hethfeld*,' while Bede says the battle was fought ' in the plain that is called *Hæthfeld*.' Continuing, Bede tells us that Cadwalla, though he bore the name and professed himself a Christian, spared neither women nor children, but ravaged the country for a long time, ' resolving to cut off all the race of the English within the borders of Britain.' Later forms of the name include the following:

DB	1086 *Hedfeld*	YS	1297 *Haytefeuld*
YR	1227 *Hetfeld*	DC	1314 *Haytefelde*
HR	1276 *Heitfeld*	NV	1316 *Haytefeld*
KI	1285 *Haitfeld*	DC	1326 *Haytefelde*

The Domesday form shows *d* for *th*, and the forms in *ei, ai, ay*, show the influence of ON *heiðr*, a heath ; but the name may fairly be explained as ' heath field,' from OE *hæth* and *feld*. In Nennius the name is *Meicen*, and in the Welsh Chronicle *Meigen* and *Meiceren*.

HATHERSHELF, Mytholmroyd, occupies the steep hillside by which the uplands of Sowerby descend to the Calder. WCR has *Haderschelf* in 1274, *Hadirchelf* in 1275, *Haderschelf* in 1307, and *Hadreshelf* in 1326, while HW 1554 gives *Hathershelf*. The termination is from OE *scylf*, a shelf or ledge of land, and the prefix may be a personal name as in the case of Hunshelf, Tanshelf, and Waldershelf. More probably, however, it is ME *hadder*, *haddyr*, heath or ling; compare the dialect-word *hadder*, *hedder*, explained in EDD as applied to various kinds of heather or ling.

HAUGH.—This name, which occurs in Sowerby and Rawmarsh, is derived from OE *halh*; see Hale.

HAWORTH, HAINWORTH, Keighley.—The DB name *Hageneuuorde* has usually been assigned to Haworth; it seems very improbable, however, that this form could have become *Hauewrth* as early as 1209; and, further, in a deed prior to 1230 both *Hawrth* and *Hagenwrthe* occur, referring apparently to different places. Early spellings are as follows:

DN	1210	*Haneworth*	PF	1209	*Hauewrth*
YD	†1230	*Hagenwrthe*	YD	†1230	*Hawrth*
CR	1252	*Hagnewurthe*	YI	1246	*Howrde*
YI	1273	*Hannewrthe*	WCR	1275	*Houwrth*
DN	1294	*Hagenworth*	KF	1303	*Haworth*
BPR	1598	*Haynworth*	PT	1379	*Haworth*

Here are two series of forms, one with *n*, the other without; these appear to have existed side by side since the 12th century, and there can be little hesitation in assigning the former to Hainworth and the latter to Haworth.

One would have expected in the case of Haworth a DB form *Hageuuorde*, from OE *haga*, an enclosure, and *weorth*, a holding or farmstead. According to rule this would have become *Hauewrth* and later *Haworth*, just as *hagathorn* became first *hawethorn* and later *hawthorn*. The interpretation of Haworth, therefore, appears to be 'enclosure, farmstead,' while that of Hainworth is 'Hagena's farmstead,' from the recorded personal name.

HAYWOOD, Campsall.—Like Heywood in Lancashire, which appears in the Whalley Chartulary in 1311 as *Heywood*, this name is doubtless either 'the wood by the enclosure,' or 'the enclosed wood,' from OE *hege*, a fence or enclosed space, and *wudu*, a wood.

HAZLEHEAD, HAZLEHURST, HAZLESHAW, HESSLE, HESSLEGREAVE.—The number of place-names derived from the hazel is not great, but in addition to the above we must note the name High Hazels near Darnall, and Lighthazels in Sowerby. The OE form of the word was *hæsel*, while the ON was *hasl* or *hesli*; and according to NED the early northern forms *hesel* and *hesyl* are probably derived from the last of these.

HAZLEHEAD, Penistone, 1256 *Heselheved*, YD 1326 *Hesilheved*, YD 1372 *Hesilheved*, is 'the hazel-tree upland.'

HAZLEHURST, which occurs near Sheffield and Halifax, is 'the hazel-tree copse,' from OE *hyrst*, a copse or wood.

HAZLESHAW, Ecclesfield, is 'the hazel copse,' from OE *sceaga*, a copse.

HESSLE, Wragby, DB *Hasele* and *Asele*, CR 1226 *Hesel*, DN 1369 *Hesil*, may be compared with Oaks and Thornes.

HESSLEGREAVE, Saddleworth, YS 1297 *Hasilgref*, is 'the hazel-tree thicket,' from OE *græfa*, a thicket, grove.

HEAD, HEADFIELD, HEADLAND.—We find 'head' as a terminal in Hazlehead, Lupset, and the three Hartsheads. In OE the form is *heafod*, and in ME *heved*, while the usual meaning is the highest point of a field, stream, valley, or hill.

HEADFIELD is a rounded eminence in the middle of the valley of the Calder near Thornhill Station.

HEADLAND is a somewhat common field-name, and is doubtless the source of the Liversedge name Headlands.

HEALD, HEALDS.—Heald Head occurs in Cawthorne, Heald Wall in Barkisland, Healds in Ecclesall, Heald in Elland, and Healds Hall in Liversedge. Referring to the last named, *Healdhousecroft* occurs in 1560 and *Healds* in 1803. EDD gives two meanings of the word Heald, (1) a shelter for cattle on the

G.

moors, (2) a slope, declivity, hill. And, further, EDD suggests that in the former sense the word is connected with ON *hæli*, a shelter, refuge; for the latter sense I would suggest ON *hjalli*, a shelf or ledge on a mountain side—compare Dan. and Norw. *hæld*, an inclination, slope. In either case the final *d* in Heald would be a supporting consonant, as in the case of Backhold and Wormald. West of Windermere there is a steep tree-covered slope called Heald Wood, and the name Heald occurs also near Garstang.

HEALEY, HEATON, HEELEY.—In discussing these names it will be helpful to examine first early forms of Heaton (Bradford) and of four names where Heaton is to-day preceded by a distinctive affix.

Cleckheaton	1086 *Hetone, Hetun*	1254 *Hetun*	1316 *Heton*
Kirkheaton	1086 *Heptone*	1199 *Heton*	1297 *Heton*
Earlsheaton	1086 *Etone, Ettone*	1286 *Heton*	1316 *Heton*
Hanging Heaton	1086 *Etun*	1266 *Heton*	1276 *Heton*
Heaton	1086 —	1276 *Heton*	1303 *Heton*

In view of the unanimity of the later forms we may take it as certain that DB *Heptone* stands for *Hetone*, and that DB *Etun*, *Etone*, *Ettone*, have lost the aspirate through the fault of the Norman scribe. In the case of *Heptone* the scribe was doubtless influenced by the name he had just written, *Leptone*.

The first element in Heaton represents OE *hēah*, *hēh*, high, concerning which Dr Skeat tells us that 'in ME the final guttural was sometimes kept and sometimes lost[1].' In our Yorkshire Heatons we have obvious examples of its loss.

But further, Dr Skeat shows that when this guttural was kept it often had an effect on the preceding vowel[2], and so OE *hēh* gave not only ME *hegh, heigh, hey*, but also ME *hygh*, whence our modern word 'high.' Thus Heaton means 'the high farmstead,' from OE *tūn*, an enclosure or farmstead, and the names Heaton and Hightown are doublets. But these names, though they possess the same meaning, are different in origin, for Heaton is early and Hightown is late, and the prefix in Heaton is local while that in Hightown is borrowed from the common tongue.

---

[1] Skeat, *Principles of English Etymology*, I, 58.   [2] *Ibid.* I, 400.

Of HEALEY or HEELEY there are five examples, HEALEY in Ossett, and the four following :

HEALEY, Batley	YD	1330 *Helay*	PT	1379 *Helay*
HEALEY, Rastrick	WCR	1306 *Heyley*	WCR	1601 *Healey*
HEALEY, Shelley	DN	1359 *Helay*	DN	1381 *Helay*
HEELEY, Sheffield	HH	1366 *Heghlegh*	PT	1379 *Helay*

In every case the meaning is 'high lea,' the terminal being derived from OE *lēah*, a lea or meadow. This word is represented in ME by such forms as *legh, leigh, ley,* and has given us also the word ' lea ' where the final guttural has disappeared. In all this OE *lēah* corresponds exactly to OE *hēah*.

HEATH, Halifax and Wakefield.—In a Nostell charter of 1120 the latter is recorded as *Heth,* and the same form occurs in YS 1297 and PT 1379. In other documents, YR 1252 for example, the name *Bruera* occurs.

HEBBLE, HEBDEN, HEPSHAW, HEPTONSTALL, HEPWORTH.—Small streams called Hebble are found in Huddersfield and near Holmfirth ; but better known is the side-stream of the Calder which flows through Halifax. Hebden Bridge gets its name from a valley and stream called the Hebden, and a parish near Skipton bears the same name. Further north, between Tees and Tweed, we find the name Hepple once and Hebburn or Hepburn three times. Early records of Hepton-stall and Hebden Bridge are as follows :

WCR	1274 *Heptonstall*	HW	1508 *Hepden Bridge*
HR	1276 *Heptonestal*	HW	1510 *The bridge of Hepden*
NV	1316 *Heptonstall*	HW	1609 *Heptonbrigg*

Hebden on the Wharfe is written *Hebedene* in DB and *Hebbeden* in YI 1305, and BCS has the name *Eblesburnon.*

HEBBLE is a very difficult word for which no suitable root presents itself. I imagine, however, that it is of Celtic origin.

HEBDEN, Halifax, is probably 'the valley of the wild-rose,' from OE *hēope,* ME *hepe,* the wild-rose, the briar, and OE *denu,* a valley ; see Shibden.

HEPSHAW, Thurlstone, may with confidence be explained as 'wild-rose copse,' from OE *sceaga,* a small wood.

HEPTON, in the name Heptonstall, Halifax, is probably
'wild-rose farmstead.'

HEPWORTH, Holmfirth, DB *Heppeuuorde*, WCR 1274 *Heppe-wrth*, PT 1379 *Hepworth*, can scarcely come from OE *hēope*. It
is almost certainly 'the farmstead of Heppa.' Searle gives
Heppo; hence we may postulate the form Heppa.

HECK, Snaith, DN 1225 *Hecke*, DN 1248 *Hec*, YI 1280
Hecke, NV 1316 *Hek*, comes directly from OE *hec, hæc*, a fence,
rail, gate. This word in the South and Midland districts became
'hatch,' but there was a Northern form 'heck,' and one of the
meanings of the latter, according to NED, is a shuttle or sluice
in a drain; compare Dan. and Norw. *hæk*, a hedge.

HECKMONDWIKE, YI 1261 *Hecmundeswyk*, WCR 1275
Hecmundewyk, HR 1276 *Kecmendewyc*, KI 1285 *Hecmundwyk*,
NV 1316 *Hekmondewyk*, PT 1379 *Hepmunwyk*. The form of
the word suggests a personal name as the first element. There
is a well-known OE name Heahbeorht which appears later as
Hechbert, and the OE Heahmund appears in an early Hunting-
donshire place-name quoted by Skeat as Hecmundegrave; see
Barugh. We may therefore explain Heckmondwike as 'the
habitation of Hecmund,' that is, 'of Heahmund,' which means
'high protector.' The termination is from OE *wīc*, an enclosure,
habitation, village.

HELIWELL, Lightcliffe, WCR 1297 *Heliwall*, 1373
Heliwelle, appears to be of Scandinavian origin. The meaning
is doubtless 'holy well,' for there is an important well near at
hand; and we may, therefore, connect the word with Dan.
hellig, holy, and *væld* (for *væll*), a well. See Holywell.

HELLABY, Rotherham, DB *Helgebi*, KF 1303 *Helghby*,
YD 1318 *Helghby*, PT 1379 *Helughby*, is 'the farm of Helga or
Helgi.' These personal names are well known, and appear in
scores of Norwegian place-names, among them Helgerud,
Helgestad, Helgenes (Rygh).

HELM, HELME.—This word occurs in the counties of
Westmorland, Lincoln, and Durham, as well as in various parts

of Yorkshire. A possible meaning is the crown or summit of a hill, from OE *helm* or ON *hjālmr*, which meant first a helmet, and later the crown or top of anything. But another meaning is a shed or outhouse; compare the Norse and Swedish dialect-word *hjelm*, a screen or shelter of boards, from ON *hjālmr* (Aasen)

HELME, Meltham, was spelt *Helme* in a deed of 1421.

HELM, Kirkheaton, is given in BM 1199 as *Helm*.

HELM, Helm Lane, Sowerby, is recorded in WCR 1275 as *Helm*, while in WCR 1307 we find the expression 'in Sowerby at le *Helmebothes*.'

HELM CLOSE is a field in Rothwell.

It is unlikely that in any of these the meaning is crown or summit; the second signification appears the more probable.

HEMINGFIELD, HEMSWORTH, Wombwell and Pontefract.—In both cases early spellings are of great value:

HR	1276 *Himlingfeld*	DB	1086 *Hamelesuurde*
YI	1303 *Hymelingfeld*	PC	†1220 *Hymeliswrd*
CH	1362 *Hymlyngfeld*	YI	1245 *Himleswrde*
CH	1386 *Hymlingffeld*	WCR	1296 *Hymeleswrth*

Among later spellings we get for Hemsworth NV 1316 *Himmelsworth*, PT 1379 *Himmesworth*. The first element carries us back to the simple personal name Hama (Searle), from which a diminutive Hamele, recorded in Searle as Hemele, appears to have been formed. From this was built the patronymic Hameling or Hemeling; compare the surname *Hamlinge* found in RPR 1592. Hence we may explain Hemsworth as 'the holding of Hamele,' and Hemingfield as 'the field of Hameling.' For the change from *Hymeleswrth* and *Hymelingfeld* to Hemsworth and Hemingfield compare Wentworth.

HENSALL near Snaith, like Melton near Rotherham, shows how intervocalic *th* may disappear. Early records are

DB	1086 *Edeshale*	SC	1315 *Hethensal*
YI	1279 *Hethensale*	NV	1316 *Hethensalle*

The strange form given by DB presents three defects due to the idiosyncrasies of its Norman scribes: (1) the omission of the aspirate, as in *Odersfelt* for Huddersfield; (2) the substitution of *d* for *th*, as in *Medelai* for Methley; (3) the omission of *n*, as in *Witreburne* for Winterburn. Yet, on the other hand, the DB record alone preserves the correct terminal, *hale*, the dative of OE *healh*, a corner or meadow, unless indeed it comes from Dan. *hale*, a tongue of land. The first element is the ON personal name Hethinn which occurs in many Norwegian place-names, for example, Hedenstad and Hedensrud.

HEPSHAW, HEPTONSTALL, HEPWORTH.—These names are discussed under Hebble.

HERRINGTHORPE, Rotherham, CH 1386 *Herryng-thorppe*, YF 1553 *Heryngthorpe*, means ' Hering's village,' from ON *thorp*, a village, and the ON personal name Hæringr (Naumann). For the absence of the sign of the genitive see Alverley and Alverthorpe.

HESKETH, West Ardsley.—Hesketh in the North Riding, near Rievaulx, appears in RC as *Hesteskeid* and *Hesteskeith*, forms which agree with ON *hesta·skeið*, a racecourse, from ON *hestr*, a horse, and *skeið*, a racecourse or links.

Year by year in West Ardsley a great horse-fair is held known far and wide as Lee Fair, and it is the custom to test the paces of the horses along Hesketh Lane. Thus the name Hesketh, being Scandinavian, bears witness to the existence of these fairs from the days of the Vikings and supports the suggestions made in the note on Tingley.

HESSLE, HESSLEGREAVE.—See Hazlehead.

HEWENDEN, Bingley, is probably from OE *hiwan*, ME *hewen*, servants, and *denu*, a valley, the first element being the gen. pl. *hiwena*. May not this refer to a valley where the inhabitants were Celts, ' hewers of wood and drawers of water' to the conquering Anglians?

HEXTHORPE, Doncaster, DB *Hestorp, Estorp*, CR 1269 *Hexthorp*, PT 1379 *Hexthorp*, is probably ' Hegg's village,' from the ON personal name Heggr, and *thorp*, a village.

HEY.—This word is derived from OE *hege* which means a hedge, a fence, an enclosed place, or a definite district not enclosed. Duignan tells us that forests were usually divided into 'hays' for administrative purposes. The OE *hege*, ME *heye*, *haye*, is allied to the OE *haga*, ME *haghe*, but it must not be confounded with OE *hecg*, which gives ME *hegge* and the modern word 'hedge.'

HICKLETON, Doncaster, DB *Chicheltone, Icheltone,* PF 1201 *Hykelton,* PC 1240 *Hikilton,* KI 1285 *Hikylton.* BCS has the place-name *Hiceleswyrth* which may be explained as 'the farm of Hicel.' It will be noted, however, that the early forms of Hickleton have neither -*s* nor -*e* to represent the genitive. Probably the name must therefore be explained as 'woodpecker farm,' from OE *hicol*; see Middendorff.

HIENDLEY or **COLD HIENDLEY,** Barnsley, was *Hindeleia* and *Hindelei* in DB, and *Hyndelay* in DN 1293, but YI 1297 has *Caldhindeley* and YD 1318 has *Coldehindelay.* The meaning is 'hind meadow,' from OE *hind,* a female deer.

HIGHAM, HIGH ELLERS, HIGHFIELD, HIGH-ROAD WELL, HIGHTOWN.—In these words the prefix is from OE *hēah, hēh,* ME *hegh, hey, hye,* high; see Healey.

HIGHAM, Barnsley, 1297 *Hegham,* 1375 *Heghome,* PT 1379 *Hegham,* is 'the high home,' from OE *hām,* a home, not from OE *hamm,* an enclosure, dwelling.

HIGHAM occurs also in Erringden and Sowerby.

HIGH ELLERS, Doncaster, formerly in the possession of Kirkstall Abbey, KC 1209 *Hechelres,* YI 1280 *Heyhelleres,* PT 1379 *Heghellers,* means 'high alders' from ON *elrir,* the alder-tree.

HIGHFIELD occurs in Ecclesall, Thurgoland, and Womersley.

HIGHROAD WELL, Halifax, is perhaps the place referred to in WCR 1277 as *Heygrode,* and, if so, the meaning is 'the high royd or clearing'; see Royd.

HIGHTOWN, Liversedge; see Liversedge.

HILL.—OE *hyll* has given the modern name 'hill,' which may be used of any land elevated above the surrounding country; to earn the name, as Wyld says, a hill need not be high. Examples in South-west Yorkshire include the following compounds: Baghill, Chidswell, Ryhill, Soothill, Tickhill, Toothill.

HILLTHORPE is in Thorpe Audlin.

HINCHCLIFFE, Holmfirth, WCR 1307 *Heyncheclyf*, PT 1379 *Hyncheclyff*, is a difficult word. The Gazetteer gives only two names with a similar prefix, Hinchin Brooke in Huntingdon, and Hinchwick in Gloucester, and for the former Professor Skeat gives no definite etymology.

HIPPERHOLME, Halifax.—Up to the end of the 14th century the recorded spellings are very consistent; afterwards the termination undergoes serious change.

DB	1086	*Hyperum, Huperun*	PT	1379	*Hyprum*
PF	1202	*Yperum*	YF	1537	*Hyprom*
YI	1266	*Hiperum*	YF	1555	*Hyperome*
WCR	1286	*Hyperum*	YF	1568	*Hipperholme*

Obviously the termination is not derived from ON *holmr*, an island, its appearance being rather that of a dative plural in -um. But what is the stem of the word? Note first that there is a stream in Derbyshire called the Hipper, and another near Goathland in the North Riding called Hipper Beck, while in addition there is a second Hipperholme between Hebden Bridge and Todmorden as well as a place in the North Riding called Hippersleight. Note further that EDD has a dialect-word 'hipper' meaning osier, while in Lancashire, according to the same authority, a field in which osiers are grown is called 'hipperholm.'

On the other hand there is in Angeln a tree-name *ippern*, while Doornkaat places on record EFris. *iper* which corresponds to Germ. *iper*, the common elm; compare Du. *ijp*, found in the place-name IJpelo, 1475 *Ypeloe*, elm lea. It seems just possible that the best early form of Hipperholme is PF 1202 *Yperum*, and that this is the dat. pl. in -um of an early form *iper*. Thus Hipperholme would be interpreted 'elms,' and would be an almost exact counterpart of the Flemish place-name Ypern (French Ypres).

HIRST, HURST.—The OE *hyrst* meant scrub, brushwood, a copse, a wood. It is found as the terminal in Ashurst, Copthirst, Elmhirst, Hazlehurst, Hollinhurst, and Kilnhurst.

HITCHELLS WOOD, Doncaster.—Early charters in KC dealing with the neighbourhood of Bessacar, make frequent mention of *Echeles*, *Escheles*, or *Hecheles*. Hitchells is doubtless the modern form of this word ; but what is its meaning ?

Duignan tells us that in the 13th century a hamlet near Wolverhampton was called *Echeles* and *Escheles*, though at a later date the name became *Necheles* or *Nechells*. He goes on to point out that all the records are Middle English, and that there is no trace of the word in Old English ; and he suggests that the origin is the OFr. *escheles*, ladders, steps, stairs, explaining Nechells as derived from *atten Escheles*, later *atte Necheles*, ' at the two-storied house.' A note in Peiffer's *Noms de Lieux* appears to hint, however, that these *escheles*, though ladders, had no connection with two-storied houses. Dealing with the word *echalet* Peiffer says it is ' a fence used in the Bocage Vendéen to enclose the meadows ' ; these fences, he goes on to say, are made with branches of trees ' which serve as a ladder (*échelle*) by which to pass from one meadow to another ; hence the name echalet.' Adopting this suggestion, *Echeles* may be explained as ' fences or stiles made with branches of trees.'

In Bardsey we find the name Hetchell Wood ; and in Picardy there are two places called l'Échelle (Robinson).

HOLBECK, Leeds, YI 1258 *Holebeke*, YF 1542 *Holbekk*, RPR 1688 *Holbecke*, means ' the beck in the hollow,' from ON *hol*, a hollow, and *bekkr*, a stream. The name occurs in Scarborough and Lincolnshire, while Normandy has a Holbec, and Denmark two Holbeks.

HOLDSWORTH, HOLDWORTH.—As is shown by the early forms given below, these names have exactly the same origin :

HOLDSWORTH, Halifax		HOLDWORTH, Bradfield	
HR	1276 *Haldewrth*	DB	1086 *Haldeuurde, Aldeuuorde*
WCR	1297 *Haldewrth*	HR	1276 *Haldewrth*
YD	1383 *Haldeworth*	YS	1297 *Haldewrth*

Obviously both names should now be written Holdworth, 'the farmstead of Halda,' from OE *weorth*, a holding or farm, and the personal name recorded by Searle.

HOLLINBUSK,   HOLLINGBANK,   HOLLIN-GREAVE,   HOLLINGTHORPE,   HOLLINGWELL, HOLLINHURST, HULLENEDGE, THICKHOLLINS. —Such early records as are available suggest a derivation from OE *holen, holegn*, ME *holyn*, holly. The spelling 'holling' occurs quite naturally for the earlier 'holegn,' *ng* replacing *gn* very much as *wh* replaces *hw* in many words; compare OE *ðegn, ðeng, ðēn*, three forms of the OE word which has given our modern word 'thane.'

HOLLINBUSK, near Bolsterstone, is 'holly-bush'; compare Dan. *busk*, Sw. *buske*, a bush.

HOLLINGBANK, Heckmondwike, and HOLLINGWELL, South Kirkby, may be explained as 'holly-tree bank,' from ON *banke*, a ridge, and 'holly-tree well,' from OE *wella*, a well.

HOLLINGREAVE, Saddleworth, CH †1272 *Holyngreue*, YS 1297 *Holingref*, is 'holly thicket,' from OE *græfa*, a thicket, grove.

HOLLINGTHORPE, Crigglestone, WCR 1297 *Holynthorp*, WCR 1307 *Holinthorp*, is 'holly-tree hamlet,' from ON *thorp*.

HOLLINHURST, Shitlington, TPR 1602 *Hollinhurste*, TPR 1607 *Hollinghirst*, is 'holly-tree copse,' from OE *hyrst*.

HULLENEDGE, Elland, WH 1316 *Holyngegge*, WCR 1478 *Holynegge*, YF 1494 *Holynege*, is 'holly-tree ridge,' from OE *ecg*, edge, a declivity, a ridge.

THICKHOLLINS, Meltham, WCR 1274 *Thyckeholyns*, YF 1537 *Thyhholyns*, may be compared with Thickbroom in Staffordshire, both from OE *thicce*, thick, close.

HOLME, HOLMFIELD, HOLMFIRTH.—The word Holme occurs as a field-name with considerable frequency. Although finally of the same origin as OE *holm*, it is not derived therefrom, its source being ON *holmr, holmi*, an islet; 'even meadows on the shore with ditches behind them are in Icelandic called holms,' says Vigfusson, and Dan. *holm* means a quay as well as a small island.   In England the word is used to designate low-lying land beside a river, land subject to inundation or almost surrounded by streams or marshes, while the OE *holm* meant 'wave, ocean, water, sea.'

HOLME, near Holmfirth, in spite of its apparent simplicity, provides a difficult problem.   In the Domesday record for Yorkshire there are at least six different places described regularly as *Holm* or *Holme*; but Holme near Holmfirth, and its neighbour Yateholme, though mentioned twice each, are never described as *Holme* but always as *Holne*.   Early forms of this Holme together with Holmfirth and Holme near Skipton are particularly striking:

HOLME, Holmfirth	HOLMFIRTH	HOLME, Skipton
DB 1086 *Holne*	WCR 1274 *Holnefrith*	DB 1086 *Holme*
WCR 1274 *Holne*	WCR 1275 *Holnefrith*	HR 1276 *Holm*
WCR 1297 *Holne*	WCR 1307 *Holnefrith*	IN 1309 *Holme*
WCR 1309 *Holne*	WCR 1309 *Holne Frith*	BM 1325 *Holme*
NV 1316 *Holm*	PT 1379 *Holmfirth*	BM 1326 *Holme*

Further, the local pronunciation of Holme (Holmfirth) is to-day quite frequently 'Hown'; hence it becomes almost certain that Holme is a usurper and that Holne or Hown is the rightful name.   One additional fact must be stated, namely, that there is a place called Holne in South Devon.

HOLME, a hamlet in Owston, is recorded as *Holme* in 1274, and doubtless comes from ON *holmr, holmi*.

HOLMES is a hamlet in Kimberworth.

HOLMFIELD occurs in Halifax.

HOLMFIRTH derives its terminal from OE *frith, fyrhthe*, a wood, coppice, forest, forest-land.   The graveship of Holme included the townships of Holme, Austonley, Cartworth, Wooldale, Scholes, Fulstone, Hepworth, Thong, and was itself a part of the lordship of Wakefield.

ALMHOLME, Doncaster, SC 1237 *Almholme*, YF 1535 *Almholme*, YF 1579 *Almeholme*, is 'elm holme,' from ON *almr*, Sw. and Dan. *alm*, an elm.

CORNHOLME, Todmorden, is either 'Korni's holme,' or 'corn holme,' from ON *korn*, grain, or the ON personal name Korni.

ESKHOLME, Thorne, KC *Escholm* and *Eschholm*, is 'ash-tree holme'; compare Dan. *asketræ* and *æsketræ*, ash-wood.

LAWKHOLME, Keighley, may be compared with Lawkland, YI 1251 *Loukelandes*, PT 1379 *Laukeland, Lawkeland,* 'land of leeks.' The prefix is doubtless from ON *laukr*, leek, garlic; compare OE *lēac*, leek.

LINEHOLME, Todmorden, is probably 'the flax holme,' from ON *lin*, flax; see Lindley.

SHAFTHOLME, Doncaster, YF 1535 *Shaftholme*, YF 1579 *Shaftholme*, derives its prefix from the ON *skaptr*, Norw. *skaft*, a shaft, pole; and its meaning is 'the holme marked by a pole.' Aasen records the dialect-form *skjefta*. See Shackleton.

WROSTHOLME occurs in Bentley near Doncaster.

YATEHOLME, Holmfirth, has for its prefix a common form of the OE *geat*, a gate.

HOLYWELL, Stainland, WCR 1285 *Heliwelle*, WCR 1336 *Helliwell*, WCR 1368 *Halywell*. We know what our forefathers believed to be the meaning, for a deed given by Watson dating from the end of the 13th century speaks of 'Henry de Sacro Fonte de Staynland.' See Heliwell.

HONLEY, Huddersfield, DB *Hanelei*, WCR 1274 *Honeley*, KI 1285 *Honlay*, WCR 1286 *Honneley*, YS 1297 *Honelay*, PT 1379 *Haunelay*, is 'Hana's lea,' from the known OE name Hana and OE *lēah*, a lea or meadow.

HOOBER is the name of a hamlet near Wentworth Woodhouse which derives its name from the conspicuous hill beneath which it lies. Early forms are YF 1569 *Hober*, and from the Wath Parish Registers, 1600 *Houber*, 1608 *Howber*. The sources of the name are OE *hōh, hō*, a heel, point of land, projecting ridge, and OE *beorg, beorh*, a hill. Compare Hooton.

HOOK.—Between Airmyn and Swinefleet the course of the Ouse makes a double bend in the form of an S reversed, and the village of Hook stands within the more northern of its loops. Early records of the name are

PC †1180 *Huck*	CR 1314 *Huk, Huck*
PF 1208 *Huc*	NV 1316 *Houk*
SC 1230 *Huck*	DN 1337 *Houke*

Note the sequence OE *gōs*, ME *gos*, Engl. *goose*, and the corresponding sequence OE *hōc*, ME *hok*, Engl. *hook*; and contrast with it the sequence OE *sūth*, ME *south*, and OE *mūth*, ME *mouth*. Obviously the early forms of Hook do not agree with the present form; they come, indeed, from a word where the vowel was *ū*, and MLG *huk*, a hook, corner, point of land, provides such a word; compare Fris. *huk* and Westphalian *huck*. According to Falk and Torp Dan. *huk* and Sw. *huk* are borrowed from MLG, and are cognate with the OE *hōc*, which meant a hook, bend, curve, corner. It is clear that the Yorkshire name has been influenced by the common word 'hook.'

HOOTON.—This name occurs four times in the portion of South Yorkshire lying between Doncaster and Sheffield. The meaning is 'the homestead on the projecting ridge of land,' from OE *hōh* or *hō*, a heel, point of land, a projecting ridge, and *tūn*, a homestead.

HOOTON PAGNEL was called *Hotun* and *Hotone* in DB; later KC 1204 gives *Hotun*, PC 1240 *Hoton Painel*, and HR 1276 *Hoton Paynell*. The place was owned in 1240 by William Paynel, hence the distinctive affix.

HOOTON ROBERT is recorded in DB as *Hotun*, by KI 1285 as *Hoton Robert*, and by NV 1316 as *Hoton sub Haia*.

HOOTON LEVITT was *Hotone* in DB, *Hotonleuet* in HR 1276, *Hotonlivet* YI 1279, *Hoton Lyveth* KI 1285.

SLADE HOOTON is mentioned by Burton as *Sled-hoton*, and in a Fine 1565 as *Slathe Howton*; the prefix is from OE *slæd*, a valley.

HOPE, HOPETOWN, HOPSTRINES, HOPTON.—
OE has the word *hop*, a small enclosed valley, and ON has *hōp*, which meant a bay or inlet, but assumed also an inland signification, and meant a hollow in the hills, a secluded spot or sheltered valley. A termination from one of these words is frequently met with, sometimes reduced to -op or -up; compare Midhope, Oxenhope, Widdop, Blacup.

HOPE occurs in Beeston, Halifax, and Honley.

HOPETOWN, Normanton, is plainly of late formation, and

may perhaps have no connection with the OE and ON words quoted above.

HOPSTRINES, Shelley, is Scandinavian, and has for its termination the ON *strind*, a border, side; see Strines.

HOPTON, Mirfield, DB *Hoptone*, DN 1218 *Hopton*, WCR 1274 *Hopton*, is 'the farmstead in the sheltered valley,' from OE or ON *tūn*, a farmstead.

HORBURY, HORLEY, HORTON.—Only Horbury appears in the Domesday record which gives *Horberie* and *Orberie*, but later spellings are as follows:

PC	1156 *Horbiri*	WCR	1374 *Horlawegrene*	IN	1246 *Horton*
PF	1202 *Orbir'*	WCR	1434 *Horlawegrene*	KF	1303 *Horton*
WCR	1286 *Horbiry*		1577 *Horley Green*	NV	1316 *Horton*

For Horton in Worcestershire Duignan gives earlier records, BCS 972 *Horton*, DB 1086 *Hortune*, and explains the name as 'muddy town,' from OE *horh*, mud, dirt. In the same way Professor Skeat interprets Hormead in Herts as 'muddy mead,' and according to Wyld a similar interpretation must be placed on Horwich in Lancashire. The first element in Horbury, Horley, and Horton is derived from the same source; we may be sure that in olden days there was at each place a tract of swampy ground.

HORBURY, Wakefield, has for its termination *byrig*, the dat. sing. of OE *burh*, a fortified homestead, stronghold, town; see Dewsbury.

HORLEY GREEN, Halifax, shows a somewhat common phenomenon, the sliding of the terminal from -lawe to -ley. The former ending comes from OE *hlǣw*, a mound, cairn, hill.

HORTON, Bradford, has the same origin and meaning as the Horton mentioned above.

HORDRON.—See Ardron.

HORNCASTLE, HORNTHWAITE.—For the former, which is situate in Hemsworth, Dodsworth gives *Hornecastell* in 1303 and 1316; and for the latter, in Thurlstone, YF has *Hornetweyt* in 1549 and *Hornethwaite* in 1559. The prefix may perhaps be the personal name Horn; more probably it is the

OE or ON *horn*, a corner, nook. Horncastle is probably 'the stronghold in the corner' from OE *castel* or ON *kastali*, a fortress ; while Hornthwaite is 'the paddock or clearing in the corner,' from ON *thveit*. See Castleford.

HORSFALL, HORSEHOLD.—WCR 1352 speaks of 'the Horsfall,' and HW 1523 has the expression 'My fermehold called *Horshald*.' NED glosses 'fall' as a slope or declivity, and 'hold' as a place of refuge or shelter. Horsfall, Horsehold, and Stoodley, all in the valley of the upper Calder, bear witness to the use once made of the neighbourhood by the Lords of the manor ; see Sowerby.

HORTON, Bradford.—See Horbury.

HOSTINGLEY, Thornhill, SE 1634 *Hostingley*, appears to have a patronymic as its first element, but Searle has no such form.

HOUGHTON.—Altogether there are in England about thirty examples of this name, three of them in South-west Yorkshire ; yet, as we shall see, not all derived from the same source. Early spellings are as follows :

GREAT HOUGHTON	LITTLE HOUGHTON	GLASS HOUGHTON
DB 1086 *Haltune*	DB 1086 *Haltone*	DB 1086 *Hoctun*
KI 1285 *Magna Haulgton*	KI 1285 *Holgton Minor*	CR 1250 *Hoghton*
KF 1303 *Magna Halghton*	KF 1303 *Parva Halghton*	NV 1316 *Hoghton*
NV 1316 *Halgton*	NV 1316 *Parva Halton*	PT 1379 *Hoghton*

Great and Little Houghton, near Barnsley, are from OE *healh*, a corner or a meadow, and may be rendered 'homestead in the meadow' ; but they would be more correctly written Haughton. Glass Houghton, near Castleford, is from OE *hōc*, a corner, angle, nook of land, and the sense is 'the farmstead in the corner of land.'

HOW.—This is derived from ON *haugr*, a mound or cairn, a word used to denote the artificial burial-mounds of the Vikings. In Icelandic literature there are many references to these burial-mounds. A passage in the Laxdala Saga reads as follows : 'So now they drank together Olaf's bridal feast and the funeral honours of Unn. And on the last day of the feast Unn was carried to the howe that was prepared for her. She was laid in a ship in the howe; and in the howe much treasure was laid with her[1].'

In South-west Yorkshire *haugr* has given us the termination in Carlinghow, Flanshaw, and Slithero; and from the same source we probably get the prefix in Howley and the medial *o* in Grenoside, Stenocliffe, and Wincobank.

HOWCANS, LONGCANS, MANKIN HOLES.—The name Longcans occurs in the Ovenden list of Overseers for 1762, and we have the following early records of Howcans, a wood in Northowram, and of Mankin Holes, a hamlet in Langfield :

WCR	1307 *Holcan*	WCR	1275 *Mankanholes*
WCR	1329 *Holcans*	WCR	1277 *Manekaneholes*
WCR	1360 *Holkans*	WCR	1308 *Mancanholes*
HW	1556 *Holcanse*	CH	1336 *Mankanholes*

These must be compared with the early forms of Alkincoats near Colne (Lancashire), where we find 1294 *Alcancotes*, 1296 *Alcancotes*, 1325 *Alcencotes*, 1343 *Alcancotes*, as well as such forms as *Altancotes* and *Altenecotes*. And we cannot but note at the same time *Olicana*, the name given by Ptolemy to the Roman station at Ilkley. It seems probable that the words are Celtic.

HOWCANS, like the first element in Alkincoats, may perhaps be a corruption of **alican*, a possible extension of Prim. Celt. **(p)alek*, a stone ; compare Ir. *ail*, a rock (Stokes).

MANKIN is possibly, as regards its first element, a derivative of Prim. Celt. **maini*, a stone ; compare W *maen*, Corn. *men*, a stone.

HOWDEN, HOWELL, HOWLEY, HOWGATE, HOWROYD, HOWSTORTH.—The first element in these names is either OE *hol*, hollow, OE *hol*, a hole, den, cavern, ON *hol*, a hollow, or ON *haugr*, a cairn, mound, hill.

HOWDEN CLOUGH, Birstall, PF 1202 *Holeden*, YF 1546 *Holden Cloughe*, is probably ' hollow valley clough,' OE *denu*, a valley, *clōh*, a clough.

---

[1] *Origines Islandicæ*, II, 150–1.

HOWELL HOUSE, Clayton, appears to be the place referred to as *Holwell* in PT 1379 (under South Kirkby); in a 16th century deed it is called *Holewell.* The meaning is 'the well in the hollow,' OE *well*, a spring or well.

HOWLEY, Batley, is mentioned by Burton in connection with Nostell Priory under the form *Hoveleo* ($v = u$), and Dodsworth gives *Howley* in 1425 and 1461. The word may be rendered 'burial-mound lea' from ON *haugr*, and OE *lēah*. Perhaps there was here, as at Carlinghow, the sepulchral mound of a Viking leader.

HOWGATE, Southowram, WCR 1274 *Holgate*, 1308 *Hollegate*, is 'the road in the hollow,' from ON *hol*, a hollow, and *gata*, a way or road.

HOWROYD, Barkisland, PT 1379 *Holrode*, is 'the clearing in the hollow'; see Royd.

HOWSTORTH, Ecclesfield, derives its second element from ON *storð*, a young wood.

HOYLAND, HOYLANDSWAINE.—Within a radius of six miles from Barnsley, in a district which has several 'thwaites,' the name Hoyland occurs three times. In each case the township rises well above the surrounding country, the highest point being a conspicuous object for many miles. Early records are:

HIGH HOYLAND.	HOYLAND NETHER.	HOYLANDSWAINE.
DB 1086 *Holand*	DB 1086 *Holand*	DB 1086 *Holande*
DB 1086 *Holant*	DB 1086 *Hoiland*	DB 1086 *Hoiland*
YF 1329 *Heghholonde*	PR 1176 *Hoiland*	YI 1266 *Holandeswayne*
PT 1379 *Hegh Holand*	KI 1285 *Holand Anstin*	NV 1316 *Holanswayne*
CH 1412 *Hyholand*	CH 1412 *Nethyrholand*	PT 1379 *Holand Swayne*

I think the name is to be connected with ON *hār*, *hōr*, high, rather than with OE *hōh*, *hō*, a projecting ridge; compare Hoober and Hooton. At the same time the form *Hoiland* found in DB and PR must, I think, be derived from ON *hǫy*, hay; compare such early forms of Austerfield, as 1237 *Oystrefeud*, 1293 *Oysterfeld*, from ON *ǫystri*. Under dialectal influence the early form *Holand* became Hoyland, just as *Soland* became Soyland. The distinctive affix in Hoylandswaine is derived from the Scandinavian personal name Sveinn or Sveini.

HOYLE.—Near Barnsley there is Hoyle Mill, and in Linthwaite Hoyle House. The word is a dialectal variation of 'hole,' from OE *hol*, a hollow ; compare Royd, formerly *Rode*, and Hoyland, formerly *Holand*.

HUBBERTON, Cartworth and Sowerby, may perhaps be derived from a personal name Hubber formed from the recorded name Hubba (Searle). I assume the loss of the sign of the genitive, as in Alverthorpe, Attercliffe, Rodley, Skelmanthorpe, and Thurstanland. The terminal is from OE or ON *tūn*, an enclosure, farmstead.

HUDDERSFIELD.—In DB we find *Oderesfelt*, *Odersfelt*, *Odresfeld*, but later spellings are of a different character and should be compared with those of Hothersall near Preston and Huddleston near Selby.

DN	†1131 *Huderesfeld*	RC	1199 *Hudereshale*	PF	1208 *Hudeston*			
CR	1215 *Hudresfeld*	PR	1201 *Hudereshal*	PC	1214 *Hudlestona*			
WCR	1275 *Hodresfeld*	PR	1206 *Huddeshal*	YI	1268 *Hodelstone*			
YS	1297 *Huderesfeld*	LI	1257 *Hudereshale*	KF	1303 *Hudleston*			
WCR	1413 *Hodresfeld*	LF	1313 *Hodirsale*	NV	1316 *Hodleston*			

The DB forms of Huddersfield are obviously at fault ; they omit the aspirate as in *Arduuic* for Hardwick, and they give *o* for *u* as in *Podechesaie* for Pudsey, defects both due to Norman scribes. The best of the early forms are doubtless *Huderesfeld*, *Hudereshale*, and *Hudleston*, and their meaning seems to be 'the field of Huder,' from OE *feld*, 'the corner of Huder,' from OE *healh*, and 'the farmstead of Hudel,' from OE *tūn*. Though no such names as Huder and Hudel are recorded, Searle has Hud and Huda, which with the common endings -er and -el would give the required forms.

But there is a complication of some importance to be dealt with, for *Hudereshale* has become Hothersall, and on the lips of the man in the street Huddersfield is sometimes 'Uthersfield' ; compare SM 1610 *Hutherfeild*, RE 1634 *Hothersfield*. When we recall the fact that the DB scribes and their successors often wrote *d* for *th*, we are compelled to ask whether *Huderesfeld* and *Hudereshale* may not, after all, represent *Hutheresfeld* and

Huthereshale. True, it is not uncommon for an earlier *th* to become *d,* as in the case of Lingards and Cudworth, and such words as 'rudder' and 'burden'; but if such a change had taken place in Huddersfield we should expect to find here and there early spellings involving *th.* On the other hand we find occasional words where an earlier *d* has been displaced by *th,* as in 'father' and 'mother'; and it seems not unlikely that under the influence of such common words as 'other' and 'another' a similar change has taken effect in Hothersall, and occasionally shows itself in the dialectal pronunciation of Huddersfield. On the whole, therefore, it seems probable that the first element in Huddersfield was originally Huder, rather than Huthhere, a name recorded by Searle.

The neighbourhood of Huddersfield appears to have been the meeting-place of several ethnic currents. Bradley, Farnley, Honley, Lindley, Huddersfield, Almondbury, Dalton, Kirkburton, Kirkheaton, seem to mark the most westerly settlements of the Anglians, whose appearance may perhaps be placed in the 7th century. Later, probably in the 9th century, came the Danes from the East. They approached by way of the valleys of the Calder and Colne, and their advance-posts are marked by several thorpes—Rawthorpe in Dalton, Finthorpe and Nether-thorpe in Almondbury, Gawthorpe in Lepton. Later still, during the 10th century, came Norsemen from the West. At this early period they settled in Golcar and Crosland; but, either at the same time or after the Norman Conquest, they settled also at Linthwaite, Slaithwaite, and Lingards, as well as in Quarmby (witness the name Burfitts Lane), in Kirkburton (witness Linfitts), and in Kirkheaton (witness the gills enumerated by Burton). The district west of Huddersfield provides, indeed, one of the strongest of Norse settlements.

In addition to the Scandinavian names already quoted, there are Ardron, Bannister, Birkby, Blacker, Booth, Bolster Moor, Cupwith, Dirker, Fixby, Garside, Little London, Lumbank, Magdale, Newbiggin, Owlers, Quarmby, Reaps, Scholecroft, Scholes, Stopes, Thurstonland, Wooldale and Woolrow. And beyond all these there are streams called Grain, districts called Lumb, woods called Storth, moorland paths called Rake, roads or lanes called Gate, grassy slopes called Slack and Wham, fields called Carr and Holme, and prominent features in the hills called Nab, Scout, and Scar.

But, further, there are several names which appear to involve a Celtic root, namely, Allen Wood, Bogden, Colne, Cowmes, Ribble, and Sude Hill.

HUGSET WOOD, Silkstone, PC †1090 and 1122 *Huggeside*, appears to have for its first element a Scandinavian personal name connected with ON *hugga*, to console.

HULLENEDGE.—See Hollinbusk.

HUMBLE JUMBLE.—This strange name occurs in connection with a beck and bridge at Alverthorpe. It is referred to in WRM 1391 as *Humble Jomble*; Denmark has a *Humlebek*; and in Lincolnshire, according to Streatfeild, the corresponding name *Humbelbec* is recorded in HR. In the Gazetteer we find three places called Humble, two in Scotland and one in Surrey; Northumberland has a Humble Hill; and between Humber and Tweed there are three Humbletons.

The possible sources are three. First there is the ON *humli*, Dan. and Norw. *humle*, the hop-plant, which has given such compounds as *humlegard* and *humlehage*, a hop-garden; secondly there is the Norse dialect-word *hummel*, barley, found in *hummelsaaker*, barley-field; thirdly there is the Norse dialect-word *humul*, a stone or boulder. A derivation from the last seems the most probable.

HUNGERHILL.—In South-west Yorkshire instances of this name occur in Haworth, Queensbury, Halifax, Ardsley, Morley, Fulstone, Hoylandswaine, and Dinnington. Similar German place-names—Hungerberg and Hungerrot, for example—are referred by Förstemann to OHG *hungar*, and we may refer our own names to OE *hungor*, ME *hunger*, explaining them as applied to land which is almost or altogether unproductive.

HUNNINGLEY, HUNSHELF, HUNSLET, HUN-STER, HUNSWORTH.—It has been made abundantly clear that, both directly and indirectly, tribal names often enter into the formation of place-names. France and Franconia were named from the Franks, Friesland from the Frisians, and Jutland from the Jutes. The names Norfolk and Suffolk designated first the inhabitants and afterwards the localities; and so also did Essex, Middlesex, Sussex, and Wessex. Large numbers of villages draw their modern names simply from the name of a family, for example, Tooting from the Totingas, Wittering from the Wihtringas; others add to the family name some such terminal as -ton or -ham, as in the case of Billingham and Billington. But, beyond this, there appear to be cases where the settler from whom a village took its name was himself known by the name of the tribe from which he sprang. We find in DB the personal name Norman, that is Northman, whence Normanton, formerly Northmanton; and other pairs of words linked in the same way are to be found, Frisa and Fryston, for example. The Celts made use of similar methods, witness Dumbarton, the fort of the Britons, and Dumfries, the fort of the Frisians. It is not altogether impossible, therefore, that a connection may exist between the names below and the tribe of the Huni—an indirect connection through an individual called by the name of his tribe.

HUNNINGLEY, Barnsley, is 'the lea of the sons of Hun'; compare the Dutch place-name Huninge (NGN).

HUNSHELF, Penistone, DB *Hunescelf*, DN 1307 *Hunshelfe*, NV 1316 *Hunclyf*, PT 1379 *Hundeschelf*, IN 1558 *Hunschelfe*, is 'Hun's shelf or ledge of land,' from OE *scylf*.

HUNSLET, Leeds, shows forms of a twofold character. DB has *Hunslet*, KI 1285 *Hunslett*, KF 1303 *Hunslett*, and PT 1379 *Hunslet*; but PF 1202 gives *Hunesflet* and *Hunesflet Ker*, NV 1316 *Hunseflet* and KC 1336 *Hunseflet*. The latter form means 'Hun's stream' from OE *flēot*, a stream, river. The former is probably 'Hun's weir,' from OE *lete*, a boundary, fence, weir; compare MHG *lette*.

HUNSTER, Bawtry, is probably from ON *staðr*, a place, and the personal name Hun; compare Ulster, Leinster, and Munster, where the terminations are from the same source.

HUNSWORTH, Bradford, was spelt *Hundesworth* in KI 1285, CR 1317, and PT 1379, but the *d* may be intrusive as in one of the early spellings of Hunshelf; thus the meaning may be either the 'farmstead of Hund,' or the 'farmstead of Hun'—most probably the former.

HUNTWICK, Wakefield, PF 1202 *Huntewich,* DN 1329 *Huntwicke, Huntewicke,* YF 1555 *Huntwick,* is 'Hunta's dwelling,' from the OE personal name Hunta, and OE *wīc,* a dwelling-place, hamlet.

HURST.—See Hirst.

HUTHWAITE, Thurgoland, mentioned in a charter of 1366 as *Huthwayt,* may perhaps be ' Hugi's clearing,' from ON *thveit* and the personal name recorded by Falkman.

ICKLES occurs in Whitley Lower and Rotherham, and the latter is recorded as *Ikkels* in VE 1535, and *Ikkyls* in 1560. The word is probably of the same type as Eccles. In the Gazetteer we find such names as Ickford, Ickleford, Ickilford, as well as Ickburgh, Ickham, Ickwell, and Ickworth. It seems possible that Ick was an early stream-name; compare the Gaulish river-name *Icauna,* now the Yonne, and the OE *Iccen,* now the Itchen.

IDLE.—See Bierley, and note that in Linthwaite there is an eminence called Idle Hill, while in Notts there is a tributary of the Trent called the Idle.

ILLINGWORTH, Halifax.—Watson believed this place so called 'from the badness or roughness of the ground.' There can be little doubt, however, that the first element of the word is the patronymic Illing, which is found again in Illington, Norfolk, and in Ilingswarf, East Friesland. Early forms of the word are WCR 1297 *Hillingwrth,* WCR 1330 *Illingworth,* PT 1379 *Illyngworth,* and the meaning is 'the farm of the Illings,' OE *worth,* a farm.

ING.—There are two quite distinct words of this form, (1) the OE patronymic suffix meaning 'son of,' and (2) the field-name signifying 'meadow.'

1. The primary use of the patronymic is well seen in early genealogies. Under the date 626 the AS Chron. says 'Penda was Pybbing, Pybba Creoding, Creoda Cynewalding,' that is, Penda was the son of Pybba, Pybba the son of Creoda, Creoda the son of Cynewald. When we come to the use of the suffix in connection with place-names we find considerable variety of treatment.

(a) The most striking use of the suffix is where the name of a family is used as the name of a place. This idiom occurs elsewhere; Essex and Norfolk, for example, are literally 'East Saxons' and 'North Folk,' and Wales means 'the Foreigners.' In OE place-names the suffix was used both in the plural— Nom. -ingas, Gen. -inga, Dat. -ingum—and, as we should scarcely expect, in the singular—Nom. -ing. We find such early forms as Hallingas, Pæccingas, Chenottinga, Basyngum, Mallingum, Cilling, and Ferring. At a later period -ingas became -inges, and -inga became -inge; while, still later, all were levelled under the form -ing. The only representatives of the simple patronymic in South-west Yorkshire are Bowling and Cridling, of which early records are as follows:

DB	1086	*Bollinc*	PF	1202 *Crideling*
YI	1246	*Bollinge*	PC	†1220 *Crideling*
CC	1265	*Bollyng*	LC	1296 *Cridelinge*
KI	1285	*Bolling*	NV	1316 *Crideling*
KF	1303	*Bollyng*	PM	1327 *Cridelinge*

These point to OE genitives plural of the form **Bollinga* and **Crydelinga*, and the meaning is '(the place) of the sons of Bolla,' and '(the place) of the sons of Crydel.' For the name Bowling compare the Dutch place-name Bolinge, the Italian Bolengo, the French Bollinghem, and the English Bolingbroke and Bollington; and for the name Crydel note that Searle gives Cryda which with the common suffix -el would give the form required.

(b) But a far more widespread use of the patronymic is in such compound words as OE *Oddingalea*, 'the lea of the family of Odda,' and *Wealingaford*, 'the ford of the sons of Wealh.' These forms are duly succeeded by such as *Oddingeleye* and *Walingeford*, and finally by Oddingley and Wallingford.

South-west Yorkshire provides only four assured examples of this class:

Cottingley	DB 1086 *Cotingelei*	CR 1283 *Cotingeleye*	
Cullingworth	DB 1086 *Colingauuorde*	PF 1235 *Cullingwurth*	
Knottingley	DB 1086 *Notingeleia*	PF 1202 *Cnottinglai*	
Manningham	CR 1250 *Maningeham*	KF 1303 *Maynyngham*	

At this point we ought perhaps to note an alternative use where the patronymic is not declined; compare *Eccyncgtune* and *Teottingtun*, Eckington and Teddington in Worcs., both early forms. Sundermann gives first place to such a use, and provides a long list of examples, including Bedinghem, Bollinghusen, and Kollinghorst.

Beyond these there were two other uses of a secondary and derivative character.

(*c*) The suffix might be used with names of places instead of persons. In this way we get such examples as *Catmeringas*, 'the dwellers at Catmere,' *Woburninga* formed from Woburn, *Wæneting* from *Wænet*, and *Hertfordinge* from Hertford.

(*d*) In compound names the suffix might be used with the force of a genitive; thus the *Wieghelmestun* of BCS (97) was written *Wigelmincgtun* in the endorsement of the charter.

But the list is not yet complete, for, in addition to all these uses, there are numerous instances where -ing usurps the place of other forms, more especially genitives in -an. Abingdon, for example, stands for *Æbbandun*; Whittington, Staffs., for *Hwitantone*; and Shellington, Beds., for *Chelwintone*, the enclosure of Ceolwynne.

When we examine the names in South-west Yorkshire, in addition to (*a*) Bowling and Cridling, and (*b*) Cottingley, Cullingworth, Knottingley, and Manningham, we find a long list where it is impossible to be clear what is the original form. This list includes Addingford, Crodingley, Darrington, Drighlington, Dunningley, Dinnington, Fallingworth, Frizinghall, Girlington, Haddingley (2), Harlington, Hostingley, Hunningley, Illingworth, Kellingley, Kellington, Pollington, Rossington, Santingley, Stanningley (3), Stannington, Trimingham.

Among names where -ing has no rightful place there are Erringden, Hollingbank, and Hollingthorpe.

2. The second word 'ing' is used to designate meadowland, 'especially low-lying land beside a stream,' and is derived from ON *eng*, Dan. *eng*—unless, indeed, there is a direct connection with the Frisian *inge*. The occurrences of the word are innumerable. We find such examples as Birk Ing, Carr Ing, Hessle Ing, Owler Ing, Sour Ing, and Toft Ing, as well as such compounds as Baitings, Cocking, Gadding, Hacking, Leeming, Ozzing, Scamming, Stocking, and Stubbing; and in addition there is a small group of names where 'ing' is used attributively, namely, Ingberchworth, Ingrow, and Ingwell, explained below.

INGBIRCHWORTH, Penistone, DB *Bercewrde*, *Berceworde*, YD 1326 *Hingbircheworth*, PT 1379 *Bircheworth*, BD 1456 *Yngbircheworth*, is derived from OE *beorce* or *birce*, the birch-tree, and *weorth*, an enclosure or farmstead; and the original name meant 'the birch-tree farmstead.' The prefix Ing, added later, is evidently intended as a contrast with that in Roughbirchworth, an adjacent village. This lends support to the theory that it is the word *eng* or *ing*, a meadow, Roughbirchworth being the farmstead in the midst of uncultivated land, and Ingbirchworth that in the midst of meadows.

INGROW, Keighley, comes doubtless from ON *vrā*, or rather *rā*, a corner, and the first element is either ON *eng*, a meadow, or the ON personal name Ingi.

INGWELL, Wakefield, WRM 1698 *Ingwell*, appears to be simply 'the well in the meadow,' from OE *well*.

INTAK, INTAKE, occurs very frequently. It signifies an enclosed piece of land, and is of Scandinavian origin; compare Sw. *intaka*, an enclosed common, and the Norwegian dialect-word *inntak*, taking in.

JAGGAR, JAGGER.—We find Jaggar Lane in Honley, Jagger Green in Stainland, Jagger Hill in Kirkheaton, and Jagger Wood in Northowram and Thurgoland.

In EDD the word 'jagger' is explained as 'a travelling pedlar, a hawker, carrier, carter, packhorse driver'; in Jamieson's Scottish Dictionary a 'jagger' is described as a pedlar, and 'jags' as saddlebags or leathern bags of any kind.

JERICHO.—This Scriptural name is to be found in the township of Stainland.

JORDAN, JORDON.—There is a place called Jordon near Rotherham; in Hopton there is Jordan Wood; in Denby Jordan Beck. Outside the county we find such examples as Jordan Bank off the Lancashire coast, Jordan Hill in Weymouth and Glasgow, and Jordan Gate in Macclesfield.

EDD explains the dialect-word 'jordan' as a piece of watery ground. It is possible, however, in such names as Jordan Wood, that the word has another origin. Owing to the Crusades the personal name Jordan became extremely popular throughout Western Europe. In the Guisborough Chartulary the name 'Johanne filio Jordani' occurs early in the 13th century.

JUBB.—The name Jubb Hill occurs in Thurgoland.

JUG, JUGS.—Among the field-names of Lofthouse we meet the curious form Midlam Jugs. EDD explains Jug as 'a common pasture or meadow,' but does not state the source of the word. Can the word be connected with a 'yoke' of land, that is, with Fr. *joug*, Lat. *jugum*?

JUM, JUMBLE, JUMP.—Near Barnsley there is a village called Jump; near Todmorden a valley called Jump Clough; in Sowerby Jumm Wood; in Langfield Jumb Hill; in Erringden Great Jumps; in Illingworth Jumples Mill; in Northowram Jum Hole; in Ardsley Jump Hill; and in Ecclesfield Jumble Hole. In addition there are fields called Jumble or Jumbles in various places, including Kirkheaton, Ecclesall and Lofthouse; and a stream near Wakefield called Humble Jumble Beck is referred to in WRM 1391 in the phrase '*Humble Jomble* in Rustanes.' From EDD we learn that Jumble means 'a rough, bushy, uncultivated hollow'; but no derivation is given.

JUNCTION, evidently a modern name, occurs in the township of Quick, Saddleworth, at a point where five roads meet.

KAYE.—Kaye Lane in Almondbury and Kaye Wood in Fulstone may have received their prefixes from a personal name;

but it seems probable in any case that Kaye should be connected with the OW *cai*, W *cae*, a hedge, fence, enclosure; compare Prim. Celt. **kagi-* which is cognate with ON *hagi* and OE *haga*. If this be an accurate guess the surnames Kaye and Haigh will have the same ultimate origin, the former having come down to us by the Celtic branch of the Indo-European family, the latter by the Teutonic.

KEBCOTE, KEBROYD, Stansfield and Sowerby.—In EDD we find the following explanations: Keb house, 'the shelter erected for young lambs in the lamb season'; keb ewe, 'an ewe that has lost her lambs'; keb, 'any creature small of its kind'; and on the other hand keb, 'an old sheep.'

KEELAM, KEELHAM.—This name occurs on the moors near Todmorden, Heptonstall, Midgley, and Denholme. Its source is most probably ON *kjölr*, which originally meant the keel of a ship, but signified later a keel-shaped range of mountains, a mountain ridge. The form Keelam would be derived from the dative plural, and would mean 'the ridges.' Keele in Staffordshire, 12th century *Kiel*, and Keelby in Lincolnshire very probably spring from the same root.

KEIGHLEY stands on the Aire, about nine miles northwest of Bradford. The pronunciation, which varies between 'Keethly' and 'Keely,' is paralleled by that of Leigh near Wigan, which, though usually called 'Lee,' is occasionally pronounced 'Leyth.' Early records give the following forms:

DB	1086 *Chichelai*	KI	1285 *Kighley*
KC	1234 *Kyhhelay*	CC	1311 *Kythelay*
YR	1244 *Kikhele*	NV	1316 *Kygheley*
YI	1273 *Kihele*	PT	1379 *Kyghlay*

The symbol *ch* was used by the Domesday scribes for more than one sound. Thus (1) *ch*=*k* in *Monechetone* and *Barchestone*, now Monkton and Barkston; (2) *ch*=*tch* in *Lachenduna* and *Blachingelei*, now Latchingdon and Bletchingley; (3) *ch*=*dg* in *Sechebroc*, which represents Sedgebrook; (4) *ch*=OE *h* or *g* in *Borch*, which represents OE *burh* or *burg*. I take the second *ch* in *Chichelai* to have the last of these values, and rewrite the

name as *Cyhelai* or *Cygelai*.   As Searle gives a personal name
of the form Cyga, the interpretation may well be 'the lea of
Cyga,' and the name would come from OE *Cyganlēah*, which
would give ME *Kygheley* quite regularly.   For the strange
spelling *Kikhele* compare *Bekhala*, now Beal, p. 69.

In the neighbourhood of Keighley there are many names of
Scandinavian origin.   In addition to various Carrs and Holmes
we find Thwaites and Braithwaite, Denby Ing and Denby Hill,
Lumb Head and Lumb Foot, as well as Flask, Lawkholme,
Scholes, Slack, and Ingrow.   As there are no 'thorpes' and as
Scholes and the Thwaites are undoubtedly Norse, we shall
probably be right in assuming a Norse origin for the whole
group; indeed, seeing that the neighbourhood of Bradford is
so predominantly Anglian, we may take it as certain that the
Scandinavian element around Keighley came from the north-
west—down the valley of the Aire.

Among possible Celtic names in the neighbourhood there
are Dob, Crumack, and the first element in Sugden.

KELLINGLEY, KELLINGTON,   Knottingley.—The
Domesday records of the latter, *Chelinctone* and *Chellinctone*, give
nc for *ng*, as in the case of Bowling, DB *Bollinc*, and Tong, DB
Tuinc.   Other early spellings are

PC 1159 *Kelinglaiam*	SC 1202 *Kelington*
PC †1160 *Kelingley*	NV 1316 *Kelington*
PC †1190 *Kellinglaiam*	PT 1379 *Kelyngton*

Among our early place-names a few, clearly of Scandinavian
origin, have as their first element the personal name Kel:

1086 *Chelestuit*, 'Kel's thwaite,' found in DB.
1182 *Kelesterne*, 'Kel's tarn,' Guisborough Chartulary.

But there is another name, clearly Scandinavian, where the first
element is a patronymic formed from Kel:

1200 *Kelingthorpe*, 'Keling's thorpe,' Guisborough Chartulary

As Kellington may be of similar origin, we may explain it as
'Keling's farmstead,' from ON *tūn*.   Yet the name may be
Anglian, 'Cēoling's farmstead'—compare DB *Cellinc*—but in
that case the initial *k* is due to Scandinavian influence.

KERISFORTH, Barnsley.—Early forms are DB *Creuesford*, PC †1240 *Keverford*, BD 1344 *Kenerosford* (for *Keueresford*?), YD 1349 *Keuerisforth*, YI 1588 *Keresforth*. The Domesday spelling is faulty, the *r* being misplaced, and we may fairly assume an early form **Keveresford* which would warrant the explanation 'Kever's ford.' A personal name of corresponding form occurs in DB, namely, *Cheure*, where *ch = k* and *u = v*, and probably this goes back to a Scandinavian name cognate with OE *ceafor*, a beetle, chafer; compare Germanic **kafru*, **kefra*, a chafer (Torp).

KERSHAW, Luddenden, is frequently mentioned in the Wakefield Court Rolls; in 1307 we find *Kirkeschawe*, in 1308 *Kirkeshagh*, in 1326 *Kerkeshagh*, in 1343 *Kerkeschagh*. The meaning of the word is 'church copse,' from OE *sceaga*, a copse or wood; but what is the church referred to? See Kirk.

KETTLETHORPE, Wakefield, DN 1242 *Ketelesthorp*, WCR 1275 *Ketelesthorp*, YS 1297 *Ketilthorp*, WCR 1307 *Ketelisthorpe*, means 'the thorpe of Ketel.' There are places called Kettlethorpe in both the North and East Ridings, the DB record in each case being *Chetelestorp*. The personal name, which is recorded in DB as Chetel and in LN as Ketill, was extremely common among the Vikings.

KEXBOROUGH, Barnsley.—The Domesday spellings are *Cezeburg* and *Chizeburg*, where *z* probably represents *ts*; compare DB *Asgozbi* for **Asgotsbi*, now Osgodby, and DB *Feizbi* for **Feitsbi*, now Faceby. Other early spellings are

YI 1284 *Kexeburg*	CH 1337 *Kesceburgh*	
YS 1297 *Kesseburg*	DN 1364 *Kesseburgh*	
NV 1316 *Keseburgh*	DN 1432 *Kexburgh*	
DN 1324 *Keskeburgh* (surn.)	YF 1545 *Kesburghe*	

Compare with these the following early forms of Flasby near Skipton (1), and Flaxby near Knaresborough (2):

(1) DB 1086 *Flatebi* (for *Flatesbi*)	(2) DB 1086 *Flatesbi*
HR 1276 *Flasceby*	KF 1303 *Flasceby*
KF 1303 *Flasceby*	NV 1316 *Flasseby*
NV 1316 *Flasceby*	PT 1379 *Fflasceby*

Here obviously *ts* has become *ss*, *sc*, and similar forms in the

case of Kexborough point to a DB form with *ts*, namely,
*Ketseburg. But Flaxby shows an *x*, due apparently to the
substitution of *ks* for *ts*, as in Kexmoor (Moorman); probably,
indeed, forms from *ks* and *ts* existed side by side, the one
surviving in Flaxby and the other in Flasby. Similar considerations apply to Kexborough—DN 1324 *Keskeburgh* being due
to metathesis—and therefore the name may be explained as
' Ketsi's fortified post.' The form *Ketsi is built up from ODan.
Keti (Nielsen) on the pattern of Grimsi, Hugsi, Elfsi (Naumann,
p. 150); compare Wilsden and Wilsick.

KILHOLME, KILPIN.—Several Yorkshire place-names
have Kil- as the first element, among them the following:

KILBURN	(NR)	DB 1086 *Chileburn*	CH †1250 *Killeburne*
KILDALE	(NR)	DB 1086 *Childale*	CH †1230 *Kildale*
KILTON	(NR)	DB 1086 *Chiltun*	CH †1250 *Kylton*
KILPIN	(ER)	DB 1086 *Chelpin*	NV 1316 *Kilpyng*

The Teutonic stem **kīla*, a wedge, has given us Norw. dial. *kile*,
a wedge, a narrow triangular piece; and of cognate origin are
Dan. *kil*, a narrow strip of land, and ON *kīll*, a narrow bay.
There is also according to Middendorff an OE word *cīll* meaning
a stream flowing in a deep bed, but this could not give Kil-, and
we have to assume either Scandinavian origin or Scandinavian
influence.

KILHOLME, Cantley, SM 1610 *Kilholme*, has for its second
element the ON *holmr*, an island, low-lying land beside a river,
and its first element is doubtless Scandinavian also.

KILPIN, Heckmondwike, may mean ' the point (or summit)
of the triangular piece of land,' from ON *pinne*, a point; compare
OE *pinn*, a point.

KILNHURST, KILNSHAW.—In the case of Kilnhurst,
Rotherham, YS 1297 has *Kilhenhirst*, and PT 1379 *Kilnehirst*;
in the case of Kilnhurst, Langfield, HW 1521 has *Kilnehirst*.
The first element is derived from OE *cylen*, a furnace, and the
terminations come from *hyrst* and *sceaga*, each of which signifies
a copse or wood. The ancient kilns were chiefly for making
charcoal, burning lime, or baking bricks; and Kilnhurst and
Kilnshaw refer doubtless to kilns for the first of these objects.

KIMBERWORTH.—See Cumberworth.

KINSLEY, Hemsworth, DB *Chineslai, Chineslei,* DN 1244 *Kynnsley,* YD 1328 *Kynnesley,* IN 1348 *Kynnesley,* is 'the lea of Cyne,' from OE *lēah* and the recorded personal name Cyne. In 1302 the spelling *Kinnersley* occurs; this proves the contemporaneous *Kynnesley* to have been trisyllabic, and provides another example of the intrusion of the consonant *r*; compare Gildersome.

KIRK.—In NED this word is described as the Northern and Scotch form of 'church,' and OE *circe* is compared with ON *kirkja.* Skeat believes the Scandinavian forms to have been borrowed from the Old English; but in any case the ultimate source of the word is the Greek neuter plural κυριακά, from κυριακός, belonging to the Lord.

In DB only one place-name in South-west Yorkshire possessed this word as prefix, viz. South Kirkby, DB *Cherchebi*; three other names which now possess it, had no such prefix in DB:

Kirk Bramwith	DB *Branuuithe, Branuuat*
Kirkburton	DB *Bertone*
Kirkheaton	DB *Etone*

The earliest mention of Kirkheaton appears to be YD 1369 *Kirkeheton,* and for Kirkburton I have no earlier record than DN 1516 *Kyrkebyrton.* But there were churches at both places long before these dates. What seems to have happened at both Kirkburton and Kirkheaton is this: The townships received their Anglian names *Bertone* and *Etone* at an early date—probably in the 7th or the 8th century. The ancient centre of population was on the hill, that is, at the places now called Highburton and Upper Heaton; the church came later, and for the greater convenience of the parish, which included other townships, it was built in each case some distance away from the centre of the township; as years passed by, hamlets sprang up near the church, and these were naturally described as Kirkburton and Kirkheaton, the townships meanwhile continuing to bear the old names Burton and Heaton; later still, when the new hamlet became the predominant partner, the whole township received the name Kirkburton or Kirkheaton. Compare Birstall.

KIRK BRAMWITH, KIRKBY, KIRKTHORPE.—

These are all of Scandinavian descent. Early records of Kirk Bramwith, Doncaster, and South Kirkby, Pontefract, are as follows:

DB 1086 *Branuuithe, Branuuat*		DB 1086 *Cherchebi, Chirchebi*	
PF 1201 *Bramwith*		YR 1267 *Suth Kyrkeby*	
YR 1252 *Bramwit*		CR 1280 *Suth Kirkeby*	
NV 1316 *Brampwyth*		KI 1285 *South Kyrkeby*	

KIRK BRAMWITH in its DB forms designates different features of the landscape—*Branuuithe* is 'bramble wood' and *Branuuat* is 'bramble ford,' from ON *viðr*, a wood, and *vað*, a ford. Possibly the two names existed side by side for a time, Bramwith finally gaining the upper hand. See Bramley.

KIRKBY is 'church village,' from ON *bȳr*, a village or hamlet. In addition to South Kirkby there is a farmhouse in Emley called Kirkby, PT 1379 *Kirkeby*; and formerly Pontefract or a part of it was also called Kirkby. See Pontefract.

KIRKTHORPE, Warmfield, is mentioned in a Fine of 1547 as *Kyrkethorp*, and means 'church village.' There is an interesting reference to Warmfield, Kirkthorpe, and Heath, in YR 1252 where the names as recorded are *Warnefeld*, *Gukethorp*, and *Bruera*.

KIRKBURTON, Huddersfield, is recorded in DB as *Bertone*, PF 1208 *Birton*, YR 1229 *Birton*, YS 1297 *Byrton*, NV 1316 *Birton*, PT 1379 *Byrton*, DN 1516 *Kirkebyrton*. It must be noted that the DB scribes often wrote *e* for *y*; if we rewrite the DB form *Byrtone*, we shall have a quite consistent series of forms, derived probably from OE *bȳre*, a cowhouse. Compare the early spellings of Burcote, Worcestershire, DB *Bericote*, 1275 *Byrcote*.

KIRKHAMGATE, Wakefield, gives an example of the OE *hamm*, an enclosure, for Kirkham is 'church enclosure,' the suffix 'gate' being a later addition signifying 'road,' from ON *gata*, a path or road.

KIRKHEATON, Huddersfield, is recorded in DB as *Etone*, FC 1199 *Heton*, KI 1285 *Heton*, YD 1369 *Kirkeheton*, PT 1379

Heton. The original name, Heton, meant 'high farm,' from OE *hēah, hēh,* high, and *tūn,* an enclosure or farm. The prefix was a later addition; see Kirk.

KIRKLEES, Brighouse.—In the park there are the remains of 'a small house for Cistercian nuns' and 'a small rectangular Roman camp' (Morris). There is also a tomb said to be that of Robin Hood, who, according to tradition, died here through the treachery of the prioress. The Wakefield Court Rolls contain the following records of the name: 1275 *Kyrkeleys,* 1314 *Kirkeley,* 1326 *Kirkeleghes,* 1423 *Kyrkeleis,* 1573 *Kirkelees,* and CR 1236 has *Kyrkelay.* The meaning is obviously 'church leas,' from ON *kirkja,* a church, and OE *lēah,* a lea or meadow.

KIVETON, Sheffield, DB *Ciuetone,* YS 1297 *Keueton,* YI 1304 *Kyueton,* YD 1326 *Keueton,* probably derives its first element from OE *cȳfe.* This word is explained as a vessel, tub, vat, and may perhaps denote a hollow. The terminal comes from OE *tūn,* an enclosure, farmstead.

KNOT or **KNOTT,** comes from ON *knūtr,* a knot, protuberance, a word used in Norway in the names of mountains, as in Jordalsnuten and Thorsnuten. This Norse word will account for the first element in Knott Wood, Stansfield, and Knot Hill, Saddleworth, the latter recorded as *Cnouthull* and *Cnothill* in 13th century charters, and *Quotil* (*u = n*) in YS 1297.

Occasionally we find the word used as a termination, thus, in the Cockersand Chartulary at the beginning of the 13th century we meet with such names as *Haluecnot* and *Gripcnottes.* The termination in Pyenot, Liversedge and Marsden, is probably to be accounted for in this way.

KNOTTINGLEY, DB *Notingeleia, Notingelai,* PF 1202 *Cnottinglai,* PC 1219 *Knottinglay,* IN 1258 *Cnottingley,* PT 1379 *Knottynglay.* The Norman scribe found the combination *cn* as difficult as the modern Anglian finds it, and, like the modern Anglian, he gave one consonant instead of two. In Bedfordshire he overcame the difficulty in another way, transcribing the name of the village now called Knotting as *Chenotinga.* The meaning of Knottingley is 'the lea of the sons of Cnot,' from OE *lēah*, a lea or meadow; compare the modern surname Knott. Nottingham has an origin quite different; in the AS Chronicle it is spelt *Snotingaham*, and in DB it is *Snotingeham*; BCS, on the other hand, has the name *Cnottingahamm*, connected with Barkham in Berkshire.

KNOWLE, KNOWLER, KNOWLES.—The origin of these words is either the OE *cnoll*, or ON *knollr*, a gently rounded hill, the summit, the top or crown of a hill.

KNOWLE occurs in Sheffield, Emley, Mirfield, and Austonley, and KNOWLES in Fixby and Dewsbury.

KNOWLER HILL, Liversedge, was called *Hustin Knowll* in 1560; but Knowler may have been used concurrently. Hustin is from ON *hūs·thing*, a council or meeting to which a king, earl, or captain, summoned his people or guardsmen. It is clear that Knowler Hill was a place of meeting for the franklins of the district. See p. 26.

KRUMLIN.—See Crimes.

LADCAR, LADCASTLE, LAD STONE.—The first is in Emley, SE 1715 *Ladcar*, the second in Saddleworth, while the third is a prominent mass of rock on Norland Moor. In each case the first element is from OE *hlæd*, a rock, pile, heap. There are other West Riding names of similar character; for example, on the hills south of Todmorden we find Two Lads, and near Ilkley an ancient boundary stone called Lanshaw Lad. The termination in Ladcastle comes from OE *castel*.

LADYTHORPE is in the township of Fenwick.

LAITHE or **LAITHES.**—This name, derived from ON *hlaða*, a storehouse or barn, occurs frequently throughout the district. In Widdop there is New Laithe Hey; in Alverthorpe, Low Laithes; in Ardsley, Wood Laithes; in Rishworth, Cheetham Laithe; in Woodlesford, Dub Laithe; in Carlton, Laithe Close; in Holme, Wood Hey Laithe; in Austonley, New Laithe; in Thurlstone, Low Laithe.

LAND.—This termination may be either Scandinavian or Anglian; compare OE *land,* ON *land,* land, district, territory, also *land* in Danish, Swedish, and Norwegian. South-west Yorkshire has the following instances: Austerlands, Barkisland, Crosland, Elland, Friezland, Greetland, Hoyland (3), Newland (3), Norland, Soyland, Stainland, Sunderland, Thurgoland, and Thurstonland. Ten of these appear to be Scandinavian, namely, Austerlands, Barkisland, Crosland, Hoyland (3), Soyland, Stainland, Thurgoland, and Thurstonland; two others, Greetland and Norland, may come either from Anglian or Scandinavian; the remnant which may be described as certainly Anglian is, therefore, quite a minority. It should be noted that six of these places, near Halifax,—Barkisland, Elland, Greetland, Norland, Soyland, Stainland—are adjacent to each other, but Elland alone appears in the Domesday record.

LANGFIELD, LANGHAM, LANGHOLM, LANG-LEY, LANGOLD, LANGSETT, LANGTHWAITE, LONGBOTTOM, LONGLEY, LONGROYD, LONG-WOOD.—The word for 'long' was *lang* in OE and *langr* in ON.

LANGFIELD, Todmorden, DB *Langfelt,* HR 1276 *Langfeld,* WCR 1297 *Langfeud,* is 'long field,' from OE *feld.*

LANGHAM, Rawcliffe, is probably from OE *hamm,* an enclosure or dwelling, rather than OE *hām,* home.

LANGHOLM, Thorne, 1342 *Langholm,* is 'long holme,' from ON *holmr,* an island, lowlying land subject to inundation.

LANGLEY is found in Emley and Bradfield, and means 'long lea,' from OE *lēah,* a lea or meadow.

LANGOLD, Letwell, 1402 *Langhald,* YF 1540 *Langold,* YF 1571 *Langald,* appears to be 'the long shed or shelter.'

LANGSETT, Penistone, CR 1290 *Langside,* NV 1316 *Lanside,* PT 1379 *Langside,* is 'the long side or slope,' OE *sīde.*

LANGTHWAITE, Doncaster, DB *Langetouet,* PF 1167 *Langethwaite,* YI 1279 *Langethauit,* NV 1316 *Langethwayt,* is 'the long clearing or paddock,' ON *thveit.*

LONGBOTTOM, Warley, WCR 1308 *Longbothem,* is from OE *botm,* which refers to lowlying land.

LONGLEY occurs in Almondbury, PT 1379 *Longlegh*; in Ecclesfield, CH 1366 *Longeley*; in Holmfirth, WCR 1307 *Longleye*; and in Norland, WCR 1285 *Langgeley*. The meaning is 'the long lea,' OE *lēah*.

LONGROYD BRIDGE, Huddersfield, 17th century *Langrodbrig*, is 'the bridge beside the long clearing'; see Royd.

LONGWOOD, Huddersfield, PF 1202 *Langwode*, DN 1383 *Langwode*, has a meaning quite obvious, from OE *wudu*.

LASCELLES HALL, Huddersfield, YD 1462 *Lascelhal*, has for its first element a place-name of French origin which became the name of a great Yorkshire family. In Brittany there is a village called Laselle; in Touraine one called Lasselle; and Lecelles, Nord, is still another form of the word. The last, written *Cella* in 1107 and 1119, is explained by Mannier as a hermitage, from Lat. *cella*, a cell, small room, hut.

LAUGHTON-EN-LE-MORTHEN, near Rotherham, is a name curiously compounded of English, French, and Scandinavian. Early records give the following forms:

DB 1086 *Lastone*	CR 1257 *Lacton in Morthing*
YR 1224 *Lactone*	KI 1285 *Laython in Morthing*
CR 1227 *Lacton*	PT 1379 *Leghton*
CR 1256 *Lacton Imorthing*	VE 1535 *Laghton*
SM 1610 *Leighton in the Mornyng* (!)	

These must be compared with early forms of Leighton Buzzard, Beds., among which we get the following:

DB 1086 *Lestone*	1291 *Leython*
PM 1272 *Leghton*	1316 *Leythone*

The Domesday forms, which show the same scribal error as Drighlington, are quite alike except for the principal vowel; and later forms approximate in the same way. It seems clear that, apart from a dialectal difference of vowel which goes back to the earliest times, the names entirely agree. As to the interpretation of Leighton, Dr Skeat says there is no difficulty. "There are plenty of Leightons," he says, "because the word simply meant 'garden.' It is from the AS *lēah·tūn*, lit. 'leek-town,' i.e. place for cultivating leeks, which was once a general word for vegetables. The AS for leek is *lēac*; but this became *leah* on account of the phonetic law whereby almost every AS *ct* passed into *ht*." The variation in vowel would be fully accounted for by the Mercian form *lēhtūn*, as contrasted with the OE form *lāhtūn*, given by Wright in his Vocabularies and quoted by Moorman.

Apart from Laughton-en-le-Morthen and Brampton-en-le-Morthen no instance of the French particles 'en le' is to be found in South-west Yorkshire; just over the border there is, however, the well-known Derbyshire example Chapel-en-le-Frith. These particles were doubtless as a rule not taken into popular speech, but were reserved for deeds and other legal documents.

There is a farm called LAUGHTON in Soyland.

LAVER, LAVERACK, LAVEROCK. — The name Laverock Hall occurs in Keighley, Haworth, Idle, Ovenden, and Marsden, while Laverack Lane is found in Brighouse and Laverack in Kirkheaton.

The only early records are BM *Laver Bridge*, Kirkheaton, which appears to be connected with Laverack, and YD 1474 *Leuerichbroke*, which is doubtless connected with Laverack Lane, Brighouse. The word Laver or Lever comes from OE *læfer*, a rush, reed, bulrush, and it seems probable that many of the names given above are derived from this source. At the same time some may come from OE *laferce*, ME *laverock*, the lark, and perhaps all have been influenced by this word.

LAW, LOW.—The simple name occurs as Law Hill in Hepworth, Southowram, and Wakefield. As a termination we find the word to-day in Canklow, Chellow, Ringinglow, Walderslow, and Whirlow; but formerly it was to be found also in Ardsley, Blackley, Dunningley, Tingley, and Tinsley. At least one of these names is Scandinavian, viz. Tingley; and we shall probably be right if in that case we refer the terminal ultimately to Prim. Norse *hlaiv*, a grave-mound, a word which according to Torp occurs only in runic inscriptions. Where the first element is Anglian we may refer -low to OE *hlāw*, *hlǣw*, ME *lawe*, *lowe*, a mound—natural or artificial—a cairn, tumulus, hill.

LAWKHOLME, Keighley.—See Holme.

LAYCOCK occurs four times, viz. in Wortley (Sheffield), in Wintersett, in Thurstanland, and near Keighley. Early spellings of the last-named are given in the first column below, and along with them for the sake of comparison are given those of Lacock or Laycock in Wiltshire.

DB	1086 *Lacoc*		CH	854 *Lacoc*
KI	1285 *Lackac*		CH	854 *Lacok*
PT	1379 *Lacokke*		DB	1086 *Lacoc*
YF	1581 *Lacocke*		DB	1086 *Lacoch*
SM	1610 *Lacock*		NV	1316 *Lacock*

The modern spelling is obviously misleading. Perhaps the name involves the suffix found in 'hillock' and 'parrock,' the stem being derived from OE *lacu*, running water, a stream, a pond. Support for this suggestion appears in the Cheshire stream-name Wheelock, of which the source is probably OE *wǣl*, a whirlpool, eddy, pool.

LEPPING, LEPTON.—Of the former, which occurs in Wadsley, there are no early spellings, but of the latter, a township near Huddersfield, we find DB *Leptone*, DN †1225 *Lepton*, YS 1297 *Lepton*, NV 1316 *Lepton*. It is scarcely possible that the meaning should be 'Leppa's farm,' for that would require such early forms as *Leppeton*; more probably the word is from Norw. *lepp*, a patch, a strip, as in the Norwegian lake-name Lepvandet (Rygh). In that case the terminations would come from ON *eng*, meadowland, and *tūn*, an enclosure or farm.

LETWELL, Tickhill, YD 1326 *Lettewelle*, PT 1379 *Lettewell*, CH 1386 *Lettewelle*, YF 1579 *Letwell*. Among the names in -well examples where the first element is a personal name are to be found only occasionally; but I think Letwell must be explained in that way, namely as 'the well of *Letta.' Although this personal name is not recorded, the corresponding strong form, Let, occurs in the Domesday record.

LEVENTHORPE, Bradford, NV 1316 *Leuwyngthorp*, PT 1379 *Leuenthorp*, is 'the village of Liufvin,' from ON *thorp*. Nielsen gives the ODan. name in the forms Liofvin and Lefvin among others, the fem. form being Liufvina.

-LEY.—This is an exceedingly common termination. It is of Anglian origin, from OE *lēah*, which means a lea or field, a tract of open ground whether meadow, pasture, or arable land. Wyld says " The original and fundamental idea seems to be ' clearance,' land from which forest has been cleared away, as distinct from *feld*, which appears to be land which has always been clear and open." The dative of *lēah* was *lēage* or *lēge* (*g*=*y*), and these gave the ME forms *laye* and *leye*.

LIDGATE, LIDGET.—Three forms of the name are known, Lidgate, Lidyate, Lidget, all from OE *hlid·geat*. This word meant a swing-gate, a gate placed across a highway to prevent cattle from straying, a gate dividing common from private land or between ploughed land and meadow. Lidyate comes directly from OE *hlid·geat*, and Lidget is a further development of the word due to palatalization ; but Lidgate shows the influence of the Scandinavian word ' gate,' a road, unless, indeed, it comes from a plural form *gatu*. We find the form Lydiate in Lancashire.

LIDGATE occurs near Holmfirth, 1514 *Lidyate* ; near Saddleworth, YS 1297 *La Lyed* ; and also in Crookes, Hipperholme, and Lepton.

LIDGET is found near Doncaster, PT 1379 *Lydeyate*, as well as in Bradford, Keighley, Pudsey, Lepton, Tankersley, and Ecclesall.

LIGHTCLIFFE, Halifax.—From 1275 onwards the name is frequently recorded in WCR as *Lithclif*. The present form of the prefix is quite misleading, the name being derived from OE or ON *hlið*, a slope, hillside, and OE *clif* or ON *klif*, a cliff.

LIGHTHAZELS, Sowerby, appears frequently in WCR where we get 1274 *Lytheseles*, 1275 *Litheseles*, 1296 *Lictheseles*, 1309 *Ligheseles*. The modern name gives the meaning accurately, the word being derived from OE *lēoht*, bright, light, and *hæsel*, a hazel.

LILLANDS.—See Lindley.

LINDHOLME, LINDRICK.—The first element appears to be derived from ON *lind*, lime-tree; but, as in the case of Lindley, the *d* may be intrusive, and the derivation may be from ON *līn*, flax.

LINDHOLME, Hatfield.—See Holme.

LINDRICK, Woodsetts, is perhaps 'lime-tree enclosure,' from OE *ric*, a fence, railing, enclosure; compare Rastrick.

LINDLEY, LINFITTS, LINGARDS, LINTHWAITE, LILLANDS.—The prefix in these words is from OE *līn* or ON *līn*, flax, and the names point to the neighbourhood of Huddersfield as formerly a centre for the cultivation of flax. Linfitts, Lingards, and Linthwaite are certainly due to the Northmen, and Lillands may possibly have the same origin, but Lindley is Anglian.

LILLANDS, Rastrick, WCR 1620 *Linlands*, WH 1775 *Linlandes*, means 'flax lands.'

LINDLEY, Huddersfield, DB *Linlei, Lillai*, WCR 1275 *Lynley*, YS 1297 *Linley*, PT 1379 *Lyndelay*, is 'flax lea,' from OE *lēah*. The early spellings in Lillands and Lindley are particularly interesting, the former proving that an assimilation of consonants has taken place, and the latter proving the intrusion of *d* as a supporting consonant.

LINGARDS, Slaithwaite, WCR 1298 *Lyngarthes*, NV 1316 *Lingarthys*, is 'flax enclosure,' from ON *garðr*, a yard, garden, enclosure. The name Lingard occurs also in Bradfield.

LINTHWAITE or LINFITTS occurs four times. There is a hamlet in Kirkburton called Linfitts, PF 1208 *Linthwait*, and there is a second Linfitts in Saddleworth. A farm in Brampton Bierlow now called Linthwaite was *Lintweit* in PC 1155, *Linttveit* in CR 1160, and *Lintewait* in YS 1297. Further, there are Linthwaite and Linfitt Hall in the Colne Valley, WCR 1284 *Lynthayt*, WCR 1307 *Lintweyt*. The meaning of all these names is 'flax paddock,' from ON *thveit*.

LINEHOLME, Todmorden.—See Holme.

LINGWELL GATE, Lofthouse, appears to have obtained its name from a field called Lingwell, 'heather field,' ON *lyng*, ling, heather, and *völlr*, a field. This derivation seems much more acceptable than one from OE *ling* and *wella*, a well, more especially seeing that the district has a considerable proportion of Scandinavian names.

LITTLE LONDON, LITTLETHORPE, LITTLE-WOOD.—The OE form of the word little is *lytel*, and the ON is *litill*, while the ME is *lytel*, *litel*.

LITTLE LONDON.—See Lund.

LITTLETHORPE or LITHROP is to be found at Hartshead and Clayton West; it is ON *litill·thorp*, little hamlet.

LITTLEWOOD is Anglian, and there are several examples of the name, but only in the case near Holmfirth are early spellings found, namely, WCR 1274 *Lyttlewode*, *Litilwode*, *Litelwode*.

LIVERSEDGE, six miles south-east of Bradford, is 'remarkable as being the place where the first effectual opposition was made to the torrent of Luddism in 1812' (Clarke). For some time the Brontë family dwelt within its borders, and its first vicar, the Rev. Hammond Roberson, was the prototype of Parson Helstone in *Shirley*. Early records give the following forms :

DB	1086 *Livresec, Liuresech*	WCR	1297 *Lyvereshegge*
FC	1199 *Liversegge*	CR	1319 *Leversegg*
WCR	1284 *Lyversege*	PT	1379 *Liversig*
KI	1285 *Leversege*	YD	1530 *Liversegge*

The earliest mention of members of the township appears in a License in Mortmain dated 1375, where *Great Lyversegge, Robert Lyversegge, Little Lyversegge*, stand for Hightown, Roberttown, and Littletown; later, in YF 1564, there is mention of *Great Lyversege* and *Little Lyversage*.

Passing to other names, we find first a certain number of stream-names such as Liver in Argyll, Levern Water in Renfrew, and Levers Water in Lancashire. It is possible that these are connected with Prim. Celt. **levo*, to wash, *lavo-*, water, the termination being the common suffix found in such early river-names as *Isara, Tamara, Samara* (Kurth); but

Middendorff compares the OE stream-name *Læfer* (BCS 949) with the Bavarian stream-name Laber and connects it with OHG *labōn*, to wash. See Laverock.

There are next (1) a number of place-names involving Liver, like Liverton in the North Riding and Devonshire, Livermere in Suffolk, and Liverpool in Lancashire, and also (2) a certain number involving Lever, like Leverton in Lincolnshire and Notts., and Lever in Lancashire. We find such early forms for Lever, Liverpool, and Liverton (NR) as

| 1212 *Lefre* | 1229 *Leverepul* | 1086 *Livreton* |
| 1282 *Leuir* | 1254 *Liverpol* | 1086 *Liuretun* |

where obviously the stream-name may again be present. Apart from Liversedge, however, no name shows the -s- of the genitive, and in our search for similar names we are thrown back upon *Leoferes·haga*, which is recorded in KCD and is explained by Dr Skeat as 'Lever's haw,' from the OE personal name Leofhere, later Leofere. A reference to Searle will show that the OE personal name Leofing could take the forms Leving and Living, and we may fairly assume therefore the OE Leofhere could become Levere and Livere.

Names with the same termination as Liversedge are (1) *Wlvesege*, found in the Cockersand Chartulary in 1250, and (2) Hathersage in Derbyshire, IN 1243 *Haversege*. These may be interpreted as 'Wulf's edge' and 'Havard's edge'; and in the same way we may explain Liversedge as 'Leofhere's edge,' or 'Leofhere's ridge,' from OE *ecg*, ME *egge*, which meant among other things a cliff and a ridge. See Edge.

LOCKWOOD, Huddersfield, WCR 1286 *Lokwode*, WCR 1307 *Locwode*, PT 1379 *Lokewod*, is perhaps 'the wood beside the fold,' from OE *loc*, an enclosure, fold, pen, and *wudu*, a wood. Compare the names Lockton, Wenlock, and Porlock, as well as the OE *gāta·loc*, a pen for goats.

LOFTHOUSE, Wakefield.—Elsewhere in the county we find Lofthouse near Sedbergh, Middlesmoor, and Harewood; and Loftus, another form of the name, occurs near Saltburn

and Knaresborough. Early spellings of Lofthouse near Wakefield and Loftsome near Howden are as follows :

DB   1086 *Loftose, Locthuse*   KI   1285 *Lofthusum*
DN   1250 *Lofthus*   KF   1303 *Lofthousum*
WCR 1285 *Lofthus*   NV   1316 *Lofthousum*
KF   1303 *Lofthouse*

Thus Lofthouse and Loftsome are the singular and plural of
the same Scandinavian word, *hūse* being dat. sing. and *hūsum*
dat. plur., and the first element being from ODan. *loft* or ON
lopt, an upper chamber, an upper floor. The name Lofthus
occurs in Norway on the shores of the Hardanger.

In comment on the Domesday form *Locthuse* it should be
noted that other Yorkshire names show such Domesday spellings
as *Locthusun* and *Loctehusum*, and that Middle Low German
possesses the alternative forms *luft* and *lucht*; and in comment
on the Icelandic form *lopt* it should be noted that one of the
peculiarities of Icelandic spelling was the use of *pt* for *ft*, *lopt*
being pronounced *loft*.

LONDON.—See Lund.

**LONGBOTTOM, LONGLEY, LONGROYD, LONG-
WOOD.**—See Langfield.

LOSCOE.—This name occurs twice: (1) near Pontefract,
KC *Loft Scoh, Loftscohg, Loschough*, (2) near Mexborough,
HR 1276 '*in bosco de Lostescoth in Barneburg.*' We may
compare with these the early names *Loftlandes* (Guisborough
Chartulary) and *Loftmarais* (Rievaulx Chartulary). In the two
last-named and in LOSCOE near Pontefract the first element is
the same as that in Lofthouse, and comes from ON *lopt*, ODan.
loft, an upper floor, Loscoe being 'the wood beside the two-storied
house,' from ON *skōgr*, a wood. LOSCOE near Mexborough on
the other hand is 'Losti's wood'; compare the personal name
Lost which appears in the Chartulary of Rievaulx.

LOXLEY, Bradfield, IN 1329 *Lokkeslay*, IN 1337 *Lokkesley*,
PT 1379 *Lokeslay*. Here is an interesting illustration of the
way in which a name may obtain a wider application. The

word, which originally meant 'Loc's lea,' from the personal name Loc recorded by Searle, became in the course of years the name of the village and also of the stream which flowed past the village. In the same way the names Agden and Ewden designated in the first instance valleys only, but were later applied to streams as well.

LUDDENDEN, Halifax, is frequently mentioned in WCR, where we find 1284 *Luddingden*, 1285 *Loddingdene*, 1296 *Luddingdene*, 1307 *Luddingden*, 1309 *Luddingdene*; in addition HPR 1568 has *Lodendyn* and *Lodynden*. Though Searle records no patronymic of the form Luding or Loding, he gives Luda and Loda, and there is ample authority for such a patronymic in the Luddingtons of Lincoln, Warwick, Kent, and Surrey, and in the Dutch place-name Ludingehus (NGN). We may fairly explain the name as 'the valley of the Ludings,' from OE *denu*, a valley. The change to Luddenden is paralleled in the case of Morthen. But see Ludwell.

LUDWELL, Thornhill, SE 1634 *Ludwell*, contains an element which occurs very frequently in English place-names. In the Domesday Survey we find such names as Lude, Ludesforde, Ludebroc, and Ludwelle, as well as Ludeburg, Ludecote, and Ludewic. Among modern place-names Ludford occurs in Shropshire and Lincoln, Ludbrook in Devon and Lincoln, and Ludwell in Wiltshire and Yorkshire. In addition, there are streams called Lud near Bolton Abbey and in Lincolnshire. Perhaps the word is of Celtic origin, and connected with Prim. Celt. **lutā*, mud; compare Welsh *llaid*, mire, mud, Gael. *lod*, *lodan*, a puddle, and note that Hogan has an early place-name *lodan* represented by the modern name Ludden.

LUMB, LUMBANK, LUMBUTTS.—Instances of the name Lumb are quite numerous. It occurs in Haworth, Oakworth, Keighley, Wadsworth, Erringden, Sowerby, Hipperholme, Liversedge, Drighlington, Almondbury, Farnley Tyas, Holme, Penistone, Bradfield and Ecclesfield. Early records, WCR 1307 *Lom*, WCR 1308 *Lom*, WCR 1370 *Lum*, show that the *b* is intrusive; and hence the derivation is either from Norw. *lom*, a tree-stem, a tree-trunk, a tree lopped of its branches (Aasen), or from Norw. *lom*, the dative plural of a word *lō* which corresponds to Germ. *loh* and OE *lēah* and means a grassy flat or lowlying meadowland by the waterside.

LUMBANK, Austonley, may come either from ON *lundr*, a grove, or from Norw. *lom*; compare Lumby, near Sherburn, recorded in BCS 972 as *Lundby* but in KI 1285 and NV 1316 as *Lumby*. The termination is derived from an early Scandinavian **banke*, the original from which Icel. *bakki* is derived.

LUMBUTTS, Langfield, derives its terminal from ON *būtr*, a log, tree-stump; compare Norw. *butt*, a stump, stub, log. As to its first element Lumbutts is like Lumbank.

LUND, LONDON.—Lund occurs in Keighley and Wombwell; London Spring in Soyland; and Little London in Rishworth, Linthwaite, Mirfield, Bentley, Ecclesall, and Healey. The source is the ON *lundr*, a grove, a small wood. In the Landnama Book (III, 6. 1) we are told of a man called Thore that he dwelt at Lund, and that he sacrificed to the grove— 'hann blotaðe lundenn'; compare the Swedish name Lund, formerly Lunden. This leads directly to the chief interest of the names now under consideration, namely, their possible connection with heathen worship.

But the name London has a second interest, its termination being probably the Scandinavian suffixed article. This article is found in thousands of Norwegian place-names. In a single county, the Amt of Hedemarken, Rygh places on record six examples of Lund and seven of Lunden, as well as numerous examples of Dalen, Haugen, Sveen, and Viken.

LUPSET, Wakefield, has the following early records: WCR 1277 *Lupesheved*, WCR 1286 *Loppesheved*, WCR 1297 *Lupesheved*, YS 1297 *Lupesheved*, WCR 1361 *Luppesheved*, DN 1363 *Lupshead*, PT 1379 *Lupishede*. Hunter expressed the opinion that the meaning is 'the high headland'; but, while the termination is certainly an early form of the word 'head,' OE *heafod*, ME *heved*, the prefix appears to be a personal name. It will be helpful to examine a number of parallel cases.

Consett	Durham	Early deed	*Conekesheved*
Ormside	Westm.	IN 1273	*Ormesheved*
Gamelside	Lancs.	IN 1323	*Gameleshevid*
Arnside	Lancs.	LF 1208	*Arnulvesheved*
Conishead	Lancs.	LF 1235	*Cuningesheved*
Whasset	Lancs.	Inquisition	*Quasheved*
Swineshead	Beds.	OE charter	*Swinesheafod*
Farcet	Hunts.	OE charter	*Fearresheafod*
Hartshead	Yorks.	PF 1206	*Hertesheved*
Thicket	Yorks.	Early deed	*Tykenheved*
Hazlehead	Yorks.	Early deed	*Heselheved*

Including Lupset we have altogether twelve examples. One of them, Hazlehead, is from a tree, OE *hæsel*; and four may perhaps involve the name of an animal : Swineshead, Farcet, Hartshead, Thicket, from OE *swīn*, a pig, *fearr*, a bull, *heort*, a hart, *ticcen*, a kid. But the prefix in ten of the twelve—including Swineshead, Farcet, and Hartshead—is most probably a personal name. Consett is plainly Conec's head; Ormside is Orm's head ; Gamelside Gamel's head ; and Arnside Arnulf's head. I can find no personal name of the form Luppo, but Searle has Loppo and Nielsen has Loppi. We may fairly assume the existence of Luppo and explain Lupset as ' Luppo's headland,' from OE *heafod* which often meant the highest point of a stream, field, or hill.

MACHPELAH occurs as the name of a farm in Wadsworth.

MAG, MAGDALE, MAGDALEN.—We find Mag Dam in Rishworth, Mag Field in Ecclesall, Mag Wood in Thurstonland, Magdale in South Crosland, and Magdalen Clough in Meltham. The most probable source of Mag is an ON word *magi*, which according to Rygh occurs in the Norwegian place-name Mageli. This word, which literally signifies stomach or belly, appears to be used to denote a narrow river gorge ; compare Wombwell.

MALKROYD, MAUKROYD, Dewsbury and Langsett.— From such names as KCD *Mealcing* and HR 1279 *Malketon*— Malton in Cambridgeshire—we may postulate an OE personal name Mealc, and hence explain Malkroyd as 'the clearing of Malke.'

MALTBY, Rotherham, DB *Maltebi*, BM 1147 *Malteby*, YR 1229 *Mauteby*, NV 1316 *Malteby*, is 'Malti's farmstead,' from ON *bȳr*, and the Scandinavian personal name Malti recorded by Nielsen. In the North Riding there is a second Maltby, DB *Maltebi*, and there are others in Lincolnshire.

MANKIN HOLES, Todmorden.—See Howcans.

MANNINGHAM, Bradford, CR 1251 *Maningeham*, WCR 1298 *Maynigham*, KF 1303 *Maynyngham*, NV 1316 *Maynyngham*, KC 1342 *Manyngham*, PT 1379 *Manyngham*, is 'the home of the Mannings,' from OE *hām*, a home or house. There are in England three Manningfords, two Manningtons, and one Manningtree, all taking their origin from the same patronymic, while Sundermann records the name Manningaland. For the dialectal variation in the first vowel see Santingley.

MANSHEAD.—In Soyland there is a hill called Great Manshead — pronounced Mawnshead (mōnzed)— and a little below the summit is a farmstead mentioned in WCR 1275 and 1277 as *Mallesheved*. The name appears to be of the same kind as Lupset; and DB has the name Mal, while Searle has such names as Malgrim and Malet.

MAPPLEWELL, Barnsley. — Burton gives *Napplewel*, where the initial consonant appears to be a scribal error, and YF 1544 has *Mapellwell*. The meaning is 'the well beside the maple-tree,' from OE *mapul* and *wella*.

MARGERY HILL, MARGERY WOOD, MARGERY HOLME, are found respectively in Bradfield, Cawthorne, and Ecclesall. I cannot do more than quote an interesting note by Peiffer on the French place-name Margheria: "Margheria est le nom des étables (cattle-sheds) situées en montagne dans les Alpes-Maritimes. Ce nom ne se trouve ni dans le Dictionnaire Provençal ni dans le Dictionnaire Italien."

MARK BROOK, MARK BOTTOMS, Ecclesfield and Upperthong.—The first element in the latter is represented by that in WCR 1307 *Merkehirst*, and its origin is OE *mearc*, a limit or boundary. We find such OE compounds as *mearc·beorh*, boundary-hill, *mearc·brōc*, boundary-brook, *mearc·denu*, boundary valley; but the most interesting example is the name of the ancient kingdom of Mercia. The corresponding ON word *mòrk*, a forest, marshland, borderland, has also produced many place-names, including Denmark, *Dan·mörk*, and Finmark, *Finn·mörk*.

MARLEY, near Bingley, DB *Mardelei, Merdelai,* PF 1209 *Merdele,* BM *Matherley,* BPR 1591 *Merley,* has for its first element the genitive Marthar of the ON personal name Mórðr. In the Domesday forms the second *r* is dropped, probably through dissimilation, and in BM the first *r* appears to be dropped for the same reason. We may explain the name as 'the lea of Marth.'

MARR, Doncaster, DB *Marra,* YI 1248 *Mar,* KI 1285 *Mare,* NV 1316 *Marre,* PT 1379 *Merre.* These agree in form with ON *marr,* which however meant the sea; but the cognate OE *mere* meant a lake or pond as well as the sea, and possibly ON *marr* was used in the same way.

MARSDEN, Huddersfield.—It will be interesting to compare the early spellings with those of Marsden in Lancashire, the latter being placed in the second column:

WCR 1274 *Marchesden*	1195 *Merkesden*
WCR 1275 *Marcheden*	1216 *Merkesdene*
WCR 1277 *Marchesdene*	1292 *Merchesdene*
WRM 1342 *Marchedene*	1305 *Merkelesdene*
PT 1379 *Mersseden*	1311 *Merclesdene*
DN 1627 *Mershden*	1332 *Merlesden*

The two names have the same terminal—from OE *denu*, ME *dene*, a valley—but that is all. Between the various forms of our own Marsden there is serious conflict. The second form, *Marcheden*, appears to be 'the borderland valley' from AFr. *marche*, ME *marche*, the border of a province or district. The two last, *Mersseden* and *Mershden*, are plainly 'marsh valley,' from OE *mersc*, ME *mersche*, *mershe*, a marsh. The remaining forms, *Marchesden* and *Marchesdene*, appear to have an intrusive *s*. It seems, therefore, that under the influence of popular etymology the earlier *Marcheden* passed into the later *Mershden*, which in its turn became Marsden just as the Oxfordshire *Mershton* became Marston.

MARSHAW.—In WH 1413 Cragg Vale or some portion of it is called *Marishai·clough*; and the spot where now the village and church are placed is called *Mareshae* in WH 1408, *Marschagh* in WCR 1308, and *Mareschawe* in WCR 1275. The bridge is still called Marshaw Bridge, and the bank close by is Marshaw Bank. The word is derived from OE *mere*, a lake, and *sceaga*, a copse or wood.

MASBOROUGH, MEXBOROUGH, Rotherham.—The early records of these places can only be distinguished by taking the presence of an *r* in the first syllable as showing connection with the former. Early forms are as follows :

PF	1206	*Merkesburch*	DB	1086	*Mechesburg*
CH	1307	*Merkesburg*	PF	1206	*Mekesburg*
CH	1320	*Merksburg*	CR	1226	*Mekesburg*
YF	1555	*Marseborowe*	YR	1247	*Mekesburgh*
YF	1572	*Markesbroughe*	NV	1316	*Mekesburg*
YF	1572	*Marshebroughe*	PT	1379	*Mekesburgh*
SCR	1606	*Marshburgh*	CH	1483	*Mexburghe*

MASBOROUGH has undergone an unusual series of changes, and for a time threatened to become Marshborough. Originally it was 'Merc's fortified place,' from OE *burg* and the personal name Merc. Then the *k* between consonants was lost, *rksb* becoming *rsb*; later *rsb* was simplified to *sb*, and so the modern form was reached.

MEXBOROUGH, on the contrary, has held an even course, and means 'the fortified place belonging to Mēk.' As to the origin of the personal name note (1) that Brons gives a Frisian name Meke, and (2) that Nielsen gives the ODan. name Miuk, which is derived from ON *mjūkr*, agile, easy, meek. From this ON word is derived the common English word 'meek,' ME *meke* (Skeat).

G.   14

MAZEBROOK, Gomersal and Ingbirchworth.—There can be no doubt about the existence of an ancient stream-name of some such form as Maze. On the borders of Westmorland there is a stream called the Maize ; in Leicestershire there is the Mease ; in Staffordshire the Mees and Meese ; and in Ross and Cromarty the Falls of the Measach.

MEAL HILL occurs four times in the neighbourhood of Huddersfield, viz. in or near Slaithwaite, Meltham, Holme, and Hepworth. The source of the word Meal is ON *melr*, a sandbank, sand hill ; compare the dialect-word ' meal,' a sandbank, recorded in EDD.

MEASBOROUGH, Ardsley near Barnsley, must be recorded as one of the derivatives of OE *burg, burh*, a fortified place.

MELLOR HILL, MELLOR LANE, Whitley Lower and Austonley.—In South-west Yorkshire Mellor is very common as a surname, derived, doubtless, from some place-name. PT 1379 has the word *Meller*, of which the origin is doubtless the ON *melr*, a sandbank, the termination being due to the plural, ON *melar*, sandbanks.

MELTHAM, Huddersfield, a very difficult name, has the following early spellings :

DB	1086	*Meltha'*	DN	1361 *Meltham*
YS	1279	*Meltham*	PT	1379 *Meltham*
NV	1316	*Muletham*	DN	1388 *Melteham*

Perhaps the strange spelling given in NV gives the necessary clue. In Danish there is a word *multebær* and in Norwegian a word *multer* used to designate the cloudberry (Larsen), and in Swedish there is a corresponding dialect-word *mylte* (Falk). Further, the Dan. word *multemyr* signifies a bog covered with cloudberry bushes (Larsen), and hence I suggest that Meltham may mean 'the home or enclosure amidst the cloudberry bushes.' In that case the *e* and *u* in the early forms represent an early *y* as was often the case. The terminal is from OE *hām*, home or OE *hamm*, an enclosure.

MELTON.—See Middleton.

METHLEY is an ancient parish almost surrounded by the Aire and Calder which unite at its eastern extremity. The church is dedicated to Saint Oswald and a mutilated carving now built into the wall near the chancel is supposed to have represented that Saint (Morris). Records of the name since the Conquest include the following :

DB	1086 *Medelai*	NV	1316 *Metheley*
PC	†1220 *Medelay*	PT	1379 *Meydlay*
PC	†1230 *Medeley*	DN	1487 *Metheley*
PC	1251 *Methelay*	VE	1535 *Methlay, Medley*
YS	1297 *Metheley*	DN	1632 *Medley als. Metheley*

As the Norman scribes frequently wrote *d* for *th*, it is possible that the three earliest forms represent *Methelai*, *Methelay*, *Metheley*. And, as an earlier *th* may give place to a later *d*, *Meydlay* and *Medley* may well be true descendants of a Domesday form *Methelai* ; compare Cudworth, Lingards, and Rodley. In a word, it seems probable that the early forms are in agreement with one another and are fitly represented by DB *Methelai*.

Assuming that this is the correct form, the most probable explanation of Methley is ' council lea,' from OE *mæðel, meðel*, a council or meeting, and OE *lēah*, a lea or meadow; Methwold in Norfolk, DB *Methelwalde, Matelwald*, appears to come from this source. That OE *meðel·lēah* should be written *Methelai* in DB, one *l* being lost, is paralleled in the case of Ulley ; compare also Owston and Shafton.

Another interpretation, agreeing with the geographical conditions, is ' middle lea,' from ON *meðal*, middle ; compare Melton, where *Metheltone* has probably superseded *Mideltone*.

MEXBOROUGH.—See Masborough.

MICKLEBRING, MICKLETHWAITE, MICKLE-TOWN.—The first element in these words is derived from OE *micel* or ON *mikill*, great.

MICKLEBRING, Conisborough, YR 1254 *Mykelbring*, HR 1276 *Mikelbring*, IN 1335 *Mikelbrink*, IN 1375 *Mykelbrynk*, YF 1536 *Mykelbrynk*, is ' the great slope,' from ON *brinka*, Dan.

brink, a slope.   The hamlet is built at the point where a tableland passes into a long and gradual descent towards the river Don.

MICKLETHWAITE, Cawthorne, is mentioned in PT 1379 under Cawthorne as *Mickilwayte*; this means 'the great clearing,' from ON *thveit*.

MICKLETOWN, Methley, MPR 1561 *Mickletowne*, means 'the great farmstead or village.'   The name can scarcely be of early origin ; see Ton.

MIDDLETON, MIDDLESTOWN, MIDDLEWOOD, MELTON.—It will be particularly interesting to compare the early spellings of Middleton, Leeds, with those of High Melton, Doncaster.

DB	1086	*Mildetone, Mildentone*	DB	1086	*Medeltone, Mideltone*
YI	1258	*Midelton*	PF	1208	*Methelton*
KI	1285	*Midylton*	YI	1252	*Methylton*
KF	1303	*Middelton*	KI	1285	*Melton-le-Heyg*
PT	1379	*Midelton*	PT	1379	*Hegh Melton*

The strange DB spellings of Middleton are due to scribal error, and we may fairly rewrite *Mildetone* as *Mideltone*.   This gives us a definite link with the second DB form of Melton, but apart from the terminal the two names are in reality quite different, the first element in Melton coming from ON *meðal*, middle, while in Middleton it comes from OE *middel*.   Thus, each name means 'middle enclosure or farm,' Melton being Scandinavian, and Middleton Anglian ; compare the Norwegian place-name Melby from *Meðal- bȳr* (Rygh), and Melton Mowbray which in DB was written *Medeltone*.   The variation between Anglian and Scandinavian in the DB forms of our own Melton is worthy of note, and so also is the fact that in the Domesday record *d* is often written for *th*.

MIDDLESTOWN, Wakefield, is the same word, but it shows (1) an instrusive *s*, and (2) the late form -town instead of the earlier -ton.   The place is referred to in DN 1325 as *Midle Shitlington*, while in YF 1523 we find *Overton, Middleton, Netherton*, the upper, middle, and lower farms.

MIDDLEWOOD, Darfield, YS 1297 *Middelwude*, tells its story with entire frankness and simplicity.

MIDGLEY, Halifax and Wakefield.—The first appears in DB as *Micleie*, where the medial consonant fails just as the final consonant does in DB *Livresec*, now Liversedge. Later records of the two Midgleys show forms which agree quite closely with one another, e.g.

MIDGLEY, Halifax	MIDGLEY, Wakefield
WCR 1274 *Miggeley*	YR 1234 *Miggeley*
WCR 1297 *Miggeley*	DN 1241 *Migeley*
WCR 1308 *Miglay*	YI 1287 *Miggeley*

It is impossible that the two Midgleys should be derived from OE *micel* or ON *mikill*, great, the early spellings being in violent conflict with such a derivation. But there is another and more obvious source, namely, OE *mycge*, ME *migge*, a gnat or midge. Iceland had formerly such names as *Mȳ·vatn*, midge-lake, and *Mȳ·dale*, midge-valley, while in the 13th century the Cockersand Chartulary gives the place-name *Migedale* in the parish of Bland. Thus, the two Midgleys may fairly be interpreted 'midge lea,' from OE *lēah*.

MIDHOPE, near Penistone, PC †1220 *Midehope*, YR 1252 *Midhop*, BD 1290 *Midhope*, YS 1297 *Midop*, *Middop*, YD 1307 *Midehope*, is 'middle valley,' from OE *mid*, middle, and *hop*, a secluded valley or retreat. Near Barnoldswick is another instance of the name, now spelt Middop.

MILNTHORPE, Sandal, 1279 *Milnethorp*, YS 1297 *Milne-thorp*, is 'mill village,' from ON *mylna*, a mill, and *thorp*, a village.

MINSTHORPE is in North Elmsall.

-MIRE occurs in Blamires and Spinksmire ; see Mirfield.

MIRFIELD, on the Calder between Dewsbury and Brig-house, has early records as follows: DB *Mirefeld*, *Mirefelt*, PC †1170 *Mirefeld*, YI 1249 *Mirefeld*, KI 1285 *Myrfeld*. The meaning is 'the swampy field,' from ON *mȳrr*, ME *mire*, a bog or marsh, and OE *feld*, a field. In the will of Henry Sayvell, HW 1483, we find a sum of 6s. 8d. bequeathed for the repair of *Mirfield brige*.

MIXENDEN, Halifax, WCR 1274 *Mixenden*, WCR 1284 *Mixhynden*, YI 1304 *Mixendene*, is derived from OE *mixen*, a dunghill, and *denu*, a valley. Staffordshire has a village called Mixen which Duignan derives from the same source; see also Sharlston.

MOLD GREEN, MOLD ROE, MOLD ROYD.—The first is in Huddersfield, the second in Lofthouse, and the third, CC 1475 *Moldrode*, in Pudsey. The word Mold goes back to OE *molde* or ON *mold*, mould, earth. The latter appears in the Landnama Book where we read of Hrolf the Hewer that 'his homestead was at *Molda·tun*,' compare the Norwegian place-names Molde and Molden.

MOOR, MOORHOUSE, MOORTHORPE, MORLEY, MORTOMLEY, MORWOOD.—These words are all from OE *mōr*, a moor or morass, or ON *mōr*, a moor, heath. Like the series of names derived from OE *mos* or ON *mosi*, a bog or marsh, they serve as witnesses to the ancient conditions. As a termination -moor or -more occurs in Pogmoor, Ranmoor, Scholemoor, Stocksmoor, and Tranmore.

MOORHOUSE, Elmsall, YR 1230 *Morhuse*, KI 1285 *Morhus*, PT 1379 *Morehouse*, may be either Anglian or Scandinavian.

MOORTHORPE, Elmsall, YD 1322 *Morthorp*, is 'the village on the moor,' from ON *mōr* and *thorp*.

MORLEY, DB *Morelei, Moreleia*, PF 1202 *Morlay*, DN 1226 *Morle*, KI 1285 *Morlay*, NV 1316 *Morley*, is 'moor lea,' from OE *lēah*, a lea.

MORTOMLEY, Sheffield, would provide a pretty puzzle if PR 1190 had not given the spelling *Mortunelea*. This is plainly 'the lea at Morton,' that is, 'the lea of the moor-farm,' from OE *tūn*, an enclosure or farm, *lēah*, a lea.

MORWOOD, Sheffield, HH 1366 *Morwood*, PT 1379 *Morewod*, is 'the wood on the moor,' OE *mōr* and *wudu*.

MORTHEN is the name of a hamlet about five miles southeast of Rotherham; it is also the name of an undefined district in which are situated Aston-in-Morthen, Laughton-en-le-

Morthen, and Brampton-en-le-Morthen. The following are early spellings:

YD	1253 *Morhtheng*[1]	YS	1297 *Morthing*
WCR	1274 *Morthyng*	YD	1345 *Morthing*
KI	1285 *Morthyng*	YF	1558 *Morthinge*

These show that the true ending is -eng or -ing, not -thing ; and seeing that the immediate neighbourhood has a large proportion of Scandinavian place-names—Micklebring, Braithwell, Stainton, Woolthwaite, Sandbeck, Firbeck, Thwaite, Maltby, Hellaby, Carr, Thurcroft, Throapham—we shall probably be right in deriving the name from ON *morð*, slaughter, and *eng*, a meadow.

If this be the correct etymology it should be found that some great struggle between the Vikings and the English took place on the site ; and in this connection it should be remembered that the direct route from Mercia to Northumbria was along the old Roman road called Riknild Street. Leaving Derby and taking the valley of the Amber, this road passed through Wingfield and Clay Cross, and then by the valley of the Rother reached Chesterfield and Beighton. At this point the course is doubtful, but the probabilities favour the crossing of the Don at Aldwark, the further course being by way of Swinton, Nostell, Normanton, and Woodlesford, to the great Roman city of Isurium. In any case the road left ' the Morthen ' but a few miles to the east.

MOSCAR, MOSELDEN, MOSELEY, MOSS.—This series of names is derived from OE *mos*, or ON *mosi*, a bog or marsh.

MOSCAR, west of Sheffield, *Mosker* and *Moskarr* in 1574, means ' the carr on the moss,' from ON *kjarr*, copsewood, brushwood.

MOSELDEN, near Rishworth, WCR 1285 *Moseleyden*, PT 1379 *Moslenden*, SE 1715 *Mosslenden*, is probably ' the valley of the lea in the marsh,' OE *denu*, a valley, *lēah*, a lea. The name appears to have been influenced by Scammonden and Ripponden.

MOSELEY GRANGE and MOSS, near Doncaster, like the adjacent Fenwick and the various Holmes and Carrs, tell their

---

[1] *YAS Journal*, XIII, p. 72.

story in no half-hearted way. All these places are situated in the triangle between the Aire, the Don, and the ancient highway from Doncaster to Castleford. In bygone days this area was largely impassable, a swamp or moss extending over many square miles. To avoid this the Roman road deviated from its direct course and swerved to the west. Early records of the names Moss and Moseley are HR 1276 *Moselay*, IN 1331 *Moseleye*, YF 1476 *Mosse*, YF 1573 *Mosseleye in the Mosse*, YF 1580 *Mosse*.

MOUNTAIN, Thornhill and Queensbury.—It is doubtful whether either of these names should be written in this way. SE 1634 records the Thornhill Mountain as *Mounton*, a form exactly paralleled by Mounton in Pembroke and Monmouth. It seems very probable that the first element in all these names should be linked with W *mawn*, peat, turf, Ir. *moin*, a marsh, moor, common, words which go back to Prim. Celt. *makni-, *mokni-, a marsh. The common word ' mountain ' comes to us from French, and it is found in our literature as early as the beginning of the 13th century.

MYTHOLM, MYTHOLMROYD.—The former occurs in or near Haworth, Hebden Bridge, Hipperholme, Holmfirth, and Midgley (Halifax) ; the latter is at the junction of the Calder and a tributary from Cragg Vale.

MYTHOLM near Holmfirth is recorded by WCR 1307 in the name *Mithomwode*, and there is a 16th century form *Mitham*.

MYTHOLM, Midgley, appears as *Mythome* in YF 1545.

MYTHOLMROYD was *Mithomrode* in WCR 1307 and 1308, and *Mitham Royd* in WH 1775 ; see Royd.

The origin is OE *mȳðum*, dat. pl. of *ge·mȳðe*, a river-mouth, the point where two rivers meet. Under Norman influence -uṁ was written -om ; then in imitation of the Anglian -ham it became -am ; and lastly, copying a well-known Scandinavian word it became -holm. Thus Mytholm means simply ' waters-meet ' or ' confluence.'

NAB.—This word, derived from ON *nabbr* or *nabbi*, a knoll, is applied to prominent hills. It occurs with considerable frequency along the western border. There are, for example, Nab Hill in Kirkheaton, Butter Nab in Lepton and Crosland, Callis Nab in Erringden, West Nab in Meltham, Hunter's Nab in Farnley Tyas, and Nabscliffe in Shepley. The name is also to be found in Bradfield, Langsett, Holmfirth, Saddleworth, Silkstone, Mirfield, Oxenhope, Shipley, Slaithwaite, Stainland, Rishworth, and Sowerby.

NAZE.—In the hill-country near Halifax and Huddersfield this word is frequently met with. There are Naze and Naze-bottom near Heptonstall, Stannally Naze in Stansfeld, Naze Hill in Wadsworth, Booth Naze in Slaithwaite, Naze Woods in Marsden, and Hard Nese in Oxenhope. It appears to be the OE *næs*, a cape, headland, projecting cliff.

NEEPSEND, NEPSHAW, NIPSHAW.—Early records of the first, which is near Sheffield, are YS 1297 *Nipisend*, YD 1361 *Nepeshende*, HH 1366 *Nepesend*. The OE *ende* meant not only a border or limit, but also a district or quarter, and the most probable explanation of Neepsend is 'Neep's quarter'; compare the modern surnames Neep and Neeper.

In the Kirkstall Coucher Book a field in Morley is designated *Nepesatherode*, Neep-shaw-royd, and the former part of the word is reproduced to-day in Nepshaw Lane. Nipshaw Lane in Gomersal is probably of similar origin, but in view of the forms in Nip- it seems not altogether impossible that ON *gnīpa*, a peak, headland, may be involved.

NETHERFIELD, NETHERLEY, NETHERTHORPE, NETHERTON.—The prefix in these names is either from OE *neoðera*, ME *nethere*, or from ON *neðri*, nether, lower.

NETHERFIELD, Kirkburton, is ' the lower field,' OE *feld*.

NETHERLEY, Holme, is 'the lower lea,' from OE *lēah*.

NETHERTHORPE, Almondbury and Thorpe Salvin, is 'the lower hamlet,' from ON *thorp*.

NETHERTON, Shitlington and South Crosland, is 'the lower farm,' from OE or ON *tūn*, an enclosure, farmstead. In the township of Shitlington we find the three names Overton, Middlestown, Netherton, and early records give the following names : YR 1234 *Schelinton Inferior*, DN 1312 *Over Shitlington*, DN 1319 *Nether Shitlington*, YF 1523 *Overton, Middelton, Netherton.* The three names last mentioned signify the upper, the middle, and the lower farms.

NETTLETON, NETTLETONSTALL.—In WCR 1284 a certain *Peter de Nettelton* is mentioned, and in 1307 and 1308 the name *Nettelton.* These may perhaps refer to Nettleton Hill in the parish of Longwood. But we also find in WCR 1308 the name *Netteltonstall* of which the identification is unknown. The prefix is the OE *netele*, a nettle. See Heptonstall.

NEWBIGGING, NEW BRIGHTON, NEWHALL, NEWLAND, NEW SCARBOROUGH, NEWSHOLME, NEWSOME, NEWSTEAD, NEWTON.—The prefix in these words comes from OE *nīwe, nēowe*, ME *newe*, new ; compare ON *nȳr*, Dan. *ny*, Sw. *ny.*

NEWBIGGING, Sandal and Thurstonland, means ' new building,' from ON *bygging*; compare WCR 1275 *Neubigging* which refers to the former.

NEW BRIGHTON is found in Cottingley ; it is, of course, a borrowed name.

NEWHALL occurs in Darfield, DB *Niwehalla*, PC 1155 *Neuhala*, KI 1285 *Newhall*, and also in Pontefract and Shitlington. The ending is from OE *heall*, ME *halle.*

NEWLAND occurs in or near Netherthong, Normanton, and Rastrick. The word ' newland ' was used of enclosures from waste land, and is to be contrasted with ' rodeland ' which was used of enclosures from wood.

NEW SCARBOROUGH, Wakefield, is of the same type as New Brighton.

NEWSHOLME, Keighley, DB *Neuhuse*, YI 1255 *Neusum*, PT 1379 *Neusom*, means ' new houses.' The words *huse* and *husum* are the datives, singular and plural, of OE *hūs*, a house, and despite their dissimilarity Newsholme and Newsome have the same meaning. Compare Gildersome and Woodsome.

NEWSOME, Huddersfield, WCR 1275 *Neusom*, PT 1379 *Neusom*, DN 1386 *Newsom*, like Newsholme, means ' new houses.'

NEWSTEAD, Hemsworth, DN 1427 *Newstede*, YF 1504 *Newstede*, means ' new place,' from OE *stede*, a site, place, station.

NEWTON, Wakefield, PR 1190 *Niweton*, WCR 1275 *Neuton*, signifies ' new farmstead or enclosure,' from OE *tūn*.

NEWTON, Doncaster, PT 1379 *Neweton*, YF 1525 *Newton*, has the same origin and meaning.

NEWTON WALLIS, DB *Niuueton*, KI 1285 *Neuton Waleys*, NV 1316 *Neuton Waleys*, was held in 1285 by Stephenus de Waleys. See Burghwallis.

NOBLETHORPE is a residence in Silkstone.

NORLAND, NORTHORPE, NORTON, NORWOOD.— The first element in these names may be either Anglian or Scandinavian, from OE *norð*, or the ON word of similar form. Between two consonants the *th* has been lost, exactly as in Norfolk, Norwich, and Normanton.

NORLAND, Halifax, occurs in WCR 1274 and YD 1322 as *Northland*, the termination being from OE or ON *land*, an estate ; see Dirtcar.

NORTHORPE occurs in Wortley (Sheffield) and Mirfield ; the latter is spelt *Norththorpe* in WCR 1297, and *Northorp* in YD 1331, and the meaning is ' north village,' from ON *thorp*.

NORTON, Frickley and Askern, is recorded in each case by DB as *Nortone*; HR 1276 has *Norton* for one of these places. The ending is from OE or ON *tūn*, an enclosure, farmstead.

NORWOOD, Hipperholme, is spelt *Northwode* in WCR 1276, and the termination is OE *wudu*, a wood.

NORMANDALE, NORMANTON, Bradfield and Wake-field.—The latter is recorded in DB as *Normatune*, in WCR 1275 and 1286 as *Northmanton*, and YS 1297 as *Normanton*. Each name is derived from a Viking settler called Northman, a personal name which is recorded in DB as Norman. See Dirtcar.

NORRISTHORPE, Liversedge, is a name of modern creation, the hamlet being formerly called Doghouse. Speaking of London in the olden days Canon Taylor says ' the hounds of the Lord Mayor's pack kennelled at Doghouse Bar in the City Road.'

NORTHOWRAM, SOUTHOWRAM, Halifax.—These names have early records as follows :

DB	1086 *Ufrun*	DB	1086 *Overe, Oure*
WCR	1274 *Northuuerum*	WCR	1286 *Southorum*
PT	1379 *Northourom*	PT	1379 *Southourom*
YF	1555 *Northourome*	YF	1546 *Southouromе*

The OE *ōfer*, an edge, brink, bank, border—cognate with Germ. *ufer*—has given many place-names : Over in Cambridge, DB *Ovre, Oure* ; Ashover in Derby, DB *Essovre* ; Edensor in Derby, DB *Ednesovre*; Okeover in Stafford, 1004 *Acofre*. It should be noted, however, that while DB *Overe* represented the dat. sing. in *e*, DB *Ufrun* represents the dat. pl. in *um*. For other examples of this inflection see Hallam, Hipperholme, Mytholm, Newsome, and Woodsome. Note also the Dutch place-name Oever, recorded in NGN III 205 as *Uvere* in 1269 and *Oeuer* in 1355.

NORTON.—See Norland.

NOTTON, Wakefield, DB *Notone, Nortone*, PC 1218 *Nottona*, YR 1234 *Notton*, NV 1316 *Notton*, goes back to ON *hnot*, a nut, and *tūn*, an enclosure or farmstead.

OAKES, OAKENSHAW, OAKWELL, OAKWORTH, OGDEN.—As a prefix the OE *āc* has assumed the forms ack, ag, augh, oak, og, and in this way has given more than a dozen ancient place-names in South-west Yorkshire ; compare Ackton, Agden, Aughton. The ME form of the simple word is *oke*.

OAKES, Huddersfield, WCR 1285 *Okes*, WCR 1297 *Okes*, should be compared with Ewes, Hessle, Popples, Thornes, and Thickhollins.

OAKENSHAW, Cleckheaton, YI 1255 *Akanescale*, YD 1355 *Okeneschagh*, shows a change of terminal, the earlier forms being from ON *skāli*, a hut, shed, shieling, while the later come from OE *sceaga*, a copse or wood. The meaning of the modern name is ' oak copse ' ; compare Birchencliffe and Birkenshaw.

OAKENSHAW, Crofton, BM *Akeneschaghe*, CR 1280 *Akenshawe*, YF 1555 *Okenshaw*, YF 1565 *Okenshawe*, has the same meaning.

OAKWELL, Birstall, PM 1333 *Okewell*, DN 1381 *Okewell*, YF

1565 *Okewell*, is 'the well or spring beside the oak-tree,' from OE *wella*. The same name occurs in Barnsley.

OAKWORTH, Keighley, DB *Acurde*, YI 1246 *Acwurde*, CR 1252 *Acwurthe*, KF 1303 *Ackeworth*, derives its ending from OE *weorth*, and means 'the homestead beside the oak-tree.'

OGDEN, Ovenden, WCR 1309 *Okedene*, comes from OE *āc* and *denu*, and means 'oak-tree valley.' For the change from oke- to og- compare Shibden.

OGDEN, Rastrick and Sowerby, comes doubtless from the same source, and has the same meaning.

ODD HILL, Dalton near Rotherham, is most probably derived from ON *oddi*, a point of land, a word which has given many Norwegian place-names such as Odde, Langodden, and Stubbodden.

ODESSA, a farm near Holmfirth, probably got its name from the famous Russian port on the Black Sea. This port was bombarded by the British fleet in April, 1854.

ODSAL, Bradford.—There is a village in Hertfordshire called Odsey, DB *Odesei*, and according to Dr Skeat this is equivalent to *Oddes·ēg*, Odd's island. The name Odd is properly a Scandinavian form, the OE being Ord. Odsal may be explained as 'Odd's corner or tongue of land'; see Hale.

OGDEN.—See Oakes.

OLDFIELD occurs in Honley, WCR 1296 *Oldefeld*, and near Oakworth, KC 1226 *Haldefeld*. The first comes from OE *eald*, old, and *feld*, a field; but there is doubt about the second which must be compared with Holdworth.

ONESACRE, Bradfield, DB *Anesacre*, YD 1432 *Onesaker*, is 'the field of An,' An being a well-known ON personal name, and *akr*, the ON for arable land.

ORGREAVE, Rotherham, DB *Nortgrave*, YD 1357 *Orgrave*, HH 1366 *Orgrave*, YD 1398 *Orgrave*, signifies 'ore pit,' from OE *ār*, ore, and *graf, græf*, a trench or pit. The DB form must be considered faulty.

OSGATHORPE, OSGOLDCROSS.—The first is near Sheffield, and the second is the wapentake in which Pontefract and Castleford are situated.   Early records are as follows :

CH	1267	*Hosgerthorp*	DB	·1086	*Osgotcros*
YS	1297	*Osgettorp*	PF	1167	*Osgodescros*
YI	1298	*Osegerthorp*	HR	1276	*Osgotecrosse*
YD	—	*Osegottorp*	NV	1316	*Osgotcrosse*
YF	1574	*Osgarthorp*	PT	1379	*Osgodcrosse*

The personal names involved appear in DB as Osgar, Osgot, and Osgod ; these represent the ON names Asgeirr and Asgautr where the first vowel is long.   Obviously Osgathorpe has oscillated between ' Osgar's thorp ' and ' Osgot's thorp,' while Osgoldcross, ' Osgot's cross,' has suffered through popular misapprehension.   For the loss of the sign of the genitive compare Alverley and Alverthorpe.

OSSETT, Wakefield, DB *Osleset*, WCR 1275 *Oselset, Ossete*, NV 1316 *Osset*, DN *Oslesete, Oselesete, Osseleset*.   We have on record the personal name Osla, and OE *set* means a seat, entrenchment, camp, so we may construe Ossett as ' Osla's seat.' The same termination is shown in the West Riding names Scissett and Wintersett and in other English names such as Elmsett, Orsett, Tattersett and Wissett.

OUCHTHORPE LANE, Wakefield, WCR 1274 *Uchethorp*, WCR 1297 *Uchethorpe*, WCR 1307 *Ouchethorpe*, but PR 1190 *Austorp*.   Austorp would mean ' east hamlet,' from ON *austr*, east, and *thorp*, a hamlet or village ; but Ouchthorpe means ' the thorpe of Uche,' Uche being a personal name found in CR.

OUGHTIBRIDGE, Sheffield.—The earliest form is DN 1161 *Uhtinabrigga* (for *Uhtingabrigga*), other early records being

YS	1297	*Wittibrig*	YD	1358	*Ughtybrygg*
YD	1323	*Uttibrig*	PT	1379	*Vghtibrig*
IN	1342	*Ughtibrigg*	YF	1488	*Ughtebryge*

The form *Uhtingabrigga* may be explained as ' the bridge of the Uhtings,' the name Uhting being recorded by Searle ; and we must look upon the later forms as developed from *Uhtigabrig, the change from -*ing* to -*ig* being not uncommon ; see Dunningley and Manningham.

OULTON, Woodlesford, CR 1251 *Olton*, WCR 1297 *Oldton*, YF 1498 *Olton*, is the 'old farm,' from OE *tūn*, an enclosure, farm; compare Oulton, Staffordshire, formerly *Oldeton*.

OUSE, OUSEFLEET.—We find *Usa* and *Use* in various early records, as well as *Useflete* in CR †1108, *Vseflet* in PF 1198, *Vsflete* in 1278, *Usflet* in 1325, and *Ossefleth* in PT 1379. Holder identifies the *Abos* of Ptolemy with the Ouse; but, although the river may have been known by both names, they are not directly connected one with the other. The origin of the word Ouse appears to be Prim. Celt. **utso-*, water, from which the Ir. *usce*, water, and the river-name Usk, are probably derivatives. The termination in Ousefleet comes from OE *flēot*, a channel.

OUSELTHWAITE, OUZELWELL.—The first element comes either from the personal name Osulf, of which the ON was Asulfr, or from OE *ōsle*, the ousel or blackbird.

OUSELTHWAITE, Barnsley, 1715 *Ouslethwaite*, may be 'Osulf's paddock or clearing,' from ON *thveit*; compare Alverthorpe and Skelmanthorpe.

OUZELWELL, Lofthouse, RPR 1589 *Ouslewell*, seems to be 'ousel well'; compare Birdwell and Spinkwell.

OUZELWELL, Thornhill, has doubtless the same meaning.

OUTWOOD was called 'The Outwood' up to the 18th century. It was part of the great demesne wood of Wakefield. This was on the north side of the town, and at the time of the Enclosure Acts amounted to some 2300 acres. In the Domesday Survey woodland to the extent of six leagues by four is said to have appertained to Wakefield.

OVENDEN, Halifax, is not mentioned in DB, but later we find YI 1266 *Ovendene*, HR 1276 *Ovenden*, WCR 1277 *Ovendene*, PT 1379 *Ouenden*. The meaning is 'upper valley,' from OE *ufan*, above; compare the Swedish place-name Ofvantorp, which Falkman derives from the corresponding ON *ofan*, OSw. *ovan*.

OVERTHORPE, OVERTON, the 'upper hamlet' and 'upper farm,' are situate respectively in Thornhill and Shitlington.

For the latter see Netherton ; for Overthorpe note YF 1564 *Overthorppe*.

OWLCOTES, OWLET HILL, OWLET HURST, OWLS HEAD.—The OE name for the owl was *ūle*, which became *oule* in ME.

OWLCOTES, Pudsey, CC *Ulecotes, Ulekotis, Oulecotes*, is obviously from OE *ūle* and *cot*, a cot or cottage.

OWLET HILL occurs in Warley (Halifax), as well as near Bolton by Bradford.

OWLET HURST, Liversedge, was written *Hullet Hirste* early in the 17th century ; its meaning is 'Owlet copse.'

OWLS HEAD, a hill in Saddleworth, was called *Hawels hede* in 1468, and has therefore no connection with the owl.

OWLER, OWLERS, OWLERTON.—Only in the case of Owlerton near Sheffield are there early records, namely, CR 1311 *Olerton*, YD 1375 *Ollerton*, SCR 1380 *Ollerton*, YD 1398 *Ollyrthon*. The signification is 'alder farmstead,' from ON *ŏlr* the alder, and *tūn*, a farmstead. From the same source comes the name Owlers which occurs in Birstall, Chevet, Marsden, and Morley, as well as in Owler Carr, Bradfield, and Owler Carr Wood, near Sheffield. The common surname Lightowler is doubtless from the same source.

OWSTON, Doncaster, DB *Austun*, DN 1284 *Owston*, CR 1294 *Ouston*, NV 1316 *Ouston*, PT 1379 *Auston*, is 'east farm,' from ON *austr*, east, and *tūn*, an enclosure, farmstead. For the early coalescing of the final *t* in *aust-* and the initial *t* in *-tun* compare Shafton and Methley.

OXENHOPE, Keighley, has passed through many phases, for example, PC †1246 *Oxenope*, PC †1250 *Oxneap*, WCR 1285 *Oxhynhope*, YD 1325 *Oxsnop*, PT 1379 *Oxenhop* ; but the meaning seems clearly 'the sheltered valley of the oxen,' from the genitive plural *oxna* of the OE *oxa*, an ox, and the OE *hop*, a sheltered valley.

OX LEE, Hepworth, PT 1379 *Oxlegh*, comes from OE *oxa*, an ox, and *lēah*, a lea or meadow. A more satisfactory spelling would be Oxley.

OXSPRING, Penistone.—Early records of the name, side by side with those of Oxton and Saxton, are as follows:

DB 1086 *Ospring -inc*	DB 1086 *Oxetone, Osse-*	DB 1086 *Saxtun*
YI 1305 *Ospring -eng*	KI 1285 *Oxton*	PF 1207 *Saxton*
NV 1316 *Ospring*	KF 1303 *Oxton*	NV 1316 *Saxton*
PT 1379 *Oxpryng*	PT 1379 *Oxton*	PT 1359 *Saxton*

The termination in Oxspring appears to come from OE *spring*, a fountain, though YF 1559 gives the name as *Oxbrynge als. Osbrynge*; compare Micklebring. But the first element is difficult to define; yet it can scarcely be OE *oxa*, an ox, witness the earliest forms as well as YF 1559 *Osbrynge*; possibly it is the personal name Osa.

OZZING, Shelley, DN 1381 *Osanz*, is a patronymic like Bowling and Cridling; compare the Frisian *Osenga* (Brons), and the German *Osanga* and *Osinga* (Förstemann).

PADAN ARAM.—Farms with this Biblical name are to be found in Kirkheaton and Old Lindley.

PADDOCK.—See Paris.

PAINTHORPE, Crigglestone, DN 1203 *Paynesthorp*, 1342 *Paynthorp*, is 'Pagen's village'; compare Ainley and Hainworth. The personal name appears in DB as Pagen and in BCS as Pagan, and an extended form appears in DB as Pagenel; compare Hooton Pagnell, formerly *Hoton Paynell*.

PARIS, PARK, PARROCK, PADDOCK.—The word 'park' is a contraction of ME *parrok*, which comes from OE *pearroc*; it is therefore of English origin, though its present form shows the influence of OFr. *parc*. Further, OE *pearroc* is derived from an older form **parr*, an enclosure; compare the dialect-word *par*, an enclosed place for domestic animals, and the verb *parren*, to enclose or bar in (Skeat). Strangely enough, as NED points out, the word 'paddock' is merely a phonetic alteration of 'parrock.'

PARIS occurs near Warley in the name Paris Gates; it occurs also, in the simple form Paris, as the name of a hamlet near

Holmfirth. Early documents show the word used as a surname in the phrase *de Paris* or *de Parys* ; it is found thus in PT 1379 in connection with Ovenden, Bingley, Hatfield, and Carlton (Barnsley). It is also recorded by YD in the form *Parysrod* in connection with Rawtonstall and Neepsend ; and it appears in YD 1502 in the phrase 'a walk milne at *Perys*.' The word seems to be the plural of *par*, an enclosure, influenced by the name of the French metropolis.

PARK, 1560 *Perocke*, is the name of a portion of Liversedge adjacent to Mirfield.

PARROCK is to be found in Rishworth and Sowerby as well as elsewhere.

PADDOCK, Huddersfield, is recorded in RE 1760 as *Parrack*, and in RE 1780 as *Paddock*.

PENDLE, PENHILL, PENNANT.—The last of these is probably referred to in WCR 1274 and 1275 as *Pendant*, but otherwise I can give no early spellings. It will be interesting, therefore, to see early forms of the Lancashire names Pendlebury, Pendleton, and Pendle Hill.

†1206 *Penlebire*	†1141 *Penelton*	1294 *Pennehille*
†1212 *Penulbery*	1246 *Penelton*	1305 *Penhul*
1300 *Penilburi*	1305 *Penhiltone*	1305 *Penhil*
1337 *Penhulbury*	1321 *Penhulton*	

Speaking of the last Dr Wyld says ' *Pen* looks like the Celtic word for "hill," etc., so that the name is pleonastic—not an uncommon thing in names which preserve a Celtic element.'

PENDLE HILL occurs in Longwood and Whitley Lower, and probably owes its first element to Welsh *pen*, the head or summit, the highest part of a field or mountain. As Pendle appears to represent Pen-hill, the whole name may mean ' hill-hill-hill ' !

PENHILL, Warmfield, may be either ' fold-hill,' from OE *penn*, a pen or fold, ' hill-hill,' from Welsh *pen*.

PENNANT, in Pennant Clough, Todmorden — formerly *Pendant*, where there is an intrusive *d*—seems quite definitely Celtic. The terminal corresponds to Welsh *nant*, a dingle or valley, and the added word ' clough ' comes from OE *clōh*, a ravine. There are several Pennants in Wales.

PENISTONE, on the western border, may be compared with Penisale, an obsolete name connected with the adjacent township of Langsett. Early records of Penistone include DB 1086 *Pengestone, Pengeston, Pangeston,* and YR 1228 *Penegelston, Penegeston.* These forms are not in agreement with the following later forms, side by side with which those of Penisale are recorded:

YI	1227 *Penigheston*	CR	1290 *Peningeshale*
YR	1232 *Peningeston*	CR	1307 *Peningesale*
YI	1258 *Peningstone*	CR	1308 *Penyngesale*
WCR	1284 *Penyngston*	CH	1358 *Penesale*

The closeness of the parallel between the forms of Penistone and those of Penisale is obvious, and we are fully warranted in explaining Penistone as 'Pening's farm'; compare CR 1252 *Peningeshalge,* Lincolnshire, and the Frisian patronymic Penninga (Brons). It is not easy to account for the earliest of the forms of Penistone. Possibly *Pengestone* stands for *Penigestone,* where Penig is an alternative to Pening; but *Penegelston* on the other hand finds no support and must be rejected.

PHIPPIN PARK lies south of Snaith, and is mentioned as early as SC †1237 in the phrase 'bosco de Fippin.'

PICKBURN, PICKNESS HILL, PICKWOOD SCAR.
—The first occurs near Doncaster, and early records give DB *Picheburne,* PF 1202 *Pikeburn,* YI 1248 *Pikebourne,* KI 1285 *Pikeburne,* KF 1303 *Pykeburn,* NV 1316 *Pickburn.* Pickness Hill is in Hoylandswaine, and Pickwood Scar in Norland. The same prefix occurs in Pickford, Warwick ; Pickhill, North Riding ; Pickmere, Cheshire ; Pickton, North Riding ; Pickwell, Sussex and Leicester; Pickwick, Wilts.; Pickworth, Lincoln and Rutland; Picton, Chester and Pembroke.

Jellinghaus records a name Pixel which in 1088 was written *Picsedila* and *Picsidila* ; but although he explains the second element as 'sedel,' a seat, he gives no explanation of the first element.

PICKLE, PIGHELL, PIGHILL.—Of the word Pickle or Pickles there are several examples. In Erringden there is

15—2

Sandy Pickle; in Denby Romb Pickle; in Oxenhope Pickles Rough; in Keighley and North Bierley Pickles Hill. But Pighell or Pighill is still more frequent; it occurs, for example, in Elland, Fixby, Skircoat, Southowram, Liversedge, Lofthouse, and Longwood; indeed, if field-names are examined, a very large number of townships will be found in which this name occurs. Passing to the history of the word, let us see what early spellings are available.

1. Connected with Elland Burton gives *Pihel.*
2. In KC there is an early form *Pictel.*
3. About 1220 the Selby Coucher Book speaks of ' Unum essartum ..., quod vocatur *Pichel.*'
4. About 1250 the Furness Coucher Book speaks of ' Totam terram ... in loco qui vocatur *Pichtil.*'

The word has, in fact, two typical forms, represented by ' pighel ' and ' pightel,' the former often hardened into ' pickel,' the latter frequently softened to ' pytle.'

When we ask what is the meaning of the word we find practical agreement between NED and EDD, for while the former explains the word as ' a small field or enclosure, a close or croft,' the latter explains it as ' a small field or enclosure, especially one near a house '; and DCR gives interesting corroboration in the phrase ' unum croftum sive toftum vocatum a pighell.' Neither NED nor EDD ventures upon a derivation.

PIKE, PIKE LAW.—The word Pike may perhaps come from OE *pīc*, a point or pike; or it may be of Norse origin and connected with the Norwegian dialect word *pīk*, a pointed mountain, *pīktind*, a peaked summit. It occurs in Pike Law, Rishworth and Golcar; Pike Low, Bradfield; and in Warlow Pike and Alphin Pike, Saddleworth.

PILDACRE, Ossett.—The termination is properly -car not -acre, witness the early spelling *Pildeker*, and its origin is the ON *kjarr*, copsewood, brushwood.

PILLEY, Barnsley, DB *Pillei*, PT 1379 *Pilley*, *Pillay*, may be ' pool meadow,' from OE *pyll*, a pool, and *lēah*, a lea. Compare DB *Pileford*, now Pilwood, near Hull.

PILLING, Skelmanthorpe, is 'willow meadow,' from ON *pill* or *pil*, a willow, and *eng*, a meadow.

PINGLE, PINGOT.—Pingle Lane occurs in Ravenfield ; but the two words are usually met with as field-names. The former is explained in NED as a small enclosed piece of land, a paddock or close ; the latter is explained in EDD as a small croft or enclosure.

PLEDWICK, Wakefield.—WCR contains many references, including 1275 *Plegwyke*, 1284 *Pleggewyk*, 1296 *Plegewyk*, 1307 *Plegwik*. In 1379 PT has *Pleghwyk*, and it is not until the 15th century that forms corresponding to that of to-day appear, namely YF 1534 *Pledewyk*, YF 1542 *Pledwyk*. Probably the meaning is 'Plecga's habitation,' Plecga being a known OE personal name.

POG, POGMOOR.—We find the spelling *Poggemore* in PT 1379, and *Poggemor* in a 13th century document in the Pontefract Chartulary, the expression being ' In territorio de Bernesleya in loco qui vocatur Poggemor.' In Wooldale there is Pog Ing ; in Liversedge Pogg Myres, 1799 *Pogmires* ; and the word Pog occurs also in Stanningley, Bradfield, and Denby. Jellinghaus records a place-name Poggenpoel in 1540; compare EFris. *pogge*, a frog (Koolman). On the other hand EDD explains *pog* as a bog, but gives no derivation.

PONTEFRACT is quite unique among the names in South-west Yorkshire, being the favoured survivor from among five rivals. A marginal note in the MS of Symeon of Durham reads as follows : ' Taddenesscylf erat tunc villa regia quæ nunc vocatur Puntfraite Romane, Anglice vero Kirkebi,' thus bearing witness to the three forms *Taddenesscylf*, *Puntfraite*, and *Kirkebi*.
    The earliest name, given in the Anglo-Saxon Chronicle under the date 947, was *Taddenesscylfe*, ' Tadden's shelf of land ' ; but the DB name was *Tateshale* or *Tateshalla*, ' Tate's corner ' from OE *healh*, or ' Tate's hall ' from OE *heall*. Concurrent with these there was a Danish name *Kirkebi*, ' church village,' and in consequence of this the distinctive prefix in South Kirkby

became necessary.   But in 1069, when the Conqueror was hastening to York to take vengeance on the Northumbrians, he was delayed, we are told, three weeks 'ad Fractipontis aquam,' at the water of the broken bridge ; and after 1090 the Latin name *Pontefracto* is to be found quite regularly in ancient charters.   Subsequently the name is found in an Anglo-French guise as *Pontefreit*, a form we may fairly attribute to the Norman retainers of the Lords of Pontefract.

It seems clear that at a very early date the name Tateshale disappeared, while Kirkebi was also lost before many generations had passed, and Taddenesscylf became Tanshelf, a subordinate member of the borough.   Thus two names remained, Pontefract and Pomfret, of which the early records are as follows :

PC	†1190	*Pontefracto*	DN 1226	*Pontefret*
PC	1135	*Pontefracto*	YI 1287	*Pomfreit*
CR	1230	*Pontefracto*	DN 1372	*Pontfret*
HR	1276	*Pontisfracti*	DN 1385	*Pomfrett*
PT	1379	*Pontiffract*	DN 1424	*Pountfreit*

The meaning of both is ' broken bridge,' Pontefract being directly from Latin, and Pomfret from Norman-French.   The Anglian and Danish names have long since disappeared, but Pontefract and Pomfret have continued side by side up to the present time, the former chiefly perhaps as the legal name, the latter as the folk-name.   Doubtless the day is not far distant when Pomfret also will be gone and Pontefract left the sole survivor.

Curiously enough, there is on record another Pontefract or Pountfreit.   A writ dated 1321 speaks of ' Pontefractum super Thamis,' and a document dated 1432 shows that this broken bridge was in Stepney, ' Pountfreit in Stepheneth Marsh.'

In olden days bridges were maintained in a variety of ways. Sometimes the funds were provided by guilds ; sometimes by endowments ; sometimes by charges on the adjoining estates. Often they came from purely voluntary sources encouraged by the promise of indulgences.   We read for instance in YR 1233 that an indulgence of ten days was to be granted to all who contributed ' to the construction of the bridge at Werreby (Wetherby),' and a 14th century document quoted by Jusserand offered the remission of forty days of penance to those who assisted ' in the building or in the maintenance of the causeway between Brotherton and Ferrybridge where a great many people pass by.' Nevertheless the history of ancient bridges is little more than a record of gradual decadence and final catastrophe, followed, after a longer or shorter interval, by restoration. Piers Plowman refers to this state of things when he speaks of ' brygges to broke by the heye weyes.' Such broken bridges—they were mostly wooden—would divert all traffic for years, and neglect attained dimensions which are now altogether inconceivable.

PONTY.—This name occurs in Honley and Kirkburton.

POPELEY, Birstall, YD 1288 *Popelay,* YD 1375 *Popelay,* is ' the lea of Pope,' from OE *lēah,* a lea or meadow. The personal name Pope would arise naturally from an earlier form Pāpa, a form which may be postulated seeing that Searle gives the name Papo ; compare OE *pāpa,* pope.

POPPLES, POPPLEWELL, POPPLEWELLS, occur respectively in Ovenden, Cleckheaton, and Warley. Early records of the second are WCR 1338 *Popilwell,* YF 1561 *Popyllwell.* Skeat gives a dialect-word *popple* signifying poplar, and Stratmann has ME *popul* with the same meaning ; hence Popplewell means ' poplar-tree well.'

PORTOBELLO, Sandal, Wakefield, is mentioned neither by Banks nor by Taylor (WRM). It probably owes its name to the events of the year 1739 when Puerto Bello, on the northern shore of the isthmus of Panama, was stormed by Admiral Vernon. The Scotch watering-place of the same name, three miles from Edinburgh, obtained its name because its first house was built by one of the seamen who served in Admiral Vernon's expedition. (*Chambers' Encyclopedia.*)

POTOVENS, a hamlet in Wrenthorpe, is recorded in the Horbury Registers in 1734 as *Potovens.* ' Thoresby in his diary, 1702, says he walked from Flanshill, Alverthorp, and Silkhouse, to the Pott-ovens (Little London in the dialect of the poor people), where he stayed a little to observe the manner of forming the earthenware and the manner of building the furnaces.'

Thus Banks in his *Walks about Wakefield*, where he states further that 'the village is legally and politely called Wrenthorpe; but whether so-named from the de Warrens, or the rabbit warren, has been doubted.'   See Little London and Wrenthorpe.

POTT, POTTER.—In Ecclesfield we find Potter Hill, HS 1637 *Potter hill*; near Goole Potter Grange; in Wadsworth Potter Cliff, YF 1557 *Potter Cliff*; near Doncaster Potteric Carr, YF 1550 *Potrecare*; and in addition to these there are Potwells, East Hardwick, and Potts Moor.   Connected with other places we find such early spellings as *Potertun*, the DB record of Potterton near Leeds; *Pottergh* in 1301 for Potter near Kendal; and *Potterlagh* and *Potterridding* in the Selby Coucher Book.

Taking all the circumstances into account it seems doubtful whether these names have any connection with earthenware or its manufacture.   On the other hand NED records a word, spelt *pot*, which signifies a natural hole or pit in the ground, a hole from which peat has been dug.   This word is described as used only in the North, and especially in districts where Scandinavian influence prevails, and it is compared with the Swedish dialect-word *putt, pott, pit*, a water-hole or abyss.   I imagine we have here the correct source of the word, and I look upon Potter as the Scandinavian plural signifying 'holes, peat-holes, or water-holes.'   This interpretation will satisfy every instance of the name quoted at the beginning of this note; and, moreover, it is in agreement with the fact that the termination in *Pottergh* and *Potrecare* are distinctively Scandinavian.

PRESTON JAGLIN, Pontefract.—See Purston Jaglin.

PRIESTLEY, PRIESTTHORPE, Hipperholme and Calverley.—The former, WCR 1275 *Presteley* and *Prestlay*, 1286 *Prestelay*, 1296 *Presteley*, is 'the priest's meadow,' from OE *prēost*, priest, and *lēah*, a meadow.   The latter is 'the priest's hamlet,' from ON *prestr*, and *thorp*, a hamlet or village.

PRIMROSE HILL.—See Rose Hill.

PUDDLEDOCK, Heckmondwike, probably derives its termination from ON *dokk*, a pool.   Compare with this the

Norse *dok*, a valley, the Scottish *donk*, a moist place, and the Sw. dialect-word *dank*, a moist marshy place or small valley. The first element is a word found alike in England, Scotland, and Ireland; but EDD gives no derivation.

PUDSEY, Leeds.—We cannot do better than compare early forms with those of Bottisham near Cambridge.

DB	1086 *Podechesaie*	1086 *Bodichesham*
KC	1198 *Pudekesseya*	1210 *Bodekesham*
CC	†1246 *Pudekesay*	1372 *Bodkesham*
KC	1250 *Pudkesay*	1400 *Botkesham*
HR	1276 *Pudesay*	1428 *Bottesham*

Prof. Skeat explains Bottisham as ' Bodec's enclosure,' and we may follow him by interpreting Pudsey as ' Pudec's island or water-meadow,' from OE *ēg*, which had both meanings. It is true that the name Pudec is not recorded by Searle ; but he gives Puda and Podda, and we may fairly assume the existence of corresponding forms with the suffix -ec, namely, Pudec and Podec. Compare Tewkesbury, formerly *Teodeces·byrig* ; Teddesley, formerly *Teodeces·leage* ; and Consett, formerly *Conekesheved.*

PUGNEYS.—This is the name of an alluvial tract of land lying between the Calder and Sandal Castle. In consequence of its nearness to the site of the battle of Wakefield, the word has sometimes been claimed as a derivative of Lat. *pugna*, a fight. We find, however, in WRM such forms as 1342 *Pokenhale*, 1577 *Pugnal*, 1713 *Pugnal*, forms which may fairly be explained as ' goblin's corner,' from OE *pūca*, a goblin, and *healh*, a corner or meadow. The transition from Pugnal to Pugney is quite irregular ; but possibly the truth is that we are in the presence of two forms, (1) *Pokenhale*, ' goblin's corner,' and (2) *Pokeney*, ' goblin's water-meadow,' from OE *ēg*, an island or water-meadow.

The Hertfordshire name Puckeridge is explained in a similar way by Dr Skeat, who gives the OE form as *Pūcan·hrycg*, and points out that the vowel in *pūca* has been shortened just as in OE *dūca*, which instead of producing ' douk ' has given ' duck ' as though from an early form *dŭca.*

PURLWELL, Batley, is to be connected with the Norw. *purla*, to gush, bubble, well up, and the Dan. *væld* (for *væll*) a spring.

PURPRISE, a farm near Hebden Bridge, is mentioned in HW 1553 as *Purprice* and *Purprise Bent*. The word is one of the few we owe to the Normans, being derived from OFr. *pourpris*, which is itself from *pourprendre*, to take away entirely ; its signification is ' a close or enclosure.' The legal term *purpresture* was used in the case of encroachments upon the property of the community or the crown.

PURSTON or **PRESTON JAGLIN,** Pontefract, DB *Prestone*, YI 1244 *Preston*, 1339 *Preston Jakelin*, PT 1379 *Preston Jakelyn*. The modern name would be more accurately Preston, ' the priest's farmstead,' OE *prēost*, a priest, and *tūn*, an enclosure, homestead.

PYE.—In South-west Yorkshire we find Pye Bank and Pye Greave near Sheffield, Pye Field near Golcar, Pye Nest near Halifax, Pyenot near Cleckheaton, and Pye Wood near Darton. Duignan records a Pye Clough in Staffordshire and a Pyehill Farm in Worcestershire, and north of Brighton the name Pyecombe occurs.

QUARMBY, Huddersfield, DB *Cornebi, Cornesbi, Cornelbi*, WCR 1274 *Querneby*, YS 1297 *Querneby*, DN 1306 *Quarnby*, later frequently *Wherneby* and *Wharneby*, is the ' mill village,' from ON *kvern*, Dan. and Norw. *kværn*, a mill, and the ON *bȳr*, a village. In Norway such compounds occur as *kværn·fos*, a mill-race, and *kværn·sten*, a mill-stone.

QUEBEC, Rishworth and Slaithwaite.—It is of course not improbable that these names owe their origin either to the capture of Quebec in 1757, or to some other connection with the Canadian city ; but as the district is strongly Scandinavian it is possible they are due to the Vikings, derived from ON *kvī*, a pen or fold, and *bekkr*, a stream. The Landnama Book has the name *Kvīa·bekkr* ; near Whitby we find Quebec ; and in the north of Scotland the name occurs again, explained by

Watson, however, as arising 'from the fact that a gentleman who had made money in Quebec settled near.'

QUEENSBURY, Halifax.—The name of this village was changed from Queenshead to Queensbury by general consent at a public meeting held May 8th, 1863 (Kelly) ; but formerly the name was Causewayend (Cudworth).

QUICK, Saddleworth, is spelt *Whyk* in DN 1313, but early forms usually show initial *qu* or *qw*, witness the following :

DB	1086 *Tohac, Thoac*	NV	1316 *Quik*
CR	1232 *Quike*	PT	1379 *Qwyk*
YS	1297 *Quyk*	DN	1629 *Quicke*

Förstemann places the word Quik on record as an element in German place-names, and connects it with ON *qvikr* and OHG *quek*, living ; compare OE *cwic*, ME *quik*, living, moving. Among the examples quoted we find the stream-names *Quekaha* and *Quecbrunn*, quick water and lively spring. That rapid streams should be called 'quick water' is in accordance with the fitness of things, and difficulty only arises when Quick is used alone as in our Yorkshire name and in the corresponding Dutch example Kuik.   Yet the origin seems clearly the same, and the meaning may well be a boggy place, a place where the ground is 'quick.'

RAINBOROUGH, RAINCLIFFE, RAINSTORTH.— Rainborough, Barnsley, PC 1155 *Reinesberga*, YI 1298 *Reynebergh*, PT 1379 *Raynebergh*, CH 1411 *Raynbargh*, is the only one of the three names of which there are early records.   Its prefix appears to be a personal name, perhaps a short form of Regenweald, while the ending is from OE *beorh*, a hill, mound.   In Raincliffe and Rainstorth, Ecclesfield, the prefix is probably the ON *rein*, a balk in a field or a steep hillside, while the terminations are from ON *klif*, a cliff, and ON *storð*, a wood.

RAISEN HALL, Brightside.—The word Raisen is doubtless to be referred to OE *ræsn*, a ceiling, a house ; compare Germanic **razna*, house.

RAKE, RAKES, RAIKES.—This word is of frequent occurrence. We find Rake Dyke and Rake Mill in Holme, Foul Rakes in Saddleworth, Cowrakes in Lindley, Rake Head in Erringden, and Raikes Lane in Birstall and Tong. The word is connected with ON *rāk*, a stripe or streak, and *reik*, a strolling, wandering ; and its later significations include (1) a rough path over a hill, (2) pasture-ground. The ME form was *rake, reike*, or *raike*. In the *Wars of Alexander* the path of righteousness is rendered 'the rake of rightwysnes,' and in *Gawayne and the Grene Knight* we find the sentence ' Ryde me doun this ilk rake,' where the word clearly means a track or road.

RAMSDEN, RAMSHOLME, RAVENFIELD, RAV-ENSTHORPE.—These are all derived from a personal name meaning 'raven,' the OE Hræfn, later Hræmn, or the ON form Hrafn. Doubtless there is here a remnant of the primitive animal worship common to all branches of the Aryan race. The raven, like the wolf, was sacred, and its name was frequently taken as their own both by Saxons and Vikings.

RAMSDEN, Holmfirth, is mentioned in PT 1379 as *Romsdeyn*, that is ' Ram's valley,' from OE *denu*, a valley. Compare Ramsey, Hunts., formerly *Hrames·ege*, explained by Professor Skeat as ' Raven's isle.'.

RAMSHOLME, Snaith, DN 1208 and 1230 *Ramesholme*, is 'the island of Raven,' from ON *holmr*, an island.

RAVENFIELD, Rotherham, DB *Rauenesfeld*, WCR 1274 *Ravenesfeud*, KI 1285 *Rafnefeld*, NV 1316 *Ravenfield*, is 'the field of Raven.'

RAVENSTHORPE, Dewsbury, is, I believe, a modern name ; but in a 13th century account of the origin of the parish church of Mirfield the stream which joins the Calder at Ravensthorpe is called *Rafenysbroke*.

RANMOOR, Sheffield, HS 1637 *Rann Moore*, is probably from ON *rann*, a house, and *mōr*, a moor. The former word has given us ' ransack,' which originally meant a house-search ; and t was frequently used in compounds such as *dverg·rann*, dwarf's house, i.e. the rocks.

RASTRICK, Halifax, has the following records : DB *Rastric,*
WCR 1274 *Rastrik,* HR 1276 *Rastrik,* PM 1327 *Raskryke,* PT
1379 *Rasterik.*   Other Yorkshire names with the same terminal
are

Escrick,	DB *Ascri*	KI 1285 *Eskeryk*	YR 1284 *Eskericke*
Wheldrake,	DB *Coldrid*	KI 1285 *Queldryk*	DM 1386 *Queldrik*

Middendorff records an OE *ric* meaning a fence or railing, an
enclosure of laths or stakes ; compare MLG *ricke, recke,* a fence,
a place fenced in, a thicket, with which many Dutch and Flemish
place-names are connected, e.g. Maurik 997 *Maldericke,* Varik
997 *Veldericke,* Herike 1456 *Hederick.*   On the other hand there
are OE *rīce* and ON *rīki,* a kingdom, and in Norway we find
such place-names as Ringerike and Raumarike.   Possibly all
the Yorkshire names above mentioned are Scandinavian :
Escrick from Dan. *æsk,* the ash; Wheldrake from the ON
personal name Kveld recorded in LN ; Rastrick from Dan. *rast,*
rest—compare Dan. *raststed,* a halting-place.   The terminal in
PM 1327 *Raskryke,* which is probably from ON *krīkr,* a nook or
corner, adds weight to the theory of a Scandinavian origin ; yet
the name may be Anglian, from OE *rǽst,* rest.

RATTEN ROW, ROTTEN ROW.—It would be difficult
to find a more widespread name.   It occurs in London and
Glasgow, Aberdeen and Shrewsbury, Newcastle and Norwich,
Spalding and Paisley, Derby and Ipswich, York and Kendal
and Durham.   In the West Riding there are instances in Halifax,
Sowerby, Lepton, Dodworth, Denby, Sedbergh, Morley, and the
Forest of Knaresborough.   The Sowerby example is mentioned
in HW 1545 as *Ratton Rawe*; the Knaresborough example in
KCR 1621 as *Ratanraw*; in the Norwich case the spelling in
the 13th century was *Ratune Rowe*; and in the Kendal example
an early form was *Rattonrawe.*
  Two explanations of Ratten present themselves.   There is
first the word *raton,* a rat, derived from OFr. *raton,* a rat, a word
which according to EDD occurs as *ratten, ratton, rattan.*   On
the other hand EDD gives a word *rotten,* also spelt *ratten,* which
means damp, boggy, saturated with rain.   This gives an explanation of the prefix at once simple and acceptable.   The

whole word may therefore be interpreted 'the swampy corner,' from ON *vrā* or *rā*. In Langsett there is Ratten Gutter, and in Oulton there is a field-name Ratten Royd.

RAVENFIELD, RAVENSTHORPE.—See Ramsden.

RAW, ROW.—The ON *vrā*, a nook or corner, lost the initial consonant in Icelandic and became *rā*, the corresponding Danish and Swedish forms being *vraa* and *vra*. In northern place-names the word appears chiefly under the forms Wray, Wrea, Ray, Raw, Row, witness the following:

Wray or Wrea, Lancs., LF 1229 *Wra*, 1513 *Wrae*
Wydra, Fewston, Yorks., KW 1571 *Widerawe*, 1577 *Witheray*
Woolrow, Brighouse, Yorks., WCR 1308 *Wollewro*, *Wolwro*
Woolrow, Shelley, Yorks., YI 1266 *Wolewra*, WCR 1275 *Wlvewro*

Probably this word occurs in South-west Yorkshire in a large number of minor names: Rawfolds, Gomersal; Raw Gate, Farnley Tyas; Raw Green, Cawthorne; Raw Lane, Stainton; Raw Nook, Lowmoor; Rawroyd, Birstall; Rawholme, Hebden Bridge; Rawroyd, Elland; Rawthorpe, Dalton near Huddersfield; as well as in Ingrow, Woolrow, and Ratten Row.

Seebohm tells us that corners of fields which from their shape could not be cut up into the usual acre or half-acre strips were sometimes divided into tapering strips, pointed at each end, called by various names. There was the OE name *gāra*, which has given the second element in Kensington Gore; in ME the phrase was *in le hirne*; in later English *in the corner*, or in Latin *in angulo*; but in early documents in Yorkshire it is common to find *del wro* or *en le wra* or *en le wro*, derived from the ON *vrā*. It is to this ancient practice that many of our surnames to-day owe their origin, for example, Gore, Herne, Corner, Rowe, Wray, and Wroe.

RAWCLIFFE, RAWMARSH, ROTHWELL.—It will be helpful to examine first some of the early forms of Rathmell, a name of undoubted Scandinavian origin equivalent to ON *rauða·melr*, red sandhill.

DB 1086 *Rodemele*   NV 1316 *Routhemell*
YI 1307 *Routhmel*   PT 1379 *Rauchemell* (*ch=th*)

Here, according to custom, DB gives *d* for *th* ; but we have also *o* for *au*, as in the case of Sowerby, from ON *saurr*, where DB has *Sorebi*. Later spellings of Rathmell give *ou* and *au* for the vowel, and represent the consonant quite accurately by *th*. Passing to the names now to be dealt with, we get the following early forms :

SC	1154 *Rodeclif*	DB	1086 *Rodemesc*	DB	1086 *Rodewelle*
SC	1229 *Rouclif*	PR	1190 *Roumareis*	DN	1235 *Routhewele*
YI	1291 *Routheclive*	PF	1206 *Rumareis*	YR	1242 *Rowell*
CH	1331 *Rauclyue*	KI	1285 *Romareis*	YI	1258 *Rowell*
NV	1316 *Rouclyff*	PT	1379 *Raumersche*	PT	1379 *Rothewell*

The first element in these names is either Rautha, the genitive of an ON personal name *Rauthi derived from ON *rauðr*, red, or it is the adjective itself in its weak form ; local considerations must be left to decide between the two.

RAWCLIFFE, Goole, has early forms like those of Rawcliffe near Blackpool and Roecliffe near Boroughbridge. The terminal comes from ON *klif*, a cliff.

RAWMARSH, Rotherham, is interesting because its terminal varies between OE *mersc*, ME *mersche*, a marsh, and OFr. *mareis*, a marsh. For DB *mesc* = *mersc* see Attercliffe.

ROTHWELL, Leeds, derives its ending either from OE *well*, *wella*, a well or spring, or from the corresponding Scandinavian word ; compare Dan. *væld* (for *væll*).

RAWTHORPE, Huddersfield.—YAS VIII 14 gives the alternative name *Raisethorpe*, where the prefix is from ON *hreysi*, a heap of stones, as in Dunmail Raise. But the name Rawthorpe Hall occurs as early as 1537, and the prefix is from ON *vrá*, a nook or corner.

RAWTONSTALL, ROALL, ROWLEY.—There can be little doubt that the first element in these names comes from OE *rūh*, ME *ruh*, *row*, *rogh*, rough, uncultivated. Early spellings are as follows :

WCR	1274 *Routonstall*	DB	1086 *Ruhale*	DN	†1225 *Ruley*
HR	1276 *Rutonestal*	PC	1159 *Rughala*	LC	1296 *Rogheley*
WCR	1298 *Routunstall*	PC	1200 *Rughale*	YS	1297 *Roulay*
YD	1336 *Routonstall*	PT	1379 *Rouhall*	PT	1379 *Rouley*

RAWTONSTALL, Hebden Bridge, doubtless possessed at first a name of two elements meaning 'rough enclosure,' the termination 'stall' being added later ; see Stall.

ROALL, Knottingley, is the 'rough corner' from OE *healh*, of which the dative is *hale*.

ROWLEY, Kirkburton, is the 'rough lea,' from OE *lēah*, a lea or meadow.

REAP, REAPS.—This word, which occurs in such names as Reap Hill and Reap Moor, is found in Wadsworth, Stansfield, Slaithwaite, Fixby, Holme, Crookes, and Hallam. It seems scarcely possible that the word should be connected with the ON *hreppa*, a share, which has given the Norfolk place-name Repps, written *Repes, Hrepes* in DB. More probably it should be connected with the Norw. dialect-word *ræpa*, explained by Aasen as a heap, a dunghill ; compare Heald.

REEDNESS, on the Ouse near Goole, PF 1199 *Rednesse*, DN 1209 *Rednesse*, DN 1304 *Rednesse*, NV 1316 *Rednesse*, SC 1324 *Redenesse*, is probably 'reedy headland,' from OE *hrēod*, a reed, and *ness*, a cape or headland.

REIN.—Names of this form, or of the form Rain or Rayne, occur occasionally ; for example there is the Rein in Calverley, Stanhope Rein in Eccleshill, and Reins in Honley. They are derived from ON *rein*, a strip of land, a balk.

RENATHORPE HALL, Shire Green, Sheffield, now Hatfield House, is recorded in YD 1279 as *Raynaldtorp*, YD 1303 *Raynaldethorpe*. The meaning is 'Raynald's village,' from ON *thorp* and the personal name given in ON as Ragnvaldr, and in DB as Raynald and Ragnald.

RIBBLE.—A small stream joining the Holme at Holmfirth bears this name, as well as its greater namesake further north. In Herts. there is a river Rib; Worcester has the name Ribbesford, formerly *Ribbedford* and *Ribetforde* (Duignan). Possibly we have here a primitive river-name in the three forms rib, ribel, ribet. Gonidec has *ribl*, a riverbank, riverside, and in Welsh there is a word *rhib* signifying a streak or driblet.

RICHMOND, Sheffield, HH 1366 *Richmond*, YD 1398 *Richemound*, is 'the strong hill,' from AFr. *riche* which meant strong as well as rich, and AFr. *mund*, a variant of *munt*, a hill (Skeat). The name Richemont occurs in Normandy and other parts of France, and we find also such names as Richebourg, Richecourt, Richelieu, Richeville (Robinson).

RIDDING, RIDING.—Both forms of the name are found, but in South-west Yorkshire the latter is the more frequent. It is derived from OE *hryding*, and means a patch of cleared land, a clearing. In Austonley we find the name Gibriding, and in Linthwaite Smithriding. For the latter note the spelling *Smethrydding* in 1305 connected with Colthorpe, and compare also the name Smithley.

RINGBY, RINGINGLOW. — Naumann provides such pairs of Scandinavian personal names as Hildr and Hildingr, Sveini and Sveiningr, Gauti and Gautingr. At the same time he gives the name Hringr, and from this we may postulate such forms as Hringi and Hringingr.

RINGBY, Northowram, is doubtless 'Hringi's farm,' ON *Hringa·bȳr*. There are several places in Norway called Ringstad and others called Ringsby, Ringstveit, Ringsrud, while Denmark has Ringsted and Sweden Ringhult and Ringstorp.

RINGINGLOW, Sheffield, 1574 *Ringinglawe*, is 'the cairn of Hringing,' from OE *hlāw, hlǣw*, ME *lawe, lowe*, a tumulus, hill; compare Dunningley. See Law, Low.

RIPLEYVILLE, Bradford.—This place is the result of the industrial expansion of the 19th century, and owes its name to its founder, Mr H. W. Ripley.

RIPPONDEN, RYBURN.—After passing through Ripponden the Ryburn joins the Calder at Sowerby Bridge.

RIPPONDEN, WCR 1307 *Ryburnedene*, WCR 1308 *Ryburnedene*, WCR 1326 *Rieuburnden*, WCR 1489 *Rybornedeyne chapel*, WCR 1599 *Ribonden*, SM 1610 *Ripondon*, is obviously 'the valley of the Ryburn,' from OE *denu*, a dean or valley.

RYBURN, the name of the stream, is mentioned in WCR 1308 in the phrase '*between Riburn and Calder.*'

The terminal is clearly from OE *burna*, a brook ; but the first element can scarcely be derived from OE *rīth, rīthe*, a rill or stream, for 'burn' could then only have been added after 'rīth' had lost its meaning.   More probably Ry- is of Celtic origin.   In Ayrshire there is a stream called Rye Water ; in the North Riding a tributary of the Derwent is called the Rye ; and in Ireland several streams to-day called Rye which according to Hogan were formerly called *Rige*.

RISE HILL, Hampole, derives its name from OE *hrīs*, shrubs, brushwood.   The ON word was also *hrīs* ; hence the modern Danish from *riis*, and the Swedish *ris*.

RISHWORTH, Ripponden, HR 1276 *Risewrd, Rissewrth*, WCR 1297 *Rissewrth*, PT 1379 *Rysseworth*, is derived from OE *risce* a rush, and *weorth*, a farmstead or holding.

RIVELIN.—This stream is a tributary of the Don, and is represented in the following early names : CH 1300 *Rivelingdene*, HH 1329 *Ryvelyndene*, HH 1383 *Ryvelingdene*, HS 1637 *Riveling Water*.   We need not hesitate to refer the river-name to the Norwegian Riflingr which comes from ON *refill*, a long strip of stuff, and has given the Norwegian place-names Revling and Revlingsvolden (Rygh).   A stream near Pickering is recorded in CR 1252 as *Tacriveling*.

ROALL.—See Rawtonstall.

ROCHE ABBEY, Maltby, is situated in a narrow wooded dell, 'bordered on the north by a low rocky scar, and watered by a musical rivulet' (Morris).   It was founded in 1147, and its Abbot is commonly described in Latin charters as *Abbas de Rupe* ; but we find the form *La Roche* as early as CR 1251. The source of the latter is obviously OFr. *roche*, a rock.   'The name of the house,' says Morris, 'was probably derived from the crags already mentioned ; but possibly it was borrowed more immediately from a particular rock which bore a rude resemblance to the Saviour on the Cross, and became in aftertime an object of devotion.'

ROCKLEY, Barnsley, YR 1250 *Rockelay*, WCR 1308 *Rockelay*, is 'Rocca's lea.' The personal name Roc, from which we may form Rocca, is recorded by Searle.

RODLEY, Calverley, is recorded in CC †1260 as *Rotholflay* and *Rozolflay*, and later as *Rothelay*, while YF 1568 has *Rodley*. The meaning is 'Hrothwulf's lea,' from OE Hrothwulf and *lēah*, a lea or meadow.

ROGERTHORPE, Badsworth. — The Domesday form, *Rugartorp*, presents the genitive *rūgar* of ON *rūgr*, rye ; but later forms like YD 1329 *Rogerthorp* and YF 1570 *Rogerthorpe* show the influence of the Norman name Roger.

-RON.—See Ardron.

ROOMS LANE.—In connection with Morley PF 1202 gives the name *Le Ruhm*, referring, doubtless, to the district to which Rooms Lane gives access. Probably the origin is the OE *rūm*, an empty space.

ROSE HILL, PRIMROSE HILL.—The word 'primrose,' which in the North often takes the form 'primrose,' came to us from the OF *primerose*, lit. first rose. It is not known in English literature earlier than the 15th century, and is used by neither Chaucer nor Gower. Primrose Hill occurs in Huddersfield and Wakefield, and Primrose Farm in Liversedge. Rose Hill occurs near Doncaster, Rawmarsh, and Penistone, and if ancient the word 'rose' is probably Celtic ; compare Welsh *rhos*, a moor or heath, Bret. *ros*, Irish *ros*.

ROSSINGTON, Doncaster, has early forms as follows: PF 1207 *Rosenton*, YR 1249 *Rosingtun*, CR 1269 *Rosington*, YI 1279 *Rosington*, PT 1379 *Rosyngton*, SM 1610 *Rosinton*. I take *Rosinton* to be an abraded form representing *Rosington*. But *Rosenton* looks like the genitive of a weak personal name, while *Rosington* is a patronymic derived from the same source. Seeing that no assured weak genitive in *-en* occurs in SW Yorkshire, we shall be justified in explaining the name as 'the homestead of the Rossings.'

ROTHER, ROTHERHAM.—Among early records of the river-name we find HR 1276 *Reder* and CH 1329 *Roder*, and among those of the place-name there are the following :

DB 1086 *Rodreham, Rodreha'*	YS 1297 *Roderham*
PF 1202 *Rodenham*	KF 1303 *Roderham*
PF 1204 *Rodenham*	NV 1316 *Roderham*
HR 1276 *Roderham*	PT 1379 *Rodirham*
YR 1286 *Roderham*	VE 1535 *Rotheram*

Although the early forms are somewhat difficult, Rotherham must be interpreted as 'the enclosure of the oxen,' from OE *hamm*, an enclosure, and OE *hrȳther*, ME *rother, rether*, an ox. Place-names showing the same first element are somewhat common. In Oxfordshire there is Rotherfield, DB *Redrefeld*, which represents OE *hrȳthera·feld*, and means 'the field of the oxen' (Alexander), and there are other Rotherfields in Sussex and Hants. In Yorkshire the name Rotherford, 'ford of the oxen,' occurs near Bowes (NR) and Leathley (WR); in Hereford-shire there is Rotherwas; in Hampshire Rotherwick; and in Surrey Rotherhithe. The persistent *d* in the early forms of Rotherham is due to Norman influence, an influence finally overcome by the Anglian pronunciation with *th*; and the early spellings in *Roden-* represent weak genitives in *-ena* formed from a stem without *r* such as is recorded by Sievers. The river-name must be explained as formed from the place-name.

ROTHWELL.—See Rawcliffe.

ROTTEN ROW.—See Ratten Row.

ROYD.—This word occurs as the termination in hundreds of field-names. It is also found as the name of a hamlet here and there, as in the case of Royd in Soyland, Royds in Beeston and Bradfield, Royds Green in Rothwell, and Royd Moor in Hemsworth. Occasionally it takes the form Rhodes, particularly when used as a surname. Early records of Royd and Boothroyd show the development of the word :

WCR 1297 *Rode*	WCR 1274 *Botherode*
WCR 1308 *Rode*	WCR 1307 *Botherode*
WCR 1360 *Rode*	WCR 1334 *Botheroide*
WCR 1377 *Roide*	WCR 1364 *Botheroid*
WCR 1497 *Roide*	WCR 1435 *Botheroide*

and its early meaning is made clear by an interesting passage in WCR 1307, where a certain piece of land is said to be 'called rodeland because it was cleared (*assartata*) of growing wood.' The word goes back, indeed, to the Germanic **ruda*, which means land newly cleared. For the succeeding history of the word we must examine three derivatives of this Germanic stem, namely, ON *ruð*, ON *rjóðr*, and MLG *rode*.

1. ON *ruð*, a clearing, has given the modern *rud* which appears in hundreds of Norwegian place-names, among them Ketilsrud, Grimsrud, and Steinsrud. If from ON *vað*, a ford, and *viðr*, a wood, we obtain 'wath' and 'with,' we may fairly expect from ON *ruð* to obtain a ME form 'ruth.' It is probable, indeed, that, after a lengthening of the vowel, the word has given us quite regularly the East Riding place-name Routh which in DB was written *Rutha*.

2. ON *rjóðr*, an open space in a forest, a clearing, should follow in the steps of ON *skjól*, a pail, which has produced the northern dialect-word 'skeel,' a milking-pail (Wall). The North Riding place-name Reeth may perhaps have come to us in this way, though the DB form *Rie* is not quite easy. Compare Greetland, DB *Greland*, which may possibly come from ON *grjót*.

3. MLG *rode*, a clearing, has given rise to a remarkable number of derivatives. Gallée, in a very full note on the subject, shows that the word could give such forms as 'roder,' 'roden,' 'rothe' and 'rothen,' as well as forms with *oi* like 'roide' and 'roiden' and others like 'rodel,' 'rodeland,' and 'roding' (NGN. II, 33–73).

Coming now to the later history of 'royd,' we see that the original short vowel, being in an open syllable was lengthened. Afterwards, under dialect influence, it became *oi*, the history being the same as that of OE *hole*, which first became ME *hole*, and later gave the modern place-name Hoyle.

The outcome of all this is obvious ; 'royd' is to be linked with LG *rode* rather than with ON *ruð* or *rjóðr*. Yet it should be noted that not a few of our local examples possess as their first element a word undoubtedly Scandinavian ; among other instances we find Boothroyd from ODan. *bóth*; Gilroyd from

ON *gil*, a valley; Mithroyd and Rawroyd from ON *miðr*, middle, and *rā*, a corner; Gambleroyd and Swainroyd from the ON personal names Gamall and Sveinn. Perhaps, indeed, we had early derivatives from both ON *ruð* and LG *rode*; but, if so, all were finally levelled out under the ME form *rode*.

Examples: Ackroyd, Anroyd, Arkinroyd, Armroyd, Blaithroyd, Brackenroyd, Brookroyd, Burkroyd, Charlroyd, Coteroyd, Crossroyd, Crowroyd, Dalroyd, Denroyd, Ellenroyd, Emroyd, Foxroyd, Greenroyd, Hanging Royd, Hillroyd, Hollinroyd, Holmroyd, Howroyd, Hudroyd, Ibbotroyd, Joanroyd, Longroyd, Lumbroyd, Malkroyd, Maukroyd, Murgatroyd, Mytholmroyd, Netherroyd, Nunroyd, Oakroyd, Oldroyd, Rattenroyd, Rawroyd, Sillroyd, Studroyd, Westroyd, Wheatroyd.

ROWLEY.—See Rawtonstall.

ROYSTON, Barnsley, DB *Rorestun, Rorestone,* YR 1233 *Roreston,* NV 1316 *Roston,* PT 1379 *Roston,* YF 1587 *Royston,* is 'Rore's farmstead,' from the ON personal name Hroarr, and ON *tūn,* an enclosure, farmstead. For the development of the first vowel compare Hoyle, Royd, Hoyland and Soyland.

According to Dr Skeat Royston in Cambridgeshire had an origin altogether different, the first element being Norman, derived from a certain Lady Roese who set up a wayside cross called after her name *Cruceroys.*

RUSBY, Denby, may perhaps be 'Hrut's farmstead,' from the ON personal name Hrut, and ON *bȳr,* a farmstead, village.

RYBURN.—See Ripponden.

RYCROFT, RYHILL, RYLEY, all derive their first element from OE *ryge,* ME *rye,* rye.

RYCROFT, Pudsey, YR 1228 *Ricrof,* is 'the rye croft,' from OE *croft,* a small field.

RYHILL, Wakefield, recorded in DB as *Rihella* and *Rihelle,* in DN 1234 as *Rihill,* in NV 1316 and PT 1379 as *Ryhill,* is simply, 'the rye hill.'

RYLEY, Kirkburton, spelt *Ryeley* in WCR 1286, and *Rylay* in YS 1297, is 'the rye lea,' from OE *lēah,* a lea or meadow.

SADDLEWORTH, WC †1230 *Sadelword, Sadelworth,* WC †1280 *Sadelword,* NV 1316 *Sadelworth,* LI 1388 *Sadilworth,* PT 1379 *Sadelworth.* No personal name of suitable form is recorded by Searle, but Förstemann gives Sadelbert and Sadalfrid, and KCD has *Sædeles·sceat.* It is not impossible, therefore, that the interpretation may be ' Sædela's holding,' from OE *weorth.* On the other hand there is an important series of names worthy of consideration in this connection, namely, Saddle Forest and Saddleback in the Lake District; Saddell, a parish in Argyll; Saddle, a mountain in Inverness ; Saddle Yoke, a mountain in Dumfries ; and Saddle Hill, an eminence in Leitrim. It seems possible, indeed, that in the case of Saddleworth the original place-name was Saddell, a word of similar form to Idle and Nostell.

It will be of interest to give a brief survey of the place-names of the whole township. From Celtic sources, carrying us back to the days of the Brigantes, there are probably the river-names Chew and Tame, the valley-name Combe, and the hill-name Featherbed. The termination of the name of the township is Anglian, and so also are Delph and Quick, Lydgate and Tunstead, Boarshurst and Micklehurst, Crawshaw and Denshaw and Castleshaw. The Scandinavian period has provided the names Austerlands, Dobcross, Foul Rakes, Knot, Slack, Slack-cote, Thurston, and Woolroad. Among these Dobcross is particularly noteworthy, bearing, as it does, definite signs of the Norse immigration from across the Irish Sea. After the Norman Conquest the de Lacy ownership led to the name Lord's Mere, while the interests of Roche Abbey provided the names Friarmere and Grange Hey, where Friar and Grange are of French origin.

SALFORD, SALTONSTALL, Huddersfield and Halifax.—For the latter WCR has *Saltonestall* in 1274, *Saltunstal* in 1275, and *Saltonstall* in 1308. Doubtless the original name was Salton, which like Salford derives its prefix from OE *salh* or *sealh,* a willow tree. Thus Salton is ' the willow enclosure,' and Salford 'the willow ford.' Compare the field-name *Salacre* found in a Survey of Almondbury dated 1548.

SALTAIRE, situate on the Aire near Bradford, was created by the enterprise of Sir Titus Salt, who established here his great worsted and alpaca factory in 1813. The sources of the name are obvious.

SALTERHEBBLE, SALTERLEE, SALTERSBROOK. —The first occurs in Halifax and Saddleworth; the second in Halifax; the third near Penistone.

In HW 1553 we find that 'John Watterhouse of Skyrcotte' left four shillings 'towards the battillyng of Sowreby brig' and a similar amount 'to the amendyng of *Salterheble.*' This apparently strange expression is quite accurate. An entry in the Almondbury Registers for 1559 tells how a certain William Brigge was drowned—a sudden tempest came upon him as he came over 'a heble or narrow bridge,' and he was blown into the water ; and EDD explains the word 'hebble' as the wooden rail of a plank bridge, or as the narrow plank bridge itself.

The first element in the three names, Salter- or Salters-, is derived from OE *sealtere,* a salter, dealer in salt, carrier of salt, gen. sing. *sealteres,* gen. pl. *sealtera.* There are many other names of similar form, including Salterford near Barnoldswick and Saltergate near Harrogate ; compare also the early name *Sealter·ford* given in BCS.

SALTONSTALL.—See Salford.

SANDAL, SANDBECK.—The first element is either OE *sand,* or ON *sandr,* Dan. *sand,* sand, gravel.

SANDAL MAGNA, Wakefield, DB *Sandala,* PF 1175 *Sandale,* PF 1202 *Sandale,* KI 1285 *Sandale,* is 'the sandy corner,' from OE *healh,* a corner or meadow, or 'the sandy tongue of land,' from Dan. *hale.* According to Kelly, the soil is loamy, while the subsoil is 'sand on sandstone rock.' It was in this parish that the battle of Wakefield was fought on Dec. 31st, 1460.

SANDAL PARVA, Doncaster, DB *Sandala,* KI 1285 *Sandhale parva,* YS 1297 *Sandale,* has the same origin and meaning.

SANDBECK, near Tickhill, CH 1241 *Sandbec,* HR 1276 *Sandebek,* YS 1297 *Sandebeck,* is Scandinavian, from ON *bekkr.*

SANTINGLEY, Wintersett, DN *Sayntingley*, appears to have for its first element a patronymic. For the first vowel compare Manningham.

SAVILE TOWN, Dewsbury.—The name has come into use since the bridge over the Calder was opened in 1863. The district is owned by Lord Savile.

SCAMMONDEN is the name of a deep secluded valley which lies west of Huddersfield and is separated from the valley of the Colne by a ridge of considerable height. Along a shoulder of this ridge ran the Roman road from Manchester (Mancunium) which passed by Castleshaw and afterwards Slack (Cambodunum). Early spellings are WCR 1275 *Scambanden*, WCR 1286 *Schambandene*, DC 1349 *Scamendene*, DC 1352 *Scammendene*, DN 1383 *Scammonden*. The first element appears to be a Scandinavian personal name *Skambani, formed from ON *skammr*, short, brief, and ON *bani*, death, a slayer; compare such names as Skamkell and Swartbani (Nielsen). The terminal comes from OE *denu*, a valley.

SCAR.—This word enters into many local names in the more hilly districts. It is found at Scarr Hill, Bradford; in Sowerby, Norland, Skircoat, Elland, Golcar, Scammonden, and Fulstone; in Grimescar, Birkby; in Nanscar, Oxenhope; and in Winscar, Wooldale. The root is ON *sker*, a rock, a precipitous cliff.

SCAUSBY, SCAWSBY, Ovenden and Doncaster.—For the latter we find early forms as follows:

DB 1086 *Scalchebi*	PF 1208 *Scauceby*	YI 1247 *Scauceby*
CH 1186 *Scalcebye*	CH 1231 *Scalceby*	NV 1316 *Scauseby*
PF 1205 *Scauceby*	CR 1232 *Scalceby*	PT 1379 *Scauseby*

The Domesday form warrants the interpretation 'Skalki's farm,' from ON *býr* and the personal name Skalki (Nielsen); but the later forms do not spring naturally from this source.

SCAWTHORPE, Adwick-le-Street, may perhaps be 'the thorpe of Skagi,' the name Skagi being recorded by Nielsen.

SCHOLES,   SCHOLEBROOK,   SCHOLE   CARR, SCHOLECROFT,   SCHOLEFIELD,   SCHOLE   HILL, SCHOLEMOOR, SCHOLEY.—Though somewhat common in South-west Yorkshire, the name Schole or Scholes is never used as the designation of a township, and finds no place in the Domesday record.   Its origin is the ON *skáli*, a shieling, log hut, shed, a word found in such early Icelandic place-names as Skala·nes, Skala·vik, and Skala·holt.

In Yorkshire the word assumes the forms Scale and Schole, the latter only being found south of the Aire.   North of that river Schole occurs but once—in Barwick in Elmete ; but Scale or Scales is found quite frequently, the simple name occurring at least eight times, while compounds number at least ten, among them Scaleber, Scalehaw, Scalemire, Southerscales, Summerscales, Winterscale and Winterscales.

But there is a further point of great interest and importance. According to Björkman the word *skáli* is West Scandinavian ; it is therefore not Danish, but Norse.   Thus the place-name Scale or Schole becomes a test by means of which we may discover the locality of Norse settlements.   Early spellings of names in South-west Yorkshire include the following :

SCHOLES, Cleckheaton, YR 1228 *Scales*, PT 1379 *Scholes*.

SCHOLES, Stainland, WCR 1308 *Skoles*.

SCHOLES, Holmfirth, WCR 1274 and 1297 *Scoles*.

SCHOLES, Keighley, PT 1379 *Scholl*, YF 1567 *Scoles*.

SCHOLES, Rotherham, YI 1284 *Scales*, YF 1544 *Scoles*.

SCHOLEBROOK, mentioned in connection with Alverthorpe, WCR 1284 *Scholbrok*.

SCHOLE CARR, Rishworth, 1593 *Scolecar*.

SCHOLECROFT, Morley, YR 1252 *Scalecroft*, YI 1264 *Scholekroft*.

SCHOLEY, Hemsworth, 1230 *Scolay*, 1379 *Scolay*.

In addition to the above we find eight other examples, namely, Scholecroft in Austonley, Scholefield in Dewsbury, Schole Hill in Penistone, Scholemoor in Bradford, West Scholes in Clayton (Bradford), West Scholes in Hoylandswaine, Scholey in Methley, and Brianscholes in Northowram.

SCISSETT, Skelmanthorpe. — No early spellings have shown themselves, but if we accept OE *ge·set*, a dwelling, or OE *set*, a seat, encampment, as the terminal, the first element will probably be a personal name as in Ossett and Winterset. A suitable name, Sisse, occurs in YS 1297, pointing to an OE form *Sissa.

-SCOE, SKEW, are derived from ON *skōgr*, a wood; compare Sw. *skog*, Dan. *skov*, and note that the word corresponds to OE *sceaga*, ME *schagh*, a copse or wood. We find the form 'scoe' in Thurnscoe, NV 1316 *Thirnescogh*, in Briscoe, and in the two Loscoes, while the form 'skew' occurs in Skew Hill, Ecclesfield; compare Askew, from ON *askr*, ash, and *skōgr*.

SCORAH WOOD, Barnsley, is tautological, the first element in Scorah being from ON *skōgr*, a wood, whilst the termination is from *vrá* or *rá*, a corner. Compare Haverah near Harrogate, spelt *Haywra* in 1311 and 1334.

SCOUT.—In EDD this word is explained as 'a high rock or hill; a projecting ridge, a precipice,' and is derived from ON *skúti*, a cave formed by jutting rocks. Among the occurrences of the word may be named East Scout, West Scout, Bald Scout in Langfield; Great Scout, Little Scout, Hathershelf Scout in Sowerby; Brown Scout in Widdop; Dill Scout in Heptonstall; Black Scout in Wadsworth; Scout Wood in Northowram; Ashday Scout in Southowram; Scout Wood and Scout Top near Marsden; Scout Hill, Ravensthorpe; Scout Bridge, Hoylandswaine; and Scout Dyke, Ingbirchworth.

SCRAITH, a hamlet in Brightside, HS 1637 *Skreth*, must be connected with ON *skrīða*, a landslip on a hillside; compare ON *skreiðr*, sliding, Dan. *skred*, slip, slide, and the modern English word 'scree' which has the same origin.

SCRAT LANE, Gomersal.—In this name we are carried back to the superstitions of the Vikings. The Norse word *skratti* meant a wizard, a warlock, a goblin. It appears in the Heimskringla Saga in the place-name *Skratta·sker*, Skratti's rock, and it has given to the folk-speech of the West Riding the name Old Scrat, used as a synonym for 't' owd Lad'—the Devil,

SCROGG, SHROGG.—Both forms are to be met with. Scrog, for example, occurs in Kirkheaton, while Shrogg is found in the adjacent township of Whitley Upper. The meaning is brushwood, shrubs, a little wood, 'scrub'; but the origin is unknown, though it appears to be Scandinavian.

SCROOBY LANE occurs in Greasborough.

SHACKLETON, SHAFTHOLME.—These are possibly examples where ON *sk* has become *sh*; compare the dialect-words 'shackle,' 'shawm' (Wall) and 'shacklet' (Flom).

SHACKLETON, Heptonstall, WCR 1274 *Schakeltonstal,* HR 1276 *Scakeltonestal,* WCR 1297 *Schakelton,* appears to come from ON *skökul,* Sw. *skakel,* a horseyard, and *tūn,* an enclosure.

SHAFTHOLME, Doncaster, YF 1535 *Shaftholme,* may perhaps come from ON *skaptr,* Norw. *skaft,* a shaft, pole; see Shafton.

SHAFTON, Royston, is interesting on account of its early forms, among which are the following:

DB	1086 *Sceptun, Sceptone*		YF	1345 *Shafton*
YR	1246 *Shefton*		PT	1379 *Schafton*
YI	1261 *Schafton*		YF	1531 *Shafton*

Just as ON *topt* stood for 'toft' and ON *gipt* for 'gift,' so here the Domesday record gives *Sceptun* for *Scefton.* Another example of the same kind, quoted by Dr Moorman, is DB *Sceptesberie* for Shaftesbury. The first element in Shafton is probably either OE *sceaft, scœft, sceft,* ME *shaft, schaft,* a shaft, pole, or a personal name derived therefrom, viz. Sceafta, which occurs in the Hertfordshire place-name Shaftenhoe (Skeat). In either case the final *t* in the first element coalesced with the initial *t* of the second element at an early date; compare Owston and Methley.

SHALEY.—See Shaw.

SHARLSTON, Wakefield, DN 1254 *Sharneston,* HR 1276 *Scarneston,* WCR 1296 *Scharneston,* PT 1379 *Sharston,* YF 1532 *Sharleston.* Obviously the first element is properly *Sharn*; yet, although the possessive *s* is present in every case, no personal name of the particular form is recorded. The word seems to be derived from OE *scearn*, dung. Professor Skeat derives Sharnbrook in Northamptonshire from the same root, and tells us that in Hampshire a dung-beetle is still called a sharn-beetle. See Mixenden.

SHAW and **SHAY** are both derived from OE *sceaga*, a copse, thicket, small wood. Early ME forms were *schagh* and *schawe*, but later the spelling *shay* frequently occurs. There is no village of the name, though as a terminal the word is quite common, witness the names Bagshaw, Birkenshaw, Blackshaw, Boshaw, Bradshaw, Castleshaw, Crawshaw, Crimshaw, Earnshaw, Fullshaw, Hepshaw, Kilnshaw, Marshaw, Murgatshaw, Nepshaw, Reddishaw, Smallshaw, Toftshaw, Walshaw, and Wilshaw.

SHAY occurs in Denholme and Austonley.

SHALEY occurs in Holmfirth, and its meaning is most probably Shay Ley, that is, 'coppice lea,' from OE *lēah*.

SHAW CROSS, Soothill, is sometimes called Shay Cross.

SHAW HOUSE, Elland, or rather the site, is recorded as *Schagh* in 1199 in Burton's account of the possessions of Fountains.

SHEAF, SHEFFIELD.—In the 12th and 13th centuries the river-name was recorded in the *Beauchief Obituarium* as *Scheth* (Addy); HS 1637 has *Sheath* quite frequently; and Hunter gives the form *Shee*. Early forms of Sheffield are:

DB 1086 *Scafeld, Escafeld*		KI 1285 *Sheffeld*	
PF 1202 *Shefeld*		NV 1316 *Sheffeld*	
PF 1208 *Sefeld*		PT 1379 *Scheffeld*	
YD 1279 *Schefeud*		VE 1535 *Sheffeld*	

The river-name has the same origin as the ordinary word 'sheath.' It comes from OE *scǣth*, *scēath*, *scēth*, ME *schethe*, *scheth*, and its original meaning was 'that which separates,' hence a boundary or limit. In Western Germany there are many place-names derived from the same Teutonic stem, **skaith-*, their usual form being *scheid*.

SHEFFIELD is obviously formed from the river-name. The initial *e* in *Escafeld* is due to the Norman scribe; compare Snaith and Stubbs. For the variation of vowel in the first element compare Emley, and note the early loss of *th* before *f.*

It is interesting to note that the river which in the early days of the Anglian settlement received its name because it was the boundary between two dominions still continues to be a boundary; it runs for several miles between the counties of York and Derby.

The place-names of the neighbourhood show that Sheffield was a centre of settlements by both Danes and Norsemen, the former largely predominating. There are nine thorpes, namely, Bassingthorpe, Herringthorpe, Grimesthorpe, Osgathorpe, Netherthorpe, Renathorpe, Silverthorpe, Skinnerthorpe, and Thorpe Hesley. There are several names containing the remnants of the ON *haugr,* namely, Sharow, Grenoside, Stenocliffe and Wincobank. But a certain number of names are distinctively Norse, for example, Gilthwaites, Gilcar, and Scholes. In addition there are such Scandinavian names as Brincliffe, Catterstorth, Crimicar, Crookes, Damflask, Little London, Moscar, Owlerton, Ranmoor, Rivelin, Storrs, and Wicker.

SHELF, SHELLEY.—OE *scylf,* ME *schelfe, shelfe,* which means a ledge or shelf of land, has given us several place-names, including Hathershelf, Hunshelf, Tanshelf, Waldershelf, and the two now in question.

SHELF, Bradford, DB *Scelf,* WCR 1275 *Schelf,* NV 1316 *Schelf,* YI 1488 *Shelf,* requires no explanation.

SHELLEY, Kirkburton, DB *Scelneleie* (*n* for *v*), *Scivelei* (*l* omitted), WCR 1275 *Schelfley,* YS 1297 *Schelflay,* PT 1379 *Schellay,* is 'the lea on the shelving land.' The position of the village is in striking agreement with the name. Shelley in Suffolk appears in KCD as *Scelflēah.*

SHEPLEY, SHIBDEN, SHIPLEY.—Early records of these names, which occur respectively near Kirkburton, Halifax, and Bradford, are as follows :

DB 1086 *Seppeleie*	WCR 1276 *Schipeden*	DB 1086 *Scipeleia*
YS 1297 *Schepelay*	WCR 1277 *Schypeden*	IN 1287 *Schippeley*
PT 1379 *Scheplay*	YI 1523 *Shipden*	CC 1328 *Schepelay*
YI 1523 *Shepley*	YI 1546 *Shybden*	YI 1554 *Shipley*

SHEPLEY and SHIPLEY have the meaning 'sheep lea,' the difference in form being due to a similar early difference; compare OE *scēp* and OE *scīp*, a sheep, gen. pl. *scēpa, scīpa*.

SHEPLEY, Mirfield, is recorded in PF 1202 under the form *Seppelae*, and has the same meaning.

SHIBDEN, which means 'sheep valley,' gives an excellent illustration of one of the commonest laws of language. A voiced consonant (*b, d, g, v, z*) and a voiceless consonant (*p, t, k, f, s*) cannot well exist side by side; both must be voiced or both voiceless. This is the reason why the *s* in 'caps' is pronounced quite differently from the *s* in 'cabs.' It is, however, usually the latter of two consonants which influences the former, thus *Shipden* has become Shibden, *Hepden* Hebden, and *Catebi* Cadeby.

SHERWOOD HALL, Knottingley, PF 1202 *Sirwud*, *Sirewud*, PT 1379 *Shyrwode* is 'the bright wood'; from OE *scīr*, bright, shining, and *wudu*, a wood.

SHIBDEN, SHIPLEY.—See Shepley.

SHIRECLIFFE, SHIRE GREEN, Sheffield. — The former is recorded as *Shirclif, Shirecliff, Shirclif*, in inquisitions of 1366, 1383, 1385. The latter, mentioned in a charter about 1220 as *Sschires*, and in a fine of 1520 as *Shier Grene*, is derived from OE *scīr*, a boundary, district, shire. But the former, though possibly from the same word, may also come from OE *scīr*, bright, shining.

SHITLINGTON, Horbury, DB *Scellintone, Schelintone*, PC 1155 *Schetlintona*, PF 1208 *Sytlington*, LC 1296 *Schitlingtone*, PT 1379 *Shytlyngton*. In his book on the place-names of Bedfordshire, Professor Skeat shows that Shillingdon was formerly *Scytlingedune*, and interprets the word as 'the down of the Scytlings,' that is, of the sons of Scytel or Scytela. He goes on to explain Scytel as a diminutive connected with Scytta, an archer. We may interpret Shitlington therefore as 'the homestead of the Scytlings,' from OE *tūn*, an enclosure, homestead.

SHUTTS, SHUTTLES, *SHUTTLEWORTH.—The larger sub-divisions of the three great fields in the common field system were called Shutts or Shots. The word is of frequent

occurrence as a field-name. In Ossett we find Shutts House;
in Batley Blew Shutt; in Wooldale and Fulstone Downshutts;
while the simple name Shutts occurs in Cawthorne. HS 1637
speaks of 'the lands called *Shuttles,'* where Shuttles may be
a diminutive of Shutts.

*SHUTTLEWORTH, Bawtry, KC 1209 *Schutleswrtha, Sutles-
wrtha*, is obviously parallel to Shuttleworth, Lancs., LF 1227
Sutttelesworth, WC 1333 *Shutelisword*, and the explanation
'Scytel's holding' may be advanced without hesitation. Compare Shitlington and Stubbs.

SIDDAL, Halifax, HW 1497 *le Sidall*, 1532 *Sidall*, 1538
Sedall, 1547 *Sydalbroke*, is probably from OE *sīd*, wide, and
healh, a corner, meadow.

SILCOATES, Wakefield, WRM 1789 *Silcotes*. The termination comes from OE *cot, cote*, or ON *kot*, a cottage; and
the first element probably refers to the particular way in which
the cottages were erected. Our modern word 'sill' is derived
through ME *sille, sylle*, from OE *syll*, which according to
Professor Skeat meant a base or support; ON has *syll*, a sill,
and Dan. *syld*, the base of a framework building.

SILKSTONE, Barnsley, is peculiarly misleading; it refers,
in fact, neither to 'silk' nor 'stone.' Early records give DB
Silchestone, PC †1090 *Sylkestona*, PF 1167 *Silcheston*, PF 1197
Silkestun, NV 1316 *Silkeston*, and the explanation is 'Sylc's
farmstead,' from OE *tūn*, a farmstead, and the personal name
Sylc recorded by Searle.

SILVERTHORPE, Braithwell. — The meaning is most
probably 'the thorpe of Silfri,' an ON personal name of that
form being on record.

SKELBROOK, SKELLOW, Doncaster.—Early records
show extraordinary variations in both cases:

DB	1086	*Scalebro, Scalebre*	DB	1086	*Scanhalle, Scanhalla*
PC	1170	*Scalebroc*	DN	1200	*Scalehale*
DN	1252	*Skelbroke*	PF	1204	*Skelehall*
YR	1253	*Skelebrok*	YI	1264	*Skelhale*
DN	1336	*Skelbroke*	PT	1379	*Skellawe*

The Domesday forms *Scanhalle* and *Scanhalla* appear to be
corrupt, but otherwise we find early forms in *scale-* followed by
later forms in *skele-* and *skel-*. Such names as come from ON
skáli, a shed or hut—Scholecroft, for example—show early forms
in *scale-*, and later in *scole-*, forms not in harmony with those
shown above. Under these circumstances the records of some
of the Yorkshire Skeltons may well be examined, and we
take (1) Skelton near Guisborough, (2) Skelton near Ripon,
(3) Skelton near Howden.

(1) DB 1086 *Sceltun*	(2) DB 1086 *Scheltone*	(3) DB 1086 *Scilton*
DB 1086 *Scheltun*	DB 1086 *Scheldone*	DB 1086 *Schilton*
CH 1180 *Scelton*	CH 1228 *Skelton*	PF 1199 *Skeltun*
CH 1239 *Skelton*	NV 1316 *Skelton*	PT 1379 *Skelton*

As the first element in these names is undoubtedly Scandinavian it will be useful to see what parallels there are in the
place-names of Denmark, Norway, and Sweden.

In the first place we find a series of names connected with
ON *skilja*, to divide, separate, and derived ultimately from Germanic **skelōn*. Among Norwegian examples from this source
Rygh gives the river-names Skilja, Skelja, and Skillebæk;
among Danish place-names Madsen gives Skjelby, Skjelbæk,
and Skjelmose; and among Swedish place-names Falkman gives
Skälhuset and Skälaholm. In these names Skjel- and Skäl-
are interpreted as meaning a boundary.

In the second place, connected with ON *skellr*, a clash,
splash, crack, and ON *skalla*, to clash, clatter, rattle, there is the
stem Skjell- found in the Norwegian Skjellaaen, and in the
plant-name Skjella.

In the third place there are the Norwegian place-names
Skallerud and Skallestad, which according to Rygh are connected
with ON *skalli*, a skull, a bald head, Norw. *skalle*, a word
sometimes applied to a barren or stony eminence (Aasen).

SKELBROOK obtains its terminal from OE *brōc*, a brook; but
the Domesday form *Scalebro* goes back to Dan. *bro*, Sw. *bro*,
a bridge. It seems extremely probable, therefore, that the
first element in Skelbrook is a stream-name. Note that a
stream—now called the Great Ings or Old Eau beck—runs
through both Skelbrook and Skellow which are adjacent to

one another ; and note further that a tributary of the Ure is called the Skell.

In SKELLOW the terminal comes from OE *hlǣw*, ME *lawe*, a cairn or burial-mound ; but the earlier forms show extraordinary divergence from the present name, and are derived from ON *hallr*, a slope, Dan. *hale*, a tongue of land, OE *heall*, a hall, or OE *healh*, a corner of land.

SKELDERGATE, Halifax.—York has a street of the same name recorded in the Whitby Chartulary as *Sceldergate* (12th century). The terminal comes from ON *gata*, a path or road ; and the first element is connected with ON *skjöld*, gen. *skjaldar*, a shield, which goes back to Germanic *skeldu*, a board, plank, shield (Torp). Probably Skeldergate means 'the road paved with planks,' just as Cluntergate means 'the road paved with logs.'

SKELMANTHORPE, Huddersfield, like Skelmersdale in Lancashire, is of Scandinavian origin, and the two names may fitly be brought together :

DB	1086 *Scelmertorp*		1086 *Schelmeresdele*
YD	1283 *Scelmarthorpe*		1202 *Skelmersdale*
WCR	1296 *Skelmarthorpe*		1202 *Skelmaresden*
NV	1316 *Skelmanthorp*		1321 *Skelmardisdale*

It seems clear that the first element is a personal name, and Nielsen presents an old Danish name Skialmar, which with a Latin ending appears also as Skielmerus and Skelmerus. Thus Skelmanthorpe is 'the village of Skelmer,' from ON *thorp*, and Skelmersdale is 'the dale of Skelmer,' from ON *dalr*. The change from Skelmarthorpe to Skelmanthorpe is probably due to the influence of such names as Normanton, Dudmanston, Copmanthorpe, and Hunmanby. A parallel case is the change from Rikmeresworth to Rickmansworth (Skeat). For the absence of the sign of the genitive compare Rogerthorpe, Renathorpe, and Herringthorpe.

SKEW.—See Scoe.

SKIERS, SKYRE WOOD.—We find PT 1379 *Skyres*, DN *Skyres*, *Skires*, for the former which is in Wentworth, and SE

1715 *Skyre Wood* for the latter which is in Golcar. Another name of similar character is Skyreholme near Burnsall, BM 1325 *Skyrom*. These words are doubtless connected with Icel. *skürr*, Sw. dial. *skur*, a shed, Skyre being a dat. sing. and Skyrom a dat. pl.

SKINNERTHORPE, Sheffield, YS 1297 *Schinartorp*, 1366 *Skynnerthorp*, is 'the hamlet of the tanner'; compare Sw. *skinnare*, a tanner, and ON *thorp*.

SKIRCOAT, Halifax, HR 1276 *Skirkotes*, WCR 1297 *Skyrecotes*, PT 1379 *Skyrcotes*, may be translated 'the bright cottages,' from ON *skirr*, clear, bright, and *kot*, a cottage.

SKITTERICK.—Small streams of this name are to be found in Wakefield, near Wath-on-Dearne, and in Emley. YR 1230 speaks of a 'duct called *Skiterik*,' apparently near Otley; and the Wath Parish Registers have the name *Skyterick* in 1640. There is a Norwegian river called Skytteren, from ON **skytra*, and an early place-name derived therefrom is *Skyttersett* (Rygh).

SLACK, SLACKCOTE.—The word 'slack' is derived from the ON *slakke*, which means a slope on a mountain edge. Places of the name occur in Barkisland, Heptonstall, Oakworth, and Quarmby. Near Bradford there is Wibsey Slack, and at Meltham Legards Slack. Ripponden and Chapelthorpe have each a Slack Lane, Lofthouse a Slack Hill, Marsden a Slack End, and Saddleworth a Slackcote and Slack Head, while the name Catherine Slack occurs in Cragg Vale, near Queensbury, and near Brighouse. But most interesting of all is the Slack in Quarmby, referred to in WCR 1275 in the description 'Thomas de *Slac* de Querneby,' and in PT 1379 as *Slak*. The breezy road along the ridge between the Colne Valley and that of Scammonden follows the line of an ancient Roman road which according to the Antonine Itinerary linked together Mamucio (Manchester), Camboduno (Slack), and Calcaria (Tadcaster); but the Ravenna geographer, dealing with the road from Mantio (Manchester) to Medibogdo (Methley), speaks of the station at Slack as Camuloduno.

SLACKCOTE, Saddleworth, is 'the cottage on the slope,' ON *kot* being a cottage or small farm.

CAMBODUNUM is derived from two ancient Celtic words, namely, *cambos*, crooked, bent, and *dunon*, a fortified place or stronghold ; hence the meaning given by Holder, 'arx curva.'

CAMULODUNUM on the other hand signifies 'the fortress of Camulos,' that is, of the god of war, Mars.

SLAITHWAITE, Huddersfield, CR 1235 *Slathweyt*, WCR 1286 *Slaghthayth*, DN 1306 *Slaghethwayte*, NV 1316 *Slaghewhait*, is a particularly interesting name. It is probably derived from the ON *slag*, slaughter, skirmish, Norw. *slag*, a blow, an action, battle, engagement, and ON *thveit*, a paddock or clearing. The Norwegian word *slagsted* is used to denote the scene of a battle or conflict, and on the same lines Slaithwaite may be interpreted 'battle clearing.'

There is a second SLAITHWAITE, situate in Thornhill Lees, Dewsbury, and pronounced like the first Slouit (slauit).

SLANTGATE occurs as the name of a road or lane in Linthwaite, Thurlstone, and Marsden. The ending is from ON *gata*, a path, and the prefix from Norw. *slenta*, to fall slanting, Sw. *slenta, slanta*, to cause to slide.

SLITHERO, Rishworth.—Watson calls the place *Slitherow*, a form which corresponds to the name *Slidrihou* found in the Cockersand Chartulary about 1213 as the name of a portion of Ainsdale, near Southport. The meaning is 'scabbard-howe,' from ON *sliðr*, a scabbard, and *haugr*, a burial-mound, or howe.

SMEATON, SMITHIES, SMITHLEY, SMITH-RIDING.—OE *smið* is a smith, and OE *smiððe*, ME *smythy*, is a smithy or forge ; but there is also an OE adjective *smeðe*, smooth, flat, level.

SMEATON, Pontefract, BCS †992 *Smithatun*, DB *Smedetone*, *Smetheton*, PC †1180 *Smithetona*, NV 1316 *Magna Smytheton*, *Parva Smytheton*, PT 1379 *Kirkesmethton*, is 'the smiths' enclosure,' from OE *smið* and *tūn*, an enclosure or farmstead.

VIII]   ALPHABETICAL LIST OF NAMES   261

SMITHIES occurs in Thornhill, TPR 1614 *Smythyes*, Barnsley, Birstall, Heckmondwike, and elsewhere; it is doubtless from OE *smiðe*, a forge.

SMITHLEY, Wombwell, recorded as *Smethelay* in IN 1307, *Smythelay* in PT 1379, and *Smythelay* in 1386, is probably 'smiths' meadow,' from OE *lēah*, a lea or meadow.

SMITHRIDING, Linthwaite, may perhaps be 'smith's clearing'; see Ridding and Smithley.

SNAILSDEN, Penistone, is most probably 'Snjall's valley,' from the ON personal name Snjall, and OE *denu*.

SNAITH, which stands on the south bank of the Aire a few miles from its junction with the Ouse, is represented in early records as follows:

DB	1086 *Esneid*	BM	1206 *Sneyd*
DB	1086 *Esnoid, Esnoit*	CH	1223 *Snaith*
PR	1154 *Sneid*	YI	1250 *Snayth*
CR	1205 *Sneyth*	NV	1316 *Snayth*

The name has considerable interest because it illustrates two of the peculiarities of the Domesday scribes, and two of the points where Old Norse differed from Old English.

The Domesday scribes wrote *esn* for *sn*; and in the same way they wrote *Escafeld* for Sheffield, *Eslingesbi* for Slingsby, and *Estretone* for Stretton. Further, the Domesday scribes wrote *d* and *t* instead of *ð*; see Bolton and Melton. In the case of Snaith *d* occurs in records later than DB, but subsequently *th* appears quite regularly.

In ON we find the consonant *ð* where OE had *d*; hence the doublets 'garth' and 'yard' from ON *garðr* and OE *geard*, an enclosure, and 'with' and 'wood' from ON *viðr* and OE *wudu*, a wood. Further, we find in ON the vowel *ei* where OE has *ā*; hence our place-names have 'stain' and 'stan' from ON *steinn* and OE *stān*, a stone, as well as 'braith' and 'brad' from ON *breiðr* and OE *brād*, broad.

Thus the earliest forms of Snaith may be read as *Sneith*, and it is plain that we owe the name to ON *sneið*, rather than OE *snād* or *snǣd*. Each of these words means 'a piece cut off,' but a secondary meaning appears to have been 'a boundary.'

Middendorff quotes such OE place-names as *Snādhyrst* and *Tattingsnād*, as well as *Snǣdhege* and *Snǣdfeld*; and Skeat gives the Bedfordshire name Whipsnade, formerly *Wibesnade.*

SNAPE, SNAPETHORPE, FOULSNAPE. — Near Darfield we find Snape Hill; in Upperthong Snape Reservoir; in Austonley Snape Clough.   But early records are available only in the case of Snapethorpe, Wakefield, and Foulsnape, Pontefract:

WCR 1275 *Snaypthorp*	CH 1220 *Fulsnap*
WCR 1277 *Snaypethorpe*	CH 1246 *Fulsnap*
WCR 1295 *Snaypethorpe*	DN 1464 *Foulesnape*

The simple name Snape occurs near Bedale in the North Riding, and near Saxmundham in Suffolk.   There is Snape Hill in Lincolnshire, and in Lancashire such names as Bullsnape, Haresnape and Kidsnape occur.

It is not easy to find the origin of the word; possibly, indeed, we have to do with different words.

1.   Most likely, perhaps, is the root which has given Danish *snabe*, a word explained in Blandinger p. 244 as having the general meaning ' odde,' a point of land, special meanings being a cape, and a strip of wood or forest.   Danish place-names derived from the word are Agersnap, Gudsnap, Kolsnap, Krogsnap, and Vandsnap.

2.   Stratmann-Bradley gives ME *snape*, a winter-pasture, and a connection with ON *snóp* is suggested.

In any case Snape, Snapethorpe, and Foulsnape, are Scandinavian in origin, and the prefix in the latter is from ON *fūl*, foul, mean.

SNODDEN HILL, SNODDLE HILL.—The former, YD 1333 *Snoden Hille*, is near Penistone; and the latter, RE 1716 *Snoddle Hill*, is in Huddersfield.   Both are doubtless connected with ON *snodinn*, Norw. *snodden*, bare, bald; compare the dialect-word *snod*, smooth.

SNYDALE, Pontefract, DB *Snitehala*, PF 1202 *Snithale*, KF 1303 *Snytall*, NV 1316 *Snytall*, PT 1379 *Snydale*, is like **Wheldale** in its suggestion that the ending is -dale; more probably, however, the terminal is OE *healh*, ME *hale*, a corner or meadow, or possibly Dan. *hale*, a tongue of land. The prefix appears to be ON *snið*, a slice ; compare Dan. *snitte*, to cut, chip.

SOOTHILL, Dewsbury, DC 1225 *Sotehill, Sothill*, YI 1251 *Sothull*, HR 1276 *Sothill*, DC 1349 *Sotehull*, PT 1379 *Sutill*, YF 1504 *Sotehill*, has probably the meaning it seems to bear, the OE word for soot being *sōt*, and the ME *sot*.

SOUTHEY, Ecclesfield, HH 1366 *Southagh*, PT 1379 *Sowthagh*, YF 1588 *Sowthay*, is simply 'south enclosure,' from OE *sūð*, south, and *haga*, an enclosure.

SOUTHOWRAM.—See Northowram.

SOWERBY, Halifax, DB *Sorebi*, WCR 1275 *Sourby*, HR 1276 *Sourebi*, WCR 1297 *Soureby*, NV 1316 *Soureby*, means 'swampy farm,' from ON *saurr*, foul, swampy, sour, and *bȳr*, a farmstead or village.

The name is of frequent occurrence in those parts of England where the Vikings settled, and in the Icelandic Book of Settlements we possess an account of the way in which more than a thousand years ago a spot in Iceland received the self-same name. Steinolf, the son of Hrolf, dwelt, we are told, in Fairdale ; one day he walked inland from Fairdale, climbed a mountain, and from thence saw a valley, great and overgrown with wood ; 'within the valley he saw a clearing, and there he raised his dwelling and called it Saurbæ, for it was very swampy ; and he called the whole dale by the same name.' In addition to *Saurbæ* we find in ON the name *Saurlith*, the swampy slope ; in the Chartulary of Cockersand Abbey names such as *Soureby*, *Sourelonde, Sourer*, and *Saurschales* are to be found ; and near Windermere there are two villages called Sawrey, swampy island.

In former days Sowerby gave its name to a district called Sowerbyshire. An old manuscript, probably of the time of James I, says ' Sowerbyshier was a several Jurisdiction or Libertie within the Mannor of Wakefielde, beinge in tymes paste accounted as a Forreste or Freechase, and replenished with deere. This Sowerbyshier was parcell of the possessions of the

Earles of Warrene and Surreye, and there were diverse vaccaries therein, and namely these, Cromptonstall, Ferneside, Oversaltonstall, Nethersaltonstall, Hadershelfe, Baitings, and Mancanholes, all knowne by meates and boundes, at the which Cattel were norished and bredd ' (WCR II, xxix). According to Ducange a Vaccaria was a cowshed or cowhouse usually constructed to hold forty cows, and situated in a pasture or woodland cleared or set apart for this head of cattle.

SOWOOD, Ossett and Stainland.—Records of the first are WCR 1277 *Soutwode*, WCR 1309 *Southwode*, DC 1573 *Sowewood*, while of the second we have the 16th century spelling *Sowewode*. The meaning is ' south wood,' from OE *sūð* and *wudu*.

SOYLAND, Sowerby, WCR 1274 *Soland*, WCR 1286 *Solande*, WCR 1297 *Solande*, YF 1553 *Soland*, YF 1572 *Soyland*, HPR *Soweland, Sowland, Soyland*, appears to mean ' sow land,' from Dan. *so*, a sow ; compare ON *sȳr*, dat. *sū*, a sow. Through dialectal influence so- became soy- as rode became royd.

SPA, SPAW, occurs with some frequency ; examples are Gunthwaite Spa, Ossett Spa, Tanhouse Spa in Ackworth, Spa Wood in Erringden, Spa Bottom in Lepton, and Spaw House in Treton.

The water of a mineral spring is often called ' spaw water,' and the spring itself a ' spaw well.' EDD explains the word in this sense, but gives no hint as to its origin, though it has been customary—at any rate in such cases as the Spa at Scarborough— to derive the name from Spa in Belgium. I venture to suggest, however, that the source of our South-west Yorkshire names is ON *spā*, prophecy ; compare ON *spā·kerling*, a prophetess, *spā·leikr*, divination, Dan. *spaa·mand*, a soothsayer, *spaa·kvinde*, a fortune-teller, *spaa·kvist*, a divining rod. That the discoloured waters of a mineral spring should in the past have been used for divination and then called ' spaw water ' seems quite within the bounds of probability, and that the name ' spaw water ' should later have lost all connection with divination is only what we should expect ; thus, there is no difficulty in regard to the sense of the word. But neither is there difficulty in regard to the phonology, for if ON *rā* could give 'raw,' as in Rawthorpe, ON *spā* could give 'spaw,' a form with which the local pronunciation fully agrees. See Raw and Ratton Row.

SPEN, Gomersal, YD 1329 *Spen*, PT 1379 *Spen*, YF 1565 *Spen*. The name is found elsewhere. Near Rochdale there is a stream called the Spenn, and York has a Spen Lane. In connection with Stalmine near Blackpool the Cockersand Chartulary has *Spen* in 1268 ; and BM records such names as *Spenneker* and *Spengate*. Further, Leithaeuser records a German place-name Spenrath, that is, Spen Royd, and Rietstap gives a Dutch place-name Spanbroek, which in form corresponds to Span Brook but means Span Marsh ; in neither case, however, is an etymology given. I suggest that the word is to be connected with the Germanic **spenan, spanan*, nipple ; compare ON *speni*, Dan. *spene*, Sw. *spene*, Fris. *spene, späne*. Thus the meaning is probably a projecting point or elevation.

SPINK, SPINKSMIRE, SPINKWELL.—The names involve either the dialect-word *spink*, a finch, or the Celtic word of the same form—compare Ir. and Gael. *spinc*, a point of rock, an overhanging cliff.

SPINK, Heptonstall, standing as it does alone, can scarcely have its origin in the bird-name.

SPINKSMIRE, Meltham, derives its terminal from ON *mȳrr*, a moor, bog, swamp.

SPINKWELL, which occurs in Dewsbury, West Ardsley, and Linthwaite, is doubtless from *spink*, a finch, and OE *wella*, a well or spring. Among early spellings we find KC †1189 *Spinkeswelle* (Aldfield), WCR 1308 *Spinkeswelle* (Holme), YF 1550 *Spynkpyghell* (Southowram).

SPITAL occurs near Pontefract, DN 1294 *Spitle Hardwicke*, as well as in Ecclesfield, Tickhill, and Wentworth. It is of French origin, from OFr. *hospital*. In Middle English the word suffered aphæresis, and became 'spital' or 'spitle.'

SPRING is found very frequently in Ordnance maps as a synonym for 'wood'; compare DC 1593 'a wood called Crakenedge Springe.'

SPROTBOROUGH, Doncaster, is the most eastern of a line of early forts or fortified places in the valley of the Don. Early spellings, DB *Sproteburg*, YR 1250 *Sprotteburg*, KI 1285 *Sprotteburg*, NV 1316 *Sprotburgh*, warrant the interpretation 'Sprot's fortified post,' from OE *burh*, and the personal name Sprot or Sprott recorded by Searle.

STAINBOROUGH, STAINCLIFFE, STAINCROSS, STAINFORTH, STAINLAND, STAINTON.—South-west Yorkshire presents six names in stain- and six in stan-, a fact which calls attention to a well-marked difference between English and Scandinavian. From a common ancestor, Teutonic *ai*, Old Scandinavian got *ei* and Old English *ā*; and in consequence the words for 'bone,' 'stone' and 'home' take the following early forms :

Teutonic	Old Norse	Old English
*baina	bein	bān
*staina	steinn	stān
*haima	heimr	hām

In Danish, as has already been shown (p. 31), the diphthong *ei* was at an early date contracted to *e*, and it thus became possible for our early place-names to present three forms :

(1) *stain-*, which is either Norse or Danish ;

(2) *stan-*, found in English names of early origin ;

(3) *sten-*, which is distinctively Danish.

But still another form is possible, viz.

(4) *stone-*, found in English names of later origin.

In the course of centuries the vowel in OE *stān* changed its pronunciation, and just as OE *hām* gave the modern word 'home,' so OE *stān* gave the modern word 'stone,' while ME has the forms *stane* and *stone* ; see Stone.

Names in *sten-* include *Stenforde*, one of the Domesday forms of Stainforth, and also Stennard and Stenocliffe.

It should be noted that some of our names show early forms of varying origin, as in the case of Stainborough, Stainforth, Stainland, Stainton.

STAINBOROUGH, Barnsley, gets its terminal from OE *burg*,

burh, a fortified place, or ON *borg*, a stronghold or castle, early forms being

DB 1086 *Stainburg, Stanburg*   CR 1252 *Steinborg*
PC †1090 *Stainburch*   NV 1316 *Staynneburgh*
PC †1160 *Steinburch*   PT 1379 *Staynburgh*

The form *Stanburg* is entirely Anglian, while *Steinborg* is entirely Scandinavian, but the signification in both cases is the same, 'the stone fort or castle.'   As there are in the neighbourhood many 'thwaites' we may conclude that the Scandinavian influence was that of Viking settlers from the west.

STAINCLIFFE, Dewsbury, is obviously 'stone cliff,' and comes from ON *steinn* and *klif*.

STAINCROSS is the name of a hamlet near Barnsley, and also of a wapentake.   We owe the name to Norsemen ; see the note on Cross.

STAINFORTH, Hatfield, is partly Anglian and partly Scandinavian ; it has the following early forms :

DB 1086 *Steinford, Stenforde*   KI 1285 *Stainford*
HR 1276 *Steynford*   NV 1316 *Staynford*

The terminal comes from OE *ford*, while the prefix is from ON *steinn*.   But the Domesday form *Stenforde* shows the Danish spelling ; compare Dan. *sten*, a stone.

STAINLAND and STAINTON, Halifax and Doncaster, both show interesting variations, witness the following :

DB 1086 *Stanland*   DB 1086 *Staintone, Stantone*
PT 1379 *Stayneland*   PF 1166 *Steinton*
CH 1276 *Staynlond*   PF 1202 *Steinton*
CH 1342 *Steynland*   NV 1316 *Staynton*

The former signifies 'the stony land,' from OE or ON *land*; and the latter is 'the stony enclosure,' from OE or ON *tūn*, an enclosure or farmstead.

STAIR.—This is the ordinary word 'stair,' and means an uphill path, an ascent ; compare OE *stīgan*, to climb, and OE *stǣger*, ME *steyer*, a stair, step.   Among examples of the use of the word as a place-name we find Stairfoot, Barnsley ; Stairs Bottom, Haworth ; and Stairs Hill, Oxenhope.

-STALL.—This termination comes from OE *steall*, a place, stall, stable ; a place for cattle.   Among dialect-meanings EDD gives cattle-shed, sheepfold, temporary shelter.

West of Halifax there are six place-names with this termination : Cruttonstall, Heptonstall, Rawtonstall, Saltonstall, Shackletonstall, and Wittonstall.   It will be noticed that each name has three elements, and it will not be unreasonable to suggest that the original names were of two elements, Crumton, Hepton, Rawton, etc.   Probably the ending -stall was added when the farmsteads became vaccaries of the Earls of Warrene.

Another name exhibiting the termination is Birstall.

STANBURY, STANDBRIDGE, STANEDGE, STAN-HOPE, STANLEY, STANNARD.—All these names are of Anglian origin, and the first element is derived from OE *stān*, a stone.   In early years—prior to the 'rounding' of the OE *ā*—the vowel in such compounds as Stanbury and Stanhope was shortened, and so the form stan- was obtained.   See the note on Stainborough, Staincliffe, etc.

STANBURY, Keighley, DN 1250 *Stanbir*, YF 1536 *Stanbury*, is 'the stone fortress,' from OE *byrig*, dat. of *burh*, a fortified place ; compare Stainborough.

STANDBRIDGE, Sandal Magna, WRM 1639 *Stan Brig*, is simply 'stone bridge,' from OE *brycg*, a bridge.   Compare Standground, Hunts., which is recorded in DB as *Stangrun* and in the Ramsey Chartulary as *Stangrunde*, 'stony ground.'   It is obvious that 'stand' was introduced by 'popular etymology' after the true meaning of 'stan' had been lost.

STANEDGE—sometimes written Standedge—a lofty moorland ridge between Marsden and Saddleworth, 1272 *Stanegge*, is obviously 'stony ridge,' from OE *ecg*, ME *egge*.

STANHOPE, Sowerby, may be explained as 'the stony valley,' from OE *hop*, a secluded valley.

STANLEY, Wakefield, DB *Stanleie*, PF 1202 *Stanleiebothum*, is 'the stony lea,' from OE *lēah*, a lea or meadow.

STANNARD, Horbury, derives its termination from OE *eard*, ME *erd*, a dwelling-place.

STANCIL,   STANNINGLEY,   STANNINGTON,
STANSFIELD.—From the Germanic stem *staina we get
not only OE stān and ON steinn, but also a large number of
personal names. Thus ON has the simple names Steinn and
Steini, as well as compounds like Steinarr, Steinulfr, Steinunnr,
Arnsteinn, Ormsteinn, while ODan. has Sten and Stenkil.
Searle gives many OE compounds, for example, Stanburh,
Stanheard, Stanwine, Stanmær, but not a single example of the
simple form Stan. That such a name existed together with
a patronymic formed therefrom is made certain by the existence
of the modern surname Stanning.

STANCIL, Doncaster, DB Steinshale, RC 1232 Stansale, KF
1303 Stansall, is 'Stan's corner,' from OE healh, a corner or
meadow.

STANNINGLEY, Bradford, Heckmondwike, and Ovenden, is
doubtless 'the lea of the Stannings,' from OE lēah, a lea or
meadow.

STANNINGTON, Ecclesfield, HH 1329 Stanyngton, is 'the
farmstead of the Stannings,' from OE tūn, an enclosure, farmstead.

STANSFIELD, Todmorden, DB Stanesfelt, HR 1276 Stanesfeld, PT 1379 Stanesfeld, is 'the field of Stan.'

STAPLETON, Pontefract, DB Stapeltone, PC 1159 Stapil-
tona, NV 1316 Stapelton, is derived from OE stapel, a post or
pillar of wood or stone, and tūn, an enclosure or farmstead.
We may explain the name as 'the farmstead marked by a
pillar.'

The word 'staple' is found in place-names under a variety of
circumstances. In Devon, for example, the name Stallbridge,
DB Staplebrige, bears witness to the use of staples or posts in
the construction of bridges. The long list of Staplefords given
in the Gazetteer shows how common it was in the olden days to
use posts in order to mark out the point at which a stream
should be crossed. Places where goods might be exposed for
sale or markets held were frequently distinguished in the same
way; and, occasionally, the meeting-place of a Hundredmoot
was marked by a staple, witness the names Barstable and
Thurstable, two of the hundreds in the county of Essex.

STATHAM, Holme, appears to come directly from the dat. pl. of ON *staðr*, a 'stead,' place, spot; compare Latham.

STAUPS, STOPES, STOUPS.—Near Hebden Bridge there is Staups Moor; in Scammonden Staups Lane; and in Northowram Staups Common, 1607 *Staupes*. The form Stopes occurs in Marsden, Bradfield and Holmfirth; and outside the West Riding the word occurs near Robin Hood's Bay in the form Stoups Brow. The source is ON *staup*, which according to Vigfusson means a knobby lump; but there is also a Norw. dialect-word *staup* which means a little deep depression or hollow.

-STEAD.—The West Riding examples of this termination are comparatively few. The name Halstead occurs in Thurgoland, Thurstonland, and Woolley; the name Newstead near Hemsworth; Barrowstead in Skelmanthorpe; and Tunstead in Saddleworth and Cleckheaton. The OE and ME is *stede*, a place, site, station, and the word corresponds to the ON *staðr*.

STENNARD, STENOCLIFFE.—The first element in these names comes from Dan. *sten*, a stone—or rather from an ODan. word of that form. See the note on Stainborough, etc.

STENNARD, Wakefield, obtains its terminal from OE *eard*, ME *erd*, a dwelling-place.

STENOCLIFFE, Ecclesfield, is recorded by Guest as *Stenocliff* in 1540 and *Stenoclyff* in 1541. Judging by such names as Grenoside and Wincobank, the second element appears to be 'how,' from ON *haugr*, a mound or cairn; thus Steno- would mean 'stone cairn.'

-STER.—This termination is usually referred to the ON *staðr*, a farm, homestead; compare Bolster Moor and Bolsterstone where the first element is obviously from ON *bolstaðr*, a farm-house. Duxter Wood in Ecclesfield, however, is written *Dukestorth* in HH 1425, and we must therefore recognise ON *storð*, as a possible source. Examples of the termination are Bannister, Meltham; Bolster Moor, Golcar; Clipster, Southowram; Copster, Thurgoland; Topster, Rishworth;

Trister, Cawthorne. In Norway the ON *staðr* has given the termination -*stad*, to be found in such names as Bolstad, Harstad, Listad, and Mustad.

STOCKBRIDGE, STOCKSBRIDGE, STOCKSMOOR, STOCKWITH, are all derived from OE *stocc* or ON *stokkr*, a stock, trunk, log.

STOCKBRIDGE, Bentley, YF 1528 *Stokbrygge*, YF 1570 *Stockbridge*, tells its story sufficiently clearly.

STOCKSBRIDGE, Sheffield, YI 1247 *Stocbrig*, PT 1379 *Stokbrig*, shows an intrusive *s*; compare Bolsterstone.

STOCKSMOOR, Thurstonland, is recorded as *le Stokes* in DN 1316, and *Stokes* in PT 1379.

STOCKWITH LANE, Hoyland Nether, is Scandinavian, its termination being derived from ON *viðr*, a wood, forest, or felled timber.

STONE, STONEROYD, STONESHAW.—Just as OE *bān* gave the modern word 'bone,' so OE *stān* gave 'stone.' But compounds formed at an early date show stan-, and place-names like Stoneroyd and Stoneshaw must have been formed at a time when OE *stān* had already become *stone*.

STONE, a hamlet in Maltby, 1324 *Stane*, 1354 *Stone*, may perhaps be so-called from a prominent rock.

STONEROYD, Kirkheaton, derives its terminal from ME *rode*, a clearing; see Royd.

STONESHAW, Heptonstall, derives its terminal from OE *sceaga*, ME *schagh*, a small wood or copse.

STONE CHAIR, Shelf.—According to an account in *Yorkshire Notes and Queries* (I, 154), this hamlet owes its name to a relic of the old coaching days. At an important junction of roads a curious double milestone existed which had a stone seat fixed between the two uprights; these uprights were placed at an angle and were held together by the flat stone which formed the seat. The inscription on the modern stone which has taken its place reads as follows: 'Stone Chair. Erected 1731. Re-erected 1891. Halifax—Bradford.'

STOODLEY, STUDFOLD, STUDROYD, have for their first element the OE *stōd*, a stud of horses.

STOODLEY PIKE, with its obelisk erected as a peace memorial after the Napoleonic wars, is well known to railway travellers between Todmorden and Hebden Bridge. The name is recorded in WCR 1275 as *Stodlay*, WCR 1296 *Stodeley*, PT 1379 *Stodlay*, and its meaning is 'stud lea,' from OE *lēah*.

STUDFOLD, Ovenden, is doubtless derived from OE *stōd·fald*, a stud-fold or paddock.

STUDROYD, Hoylandswaine, is 'stud clearing'; see Royd.

STORRS, STORTH, STORTHES.—These come from ON *storð*, a young plantation or wood. The word is of very frequent occurrence throughout the western hill-country. It is found in one or other of its forms in Elland, Birkby, Huddersfield, Linthwaite, Thurstonland, Ossett, Oxspring, Darfield, Heeley, Ecclesfield, and Bradfield. Storthes in Thurstonland is referred to in WCR 1275 as *Stordes* and 1286 as *Storthes*.

HOWSTORTH, in Ecclesfield, HS 1637 *How Storth*, may be either 'the wood marked by a cairn,' from ON *haugr*, a how or cairn, or 'the wood in the hollow,' from ON *hol*.

RAINSTORTH, in Ecclesfield, is probably 'the wood on the balk or rein,' from ON *rein*, a balk or steep hillside.

STOTFOLD, STOTLEY.—Here the first element would seem to be the ME *stot*, a horse or bullock.

STOTFOLD, Hickleton, DB *Stotfald*, *Stotfalde*, KI 1285 *Stodfold*, PT 1379 *Stodefold*, has been influenced by OE *stōd·fald*, a stud-fold or paddock.

STOTLEY, Saddleworth, is 'the lea of the horse or bullock,' from OE *lēah*, a lea.

STOUPS.—See Staups.

STRAFFORD, the wapentake in which are Sheffield and Rotherham, derives its name from an ancient ford across the Don not far from Conisborough, where to-day we find the name Strafford Sands. Among early records are the following:

DB	1086 *Straforde, Strafford*	HR	1276 *Strafford*
DB	1086 *Strafforth*	KI	1285 *Strafford*
PF	1166 *Straford*	PT	1379 *Strafford*

Although there is no sign of a second *t*, the meaning is probably 'street ford,' from OE *strǣt*, a street or highway, and *ford*, a ford. The Berkshire name Straffield, DB *Stradfeld*, is explained as 'street field' by Skeat, who says the word Street in such cases 'commonly refers to a Roman Road.'

STRAND.—On the opposite bank of the Calder from Horbury, is a stretch of land called the Strands, derived either from OE *strand*, a strand, shore, or from ON *strönd*, a border, coast, shore. This word is in frequent use in ON place-names, witness the names *Strond* and *Skarth·strond*; it is also found in Shetland as Strand, and in Cumberland as Strands.

STRANGSTRY WOOD, Elland.—WCR 1394 has *Strang-stigh Wood* and WCR 1437 has *Strangstyes*, so we may without hesitation explain the name as 'the arduous path,' from OE *strang* or ON *strangr*, strong, hard, arduous, and OE *stīg* or ON *stīgr*, a path. The modern spelling shows the assimilation of the initial consonants of the second syllable to those of the first.

STREETHOUSE, Normanton, is on the line of the Roman road which passed from Pontefract through Featherstone and Agbrigg to Wakefield, and it is believed to owe its name to that fact, OE *strǣt*, a street, highway, from Lat. *strata*, being regularly applied to Roman roads. The first element in Street Side, Ossett, is held to have the same origin, as well as the termination in Tong Street and Adwick-le-Street.

STREETTHORPE, Hatfield, DB *Stirestorp*, HR 1276 *Stirtorp*. PT 1379 *Stirestrop*, means 'Styr's village,' Styr being a well-known personal name.

STRINES.—Though not so common as Slack or Storth, this Scandinavian word is found in several parts of South-west Yorkshire. It occurs in Scammonden, Denby, Hepworth, Saddle-worth, and near Sheffield. In addition, a farm in Shelley is called Hopstrines; a brook in Erringden is recorded by Watson in 1336 as *Southstrindbroc*; a farm near Heptonstall is mentioned in HW 1521 as *Stryndes*; and there was also, it would seem, a part of Northowram called by this name, for an entry in WCR 1352 speaks of *Le Stryndes*.   One of the eight petty kingdoms in the basin of Trondhjem Fiord was called Strind, and in the Landnama Book it is recorded of Eyvindr Vapna and Refr the Red that they came to Iceland 'from Strind in Throndheime.' The word comes from ON *strind*, a border, side, and is found as a termination in the name Hopstrines.

STUBB, STUBBING, STUBLEY.—These are derived from OE *stybb*, or ON *stubbi*, a stub or stump.   The word indicates, therefore, the former existence of woodland, just as do the words 'riding' and 'royd.'

STUBBING, which is very common, means 'stump meadow,' from ON *eng*, a meadow; HS 1637 has *Stubbing* in Bradfield.

STUBLEY, Heckmondwike, YD 1373 *Stublay*, 1375 *Stubelay*, means 'stump lea,' from OE *lēah*, a lea or meadow.

STUBBS, near Hampole, DB *Eistop*, KI 1285 *Stubbes Lacy*, shows Norman influence in the initial vowel of the DB spelling.

STUBBS WALDEN, near Pontefract, DB *Eistop*, YI 1244 *Stubbes*, NV 1316 *Stubbes*, shows the same influence.

STUDFOLD, STUDROYD.—See Stoodley.

SUDE HILL, Fulstone.—Although there are no early records of the name it seems extremely probable that it is of Celtic origin, cognate with Gael. *suidhe*, OIr. *suide*, a resting-place, a seat.   There are places in Ross and Cromarty called Suidh Ma-Ruibh, Malruba's seat, places where Malruba was accustomed to rest on his journeys; and Watson also records the name Suddy, 1227 *Sudy*, 1476 *Suthy*, explaining it as seat.   Hogan places on record about thirty examples of the name; compare *suidhe finn*, now Seefin, and *suidhe gabha*, now Seagoe.

SUGDEN, SUGWORTH.—In EDD the word *sog, sug, sugg*, is explained as a morass, or soft boggy ground.   It appears to be of Celtic origin; compare Welsh *sug*, Irish *sugh*, Gaelic *sugh*, which mean sap, moisture, and note the name Sug Marsh, Timble.

SUGDEN, Haworth, PT 1379 *Sugden, Sugdeyn,* WCR 1379 *Sugden,* may be interpreted 'the swampy valley,' from OE *denu,* a valley.

SUGDEN, Bradfield, has doubtless the same meaning.

SUGWORTH, Sheffield, YF 1540 *Sugworth,* is 'the swampy farm,' from OE *weorth,* a holding, a farmstead.

SUNDERLAND, Northowram and Hebden Bridge.— Probably connected with the former, WCR 1274 has *Sondreland* and *Sundreland,* and WCR 1286 *Sonderlande.* The origin is plainly the OE *sonderland,* sundered land, private property.

SUTCLIFFE, Hipperholme, WCR 1274 *Suthclif,* WCR 1297 *Sutheclyf,* is the 'south cliff,' from OE *sūð,* south, and *clif,* a cliff.

SUTTON, Campsall, DB *Sutone,* means 'the south farm or enclosure,' from OE *sūð* and *tūn*; compare WHS †1030 *Suðtune.*

SWAITHE, Worsborough, 1284 *Swathe,* 1313 *Swath,* is derived either from ON *svæði,* an open space, or from ON *svað,* a slippery place, a slide; compare Norw. *svad,* a mountain slope, bare rock.

SWINDEN, SWINEFLEET, SWINLEY, SWIN-NOW, SWINSEY, SWINTON.—Perhaps the first element is the personal name Suin recorded in DB, but more probably it is the OE *swīn,* ON *svín,* a pig. Reference is made in LN to the way in which an Icelandic valley became known as Swine-dale: 'Steinolf,' we are told, 'lost three swine, and they were found two winters later in Svina·dale.'

SWINDEN, Penistone, in early deeds *Swyndone* and *Swyndene,* is 'swine valley,' from OE *denu,* a valley, though one of its early forms is from OE *dūn,* a hill.

SWINEFLEET, on the Ouse, 1304 *Swynflet,* 1344 *Swynflete,* YF 1541 *Swynflete,* is perhaps 'swine channel,' from OE *flēot* or ON *fliōt,* a river or channel.

SWINLEY, Cleckheaton, appears to be 'swine lea.'

SWINNOW, Pudsey, CC *Swynhagh, Swynehagh,* is exactly paralleled by the OE *swīn·haga* and the ON *Svīn·hage,* a ' swine-enclosure.'

SWINSEY, Meltham, WCR 1307 *Swynstye,* is obviously ' swine sty,' from OE *stīgo,* a sty.

SWINTON, Rotherham, DB *Suintone* and *Swintone,* CR 1227 *Swinton,* KI 1285 *Swinton,* is ' swine enclosure,' from OE or ON *tūn,* an enclosure.

SWITHEN, SWITHENS, Darton and Sowerby, come from ON *sviðinn,* which is applied to places where the copse or heather has been burnt. At Bramley (Rotherham) a place of the same name is recorded in 1318 as *le Swythen,* and in the Lake District there is Sweden How.

Scores of names are to be found in Norway where the ON *sviðinn* is represented by *sveen,* among them Bergsveen, Kvernsveen, Langsveen, Nordsveen, and Sandsveen. Rygh explains *svið, sviða,* as a place which has been cleared by burning.

SYKE, SYKES, SYKEHOUSE.—The local name Sykes is to be found in Saddleworth, PT 1379 *Sykes,* and near Keighley; Sykehouse, YF 1555 *Sykhowse,* is near Thorne; Syke Fold is in Cleckheaton and Syke Lane in Sowerby. The etymology is from OE *sīc,* a runnel, or ON *sīk,* a ditch or trench.

TAME.—This stream rises in Saddleworth, and after flowing past Staleybridge, joins the Mersey at Stockport. Records of the 13th and 14th centuries have the spelling *Thame,* but an early deed given by Dodsworth has *Tome.* Other river-names apparently from the same root are the Cornish Tamar, the Staffordshire Tame, the Worcester Teme, the stream flowing through Tempsford in Bedfordshire, and the Thames. We find for the Thames *Tæmese·muth* in the AS. Chron. 892, *Tæmese forda* in the AS. Chron. 921 for Tempsford, DB *Tamedeberie* for Tenbury, and DB *Tameworde* for Tamworth. Hogan records no Goidelic river-name of similar form.

TANKERSLEY, Barnsley, DB *Tancreslei*, PC †1238 *Tancreslay*, YR 1252 *Tankerlay*, NV 1316 *Tankeresley*, is 'Tanchere's lea,' from OE *lēah*, and the personal name recorded by Searle.

TANSHELF, Pontefract, is connected with an interesting chapter in our early history. During the first half of the 10th century there were in Northumbria Viking rulers who threatened to make York once more the head of Britain; 'it needed,' as Freeman says, 'campaign after campaign, submission after submission, revolt after revolt, before the stubborn Dane finally bowed to his West-Saxon lord.' In 924 Edward of Wessex had succeeded in obtaining their submission, and in 925 Athelstan had recognised them by giving his sister in marriage to their king; but after Athelstan's death they threw off the yoke led by Anlaf of Ireland. In 944 Edmund expelled Anlaf and once more subdued his people, but when Edmund died they again revolted and chose Eric for king. Edmund's brother and successor immediately marched into Yorkshire and at Tanshelf once more received the submission of the Danes. The account in the Saxon Chronicle under the year 947 reads as follows : 'In this year came Eadred, king, to Taddenes·scylfe, and there Wulstan, archbishop, and all the Witan of the Northumbrians pledged their faith to the king. And within a little while they belied it all, both pledge and oaths as well.' We have here the earliest record of the name, namely, *Taddenes scylfe*. Later records are CR 1257 *Tanshelf*, YI 1258 *Tanesolf*, LC 1295 *Thanschelf*, DN 1362 *Tanshelfe*. The OE *scylf* means a ledge or shelf of land, and the name Tanshelf may be interpreted as 'Tadden's ledge' or 'Tadden's shelf of land.' Although the personal name Tadden is not recorded, we possess the forms Tade and Tado. See Pontefract, Shelf, and Adwick-le-Street.

THONG, TONG.—Upperthong and Netherthong are near Holmfirth, while Tong, a picturesque rural spot surrounded by great centres of industry, is on the borders of Bradford. Early spellings of the two names, together with records of the East Riding Thwing, are as follows :

WCR	1274	*Thwong*	DB	1086	*Tuinc*	DB	1086	*Tuenc*
WCR	1286	*Hoverthong*	PF	1203	*Tanga*	KI	1285	*Tweng*
WCR	1308	*Thounge*	CR	1232	*Tange*	KF	1303	*Tweng*
YI	1366	*Thwonge*	KF	1303	*Tong*	NV	1316	*Twenge*
YF	1575	*Thonge*	NV	1316	*Tonge*	CH	1339	*Tweng*

Three points should be noted : (1) the two Domesday forms, *Tuinc* and *Tuenc*, both show *nc* for *ng*; (2) the later forms of Tong show the same change as that exhibited in long, strong, wrong, which come from OE *lang, strang, wrang*; (3) the place-name Thong, like the common noun of the same form, comes from ME *thwong*, which is itself derived from OE *thwang*.

Doubtless the Domesday names for Tong and Thwing are variations of the same word, and go back like Thong to the Teutonic type, **thvangi*, a type which has given OE *thwang*, and ON *thvengr*, a thong or strap, and which is to be referred to a verbal form meaning to constrain. The post-Domesday forms of Tong on the other hand go back to OE *tange* or ON *tangi*, Dan. *tange*, a tongue of land.

When we examine the localities we find that three of the places, Tong, Upperthong, and Netherthong, possess similar characteristics : each consists of a spur given off by the main ridge of hills, and each is flanked by streams which unite where the spur runs out. Thwing is similar in being placed on a spur of the Wolds, but different in possessing only one stream. This stream, called the Gipsey Race, changes its course near Thwing from east to south, and thus makes a half circuit of the hill on which the village stands. It appears, therefore, that the meaning of Thong and Thwing is practically the same as that of Tong, namely, a spur or tongue of land.

Additional examples are Tong Lee in Marsden, and Tong Royd in Elland.

THORNE, THORNES, THORNBURY, THORN-CLIFFE, THORNHILLS, THORNSEAT, THORNTON.

—These are derived from the thorn, OE or ON *thorn*, which in the district south of the Aire has provided many place-names. To complete the list we must add Amblerthorn, Arbourthorn, Cawthorne, Thornhill, and Thurnscoe.

There is an interesting peculiarity in the DB forms due to Norman influence, namely, the substitution of *t* for initial *th*; compare Torp and le Torp, Norman place-names from ON *thorp* (Robinson). Among the examples in South-west Yorkshire where this substitution has taken place there are three names from the thorn, two Thorpes, Throapham, Thrybergh, Thurgoland, Thurlstone, and Thurstonland.

THORNE, Doncaster, DB *Torne*, YI 1276 *Thorne*, YS 1297 *Thorn*, IN 1335 *Thorne*.

THORNES, Wakefield, WCR 1275 *Thornes* and *Spinetum*.

THORNBURY, Bradford, comes from OE *burh*, a fortified post.

THORNCLIFFE occurs in Tankersley.

THORNCLIFFE, Kirkburton, PF 1202 *Thornotelegh*, PF 1208 *Thornetele*, WCR 1275 *Thorniceley*, YS 1297 *Thornykeley*, WCR 1307 *Thorntelay*, YD 1316 *Thornecley*, DN 1517 *Thornclay*, has seen a struggle between two forms of the first element, Thornic and Thornot.

THORNHILLS, Brighouse, is from OE *thornig*, thorny, and *healh*, a corner or meadow, witness the forms *Thornyhales* in WCR 1333, *Thornyales* in WCR 1339, *Thornyals* in WCR 1419. See Hale.

THORNSEAT, Bradfield, *Thorneset* in 1329, is the 'seat beside the thorn,' from OE *sǣte*, ME *sete*.

THORNTON, Bradford, DB *Torentone*, YI 1246 *Thornton*, HR 1276 *Thorenton* is the 'farm beside the thorn,' from OE *tūn*, an enclosure, homestead.

THORNHILL, Dewsbury, DB *Tornhil*, *Tornil*, PF 1175 *Tornhill*, PR 1190 *Tornhill*, YR 1234 *Tornhill*, YD 1292 *Thornhulle*, NV 1316 *Thornhull*, PT 1379 *Thornhill*, is the OE *thorn·hyll*.

Fragments of crosses discovered here are of extreme interest, and show that an ecclesiastical establishment existed on the spot centuries before the Norman Conquest. One of these fragments is believed to be a memorial of the King Osberht who was slain in battle by the Danes in the year 867, the year when first the Northmen invaded Yorkshire in force. The account of the AS. Chronicle is as follows: 'In this year the

(Danish) army went from East Anglia over the mouth of the Humber to York in Northumbria. And there was much dissension among the people (the Northumbrians), and they had cast out Osberht their king and had taken to themselves a king, Ælla, not of royal blood. But late in the year they resolved that they would fight against the (Danish) army, and therefore they gathered together a large force and sought the (Danish) army at the town of York, and stormed the town. And some of them got within, and there was immense slaughter of the Northumbrians, some within, some without, and both the kings were slain[1].'

THORPE.—This is one of the most interesting of our place-name elements. It is derived from a Teutonic stem *thurpa*, a troop, a host, a throng of people, a village; and is cognate with Lat. *turba*, a crowd of people, and OW *treb*, a house. From the stem *thurpa* come ON *thorp* and OE *thorp*, as well as OFris. *thorp*, *therp*, OHG *dorf*, Du. *dorp*. Place-names derived from this stem are common in Norway, Sweden, Denmark, Germany and Holland; compare the Danish names Ulstrup, Qverndrup, Skallerup; the Norwegian Nordtorp, Södorp; the Frisian Olterterp, Ureterp, Wijnjeterp; and the German Allendorf, Meldorf, Warendorf.

In South-west Yorkshire either alone or in composition there are at least sixty-three examples. Reasons for describing these as distinctively Danish have already been given (Chap. III, p. 33); but an additional series of names which give strong support to the argument may here be quoted, viz. the eight names containing -thorpe which are found near Sheffield: Bassingthorpe, Herringthorpe, Grimesthorpe, Osgathorpe, Netherthorpe, Renathorpe, Silverthorpe, Skinnerthorpe. In all these instances the first element may be Scandinavian, but in only two can it possibly be Anglian; in six it is certainly Scandinavian, but in none is it certainly Anglian.

THORPE, near Leeds, was DB *Torp*, KF 1303 *Thorp*.

THORPE IN BALNE was simply *Thorp* in 1150 and 1320.

---

[1] *YAS Journal*, IV, p. 416, VIII, p. 49.

THORPE AUDLIN, Pontefract, was *Torp* in DB, *Thorp* in NV 1316, and *Thorp Audelyn* in PT 1379. Audlin is doubtless derived from the DB name Aldelin.

THORPE SALVIN, on the Notts. border, was *Rykenildethorp* in HR 1276 and *Rikenildthorp* in KI 1285, but NV 1316 has *Thorp Salvayn* and PT 1379 *Thorp Saluayne*. It was held in 1285 by Radulphus Salvayn and in 1303 by Antonius Salvayn.

THORPE HESLEY, Kimberworth, was *Thorpe* in 1307.

In addition to the five Thorpes above named, there are two Thorpes in Sowerby, one in Idle, and one in Hoylandswaine, together with the following:—Alverthorpe, Armthorpe, Astonthorpe, Bassingthorpe, Chapelthorpe, Dowsthorpe, Edenthorpe, Edderthorpe, Finthorpe, Gannerthorpe, Gawthorpe (2), Goldthorpe, Grimethorpe, Grimesthorpe, Herringthorpe, Hexthorpe, Hillthorpe, Hollingthorpe, Kettlethorpe, Kirkthorpe, Ladythorpe, Leventhorpe, Lithrop (2), Milnthorpe, Minsthorpe, Moorthorpe, Netherthorpe (2), Noblethorpe, Northorpe (2), Norristhorpe, Osgathorpe, Ouchthorpe, Overthorpe, Painthorpe, Priestthorpe (2), Ravensthorpe, Rawthorpe, Renathorpe, Rogerthorpe, Scawthorpe, Shipmanthorpe, Silverthorpe, Skelmanthorpe, Skinnerthorpe, Snapethorpe, Streetthorpe, Throapham, Upperthorpe, Wilthorpe, Woodthorpe (2), Wrenthorpe.

Many of the above are doubtless post-Conquest. Ravensthorpe and Norristhorpe appear to have been created during the last century; Chapelthorpe and Noblethorpe have French prefixes; in Hillthorpe and Woodthorpe the prefixes are obviously Anglian; and Astonthorpe is a secondary formation.

THREAP CROFT, THREAPLAND, Ovenden and Pudsey.—The Cockersand Chartulary has the 13th century spellings *Threpridding* and *Trepcroft*, and among 12th and 13th century examples in Scotland Johnston notes *Trepewode* and *Threpeland*. Quotations in EDD read as follows: (1) ‘A long tract of land stretches southward which was formerly Debateable Land, or Threap Ground’; (2) ‘Part of Wooler Common is still undivided, owing to disputes; it is called Threap Ground.’ And, further, Johnston quotes from a 15th century truce between England and Scotland the expression

'The landez callid Batable landez or Threpe landez.' The prefix is from OE *threapian*, to reprove, correct; and Threapland is 'land about which there is dispute.'

THROAPHAM, Tickhill, DB *Trapun*, YD 1499 *Thropon*, VE 1535 *Thropon*, is by no means an easy word. The DB scribe gives *t* for *th* and *n* for *m* in accordance with his usual custom, but he also appears to have given *a* for *o*. If we may take the DB form as **Thropum*, a form agreeing with the modern name, the meaning will be 'the thorpes.'

THRUM HALL, Halifax and Rishworth.—The name Thrum means a border or edge; compare ON *thrömr*, the brim, edge, verge, MHG *drum*, and the Dutch place-names Drumt, 850 *Thrumiti*, 1200 *Drumthe*, and Dreumel, 893 *Tremile*, 1117 *Trumele*, 1226 *Drumel*.

THRYBERGH, Rotherham, is an interesting name of which early spellings are as follows :

DB	1086 *Triberge, Triberga*	KI	1285 *Tryberg*
PC	†1194 *Triberge*	NV	1316 *Trebergh*
DN	1200 *Triberg*	WCR	1375 *Thrybargh*

As the plural of ON *berg*, a rock or cliff, is *berg*, while the plural of OE *beorg*, a mound or hill, is *beorgas*, it would appear most satisfactory to explain the word as Scandinavian and equivalent to 'the three cliffs,' the first element being from ON *thrīr*, three. The word *berg* is extremely common in Norwegian place-names, and Rygh places on record such examples as Nordberg, Lundberg, Sandberg, and Steinberg. McClure suggests that Thrybergh may perhaps be the *Trimontium* of Ptolemy.

THURGOLAND, THURLSTONE, THURSTON-LAND.—These names may well be taken together. They are all of Scandinavian origin, and in DB they all show initial *t* for *th*—due to Norman scribes.

DB	1086 *Turgesland*	DB	1086 *Turulfestune*	DB	1086 *Tostenland*
PF	1202 *Turgarland*	YI	1298 *Thurlestone*	PF	1202 *Thurstanland*
CH	1294 *Thorgerland*	YD	1301 *Thurleston*	WCR	1284 *Thorstanlande*
PT	1379 *Thurgerland*	PT	1379 *Dhurleston*	YS	1297 *Thurstanland*

Among the personal names in the Domesday record we find Turgar, Turulf, and Turstan. These are obviously the names

VIII] ALPHABETICAL LIST OF NAMES 283

we require to explain the three place-names, and it is clear they come from ON Thorgeirr, Thorolfr, and Thorsteinn. The terminals are ON *land*, land, an estate, territory, and ON *tūn*, an enclosure or farmstead.

THURNSCOE, Doncaster, appears to have been particularly troublesome to early scribes. Early forms are as follows:

DB 1086	*Ternusc, Ternusch*	YR 1269	*Tihirneschouth*
DB 1086	*Dermescop*	HR 1276	*Thirnnesch'*
CR 1187	*Tirnescogh*	CR 1280	*Thirnesco*
PF 1190	*Tirnesco*	NV 1316	*Thirnescogh*

The latest of these forms is quite the most accurate, CR 1187 *Tirnescogh* only failing because the initial is *t* instead of *th*. Despite the extraordinary variations in the name its meaning is quite plain, 'thorn wood,' from ON *thyrnir*, a thorn-tree, and *skōgr*, a wood.

THWAITE.—This word is characteristic of the districts settled by the Norsemen. It is derived from ON *thveit*, a parcel of land cleared of wood, an outlying cottage with its paddock, and corresponds to the Norwegian *tveit* and the Danish *tved*. Flom tells us that in Norway *tveit* is far more common than *tved* in Denmark. A small map of Denmark which shows dozens of thorpes—among them Ingstrup and Tulstrup, Skallerup and Dallerup, Tamdrup and Qverndrup—has only two thwaites, Nestved and Egtved ; and Lincolnshire, with its great mass of Danish names, has only one thwaite (Streatfeild). When used as a suffix in our English place-names the word sometimes takes upon itself quite extraordinary forms, such as -fitt as in Gumfitt (Gunthwaite) and Langfitt (Langthwaite), and -foot as in Follifoot and Moorfoot.

South of the Aire we find twenty-six examples of the word. There are four Thwaites, two Braithwaites, two Gilthwaites, two Linthwaites, and two Slaithwaites. There are also Alderthwaite, Birthwaite, Butterthwaite, Falthwaite, Gunthwaite, Hornthwaite, Huthwaite, Langthwaite, Ouselthwaite, and Woolthwaite, as well as Burfitts, Garfitt, and two Linfitts. In addition there are certain ancient names apparently not now in use, Brigthwaite, Micklethwaite, Oggethwaite, Salthwaite, and Thunnethwaite.

THWAITE, Leeds, is recorded in HR 1276 as *Rothewelletwayt*, and in RPR 1673 as *Thwaite*.

THWAITE, Ecclesfield, is *Thwayt* in PT 1379.

THWAITES, Keighley, was *Twhaytes* in YI 1303, *Thwaythes* in PT 1379, and *Thwayts* in YF 1558.

THWAITE HOUSE is mentioned in YF 1550 in connection with Firbeck as *Twaite*, and in 1576 as *Thwaite*.

TICKHILL, in the extreme south, has the remains of a priory and castle, the latter on the site of an ancient fortified mount. In DB it is called *Dadesleia*, a name still to be recognised in Dadsley Well; but PR 1130 has *Tykehull*, PR 1161 *Tichehill*, CR 1232 *Tikehull*, WCR 1309 *Tickehill*. The etymology is very doubtful. The first syllable may possibly be from ON *tīk* which gives OE *tike*, a dog; or it may be from some such personal name as OE Tica or Ticca, recorded by Searle. Tickenhall, Staffordshire, is *Ticenheale* in an early charter, and Ticknall, Derbyshire, is *Ticenheal*, 'the kid's meadow,' while Tickenhill in Worcester is explained as 'the kid's hill.' There are other village-names of similar type, for example, Tickton and Tickford, Tickenham and Tickenhurst.

TILTS, Doncaster and Thurgoland.—An undated inquisition dealing with the former speaks of *Langethauit* and *Thils*; another inquisition, dated 1304, gives the form *Tilse*; and in 1602 we find *Langfitt cum Tilse*. It seems clear that the second *t* in Tilts is intrusive, and the source of the name appears to be OE *thille*, a plank, a stake; compare Icel. *thilja*, a plank, Sw. *tilja*, a plank, floor. Thus the meaning is 'the planks'; compare the Norw. dialect-word *Skjeldtile*, a plank-way.

TINGLEY, TINSLEY.—The latter is recorded in DB as *Tineslauue, Tirneslauue*, but the former finds no place in that valuable survey. Later spellings are as follows:

WCR	1284	*Tyngelowe*	PR 1103	*Tineslei*
WCR	1296	*Thyngelawe*	YR 1230	*Tineslawe*
WCR	1308	*Thinglowe*	KI 1285	*Tinneslawe*
YF	1551	*Tynglay*	YS 1297	*Tyneslowe*
YF	1558	*Tynglawe*	NV 1316	*Tynneslawe*

The termination *lawe, lowe,* comes from OE *hlǣw,* a burial-mound, cairn, hill, while *ley* comes from OE *lēah,* a lea or meadow. Both names, like Ardsley, Blackley, Dunningley, show -ley where early records give -lawe or -lowe ; but in the case of Tinsley the facts point rather to selection than substitution.

TINSLEY, Sheffield, is 'Tynne's lea,' but it seems probable that in early days the names Tynneslei and Tynneslawe existed side by side, the latter signifying 'the burial-mound of Tynne.'

TINGLEY, Morley, means 'the lea of the Thing,' that is, 'Assembly field,' from ON *thing*; but the earlier form *Thynge-lawe* meant 'the mound of the Thing.' Everything points to the fact that Tingley was once a great meeting-place for the freemen of the neighbourhood. It is situated at the point where two great roads intersect, the road from Dewsbury to Leeds, and that from Bradford to Wakefield. The hill from which its earlier name was derived, and where doubtless the annual meetings were held, is a prominent object. But more remarkable perhaps is the existence of a notable fair, held close at hand, which in all probability owes its origin to these very meetings. Lee Fair, as it is called, is known throughout the Riding. It is described as a horse, cattle, and pleasure fair ; and it is held annually on two separate dates, the 24th of August and the 17th of September— 'the former and the latter Lee.'

At some period after the conquest by the Danes in 867 the existing divisions of Yorkshire were transformed. It was then that the county was divided into ridings, and the ridings into wapentakes. York remained outside the ridings in a position of unchallenged supremacy, but each riding had its own centre, the North as it would seem at Northallerton, the East at Beverley, and the West at Wakefield. This being so, we should expect to find the meeting-place of each riding within a short distance of its capital, and when everything is considered the suggestion that Tingley was the meeting-place for the West Riding can scarcely be seriously contested. For the East Riding it may well have been (as suggested in the *Victoria County History*) at Craikhow near Beverley, and for the North Riding I suggest Fingay Hill (RC *Thynghou*) about five miles from Northallerton[1].

---

[1] But see *Victoria County History of Yorkshire*, II, 134.

TODMORDEN, on the western border, LAR 1247 *Tot-mardene, Tottemerden,* WCR 1298 *Todmereden,* WC 1329 *Todmarden,* LI 1396 *Todmereden,* HW 1521 *Todmereden,* is a name of three elements.   First, there was the two-stem name Totmar, and afterwards came the three-stem form Totmardene. The third element is obviously from OE *denu,* a valley; and the second is most probably from OE *mere,* a pool, lake, or marsh. But the first is not so simple.   It may possibly be the OE personal name Tota, Totta; or it may be the OE *tote,* a tuft of grass, a heap, an eminence.   See Leithaeuser who gives ON *tota,* a peak, and a corresponding MLG form *tote.*   We may explain Todmorden as 'the valley of Totmar,' while Totmar is probably 'hill-marsh' or 'hill-lake.'

TOFT, TOFTSHAW, TOPCLIFFE.—The word 'toft' is of Scandinavian origin; compare ON *topt* (pron. *toft*), Dan. *toft.* It means a croft, a field, a cleared space for the site of a house, a homestead.   As the name of a village the word occurs in Cheshire, Lincoln, Norfolk, Cambridge and Warwick; in Normandy it is often found as a suffix in the form -tot, as in Yvetot, Ivo's toft, and Langetot, long toft.   On the other hand, according to Canon Taylor, it is very scarce in Norway and Westmorland, and quite unknown in Cumberland.   The word appears, therefore, to be Danish rather than Norwegian.   In the West Riding it occurs chiefly as a field-name, as for example at Pudsey, Cleckheaton, Liversedge, Morley, Lofthouse, and Hunshelf; but it occurs also in Eastoft and the two names following:

TOFTSHAW, Hunsworth, PT 1379 *Toschagh* and *Thofthagh,* has an Anglian termination, from OE *sceaga,* a copse, small wood.

TOPCLIFFE, Morley, WCR 1296 *Tofteclive,* WCR 1297 *Tofteclyve* and *Thofteclyf,* has for its termination the ON *klif,* a· cliff.

TOM HILL, Oxspring, may perhaps be derived from the Celtic *tom,* a hillock, Ir. Gael. Welsh *tom.*   Tomdow in Argyll is *tom-dubh,* the black hillock.

-TON, -TOWN.—The long vowel of *tūn* was shortened in compounds, and the word was written *ton* as in Newton, or *tun* as in Tunstead; but when it stood alone it gave ME *toun*, later *town*; hence place-names with the termination 'town' are comparatively late. In 1375 the three hamlets of Liversedge— Hightown, Roberttown, Littletown—were called Great Lyversegge, Robert Lyversegge, and Little Lyversegge, and as late as 1564 the names Great Lyversege, and Little Lyversage occur.

The original meaning of the OE *tūn* was an enclosure, a place surrounded with a bank or hedge, the word being connected with the verb *tȳnan*, to fence, to hedge in. Hence the name Barton meant an enclosure for corn, and Appleton an apple orchard. Subsequently the word denoted a homestead, a farmhouse with all its belongings; and last of all it took the signification town or village.

It is usual for place-names in -ham to have as their first element the name of the settler who first made there his home, but those in -ton are more commonly preceded by an adjectival term descriptive of the local situation or its general character, as in the case of Aston, Clayton, Newton, and Norton; yet at the same time there are many such names as Royston, Rore's homestead, and Silkstone, Sylc's homestead.

TONG, TONGUE.—See Thong.

TORNE is the name of a small stream which passes through Rossington and Auckley. In KC 1187 we find the name *Tornwad*, that is, Torn-wath; and KC makes further reference to the stream in the phrase 'in aquam magnam que vocatur Thorn.' The word must be compared with the first element in *Turnocelum*, an early Celtic name in the North of England (Williams); with *Tornolium*, an early form of the French place-name Tournoël (Williams); and with *Tornepe*, the 12th century form of the Flemish river-name Tourneppe (Kurth). In the last example the terminal comes from the Celtic *-apa*, a word cognate with Lat. *aqua* (Stokes). It seems fairly certain, therefore, that Torne is of Celtic origin. Compare the name with Balne, Colne, and Dearne.

TRANMORE, Balne, appears in CR 1305 as *Tranemore*, while BM has *Tranemoore*. The prefix represents ON *trana*, a crane, a bird formerly abundant in Great Britain, and prized as food, but now extinct; compare Sw. *trana* (for *krana*), Dan. *trane* (for *krane*) and OE *cran*, a crane. The Scandinavian 'tran' appears in many Yorkshire place-names, including several Tranmires and Trenholmes; but in Tranby the first element is probably the personal name Trani recorded by Nielsen. The English 'cran' appears in such names as Cranbrook, Cranborne, Cranfield, and Cranford.

TRETON, Rotherham, DB *Tretone, Trectune*, PF 1204 *Treton*, KI 1285 *Tretthon*, NV 1316 and PT 1379 *Treton*, is probably 'tree farmstead' from OE *trēo*, a tree, and *tūn*, an enclosure or farmstead; compare the OE names *trēow · steall* and *trēow · stede* recorded by Middendorff.

TRIANGLE, a hamlet in Sowerby, appears to have obtained its name from a triangular plot of ground situated in the acute angle where two roads meet.

TRIMINGHAM, Halifax, is recorded as *Trimingham* in WCR 1274 and 1275 and *Trymyngham* in WCR 1307. Its first element must be compared with the personal name Trimma recorded by Searle.

TRIPPEY, Liversedge.—This name may perhaps be connected with the Icel. *threp, threpi*, which meant a ledge, rising ground, an eminence. Aasen connects with this ON word the Norwegian word *trip* which has similar meanings.

TRUMFLEET, Doncaster, PF 1203 *Trumflet*, DN 1322 *Trumflet*, DN 1360 *Trunsflete*, DN 1361 *Trumflete*, is either 'border channel,' from Norw. *trum*, a border, edge, and ON *fljōt*, a channel, or 'stump channel,' from OE *trum*, a tree-stump, and OE *flēot*, compare *wyrttrum*, a word occurring in BCS. See Thrum.

TUDWORTH, Doncaster, DB *Tudeuuorde*, is 'the holding of Tuda,' Tuda being a well-known name.

TUNSTEAD, Saddleworth and Cleckheaton, corresponds to the OE *tūn·stede*, a townstead, the site of a farmstead or village.

TWISTLE, TWIZLE.—The OE *twisla* meant a confluence, the fork of a river or road. It corresponds to ON *kvisl*, a branch or fork of a tree, and it occurs in Twizle Clough, Holme, in Briestwistle near Thornhill, in Wightwizzle near Penistone, and in the name *Breretwisel* near Wath-on-Dearne.

TYERSALL, Pudsey, has been taken for a descendant of early forms like those of Teversall, Notts; but a comparison of the recorded spellings makes the matter quite plain.

PF	1203	*Tireshale*	DB	1086	*Tevreshalt*
KC	1267	*Tyrissale, Tyrsale*	YR	1275	*Thiversold*
HR	1276	*Tirsal*	YR	1280	*Tyversolde*
KF	1303	*Teresall*	VE	1535	*Teversholt*
PT	1379	*Tyrisall, Tiresall*	VE	1535	*Teversall*

The second element, which is best shown in PF 1203 *Tireshale*, comes from OE *healh*, a corner or meadow. The first element is a personal name, probably *Tȳr*; compare the modern surname Tyers.

UGHILL.—See Gilcar.

ULLEY, Rotherham.—If we remember that the Domesday scribes often wrote *o* for *u*, we shall find the early spellings very consistent with one another.

DB	1086	*Ollei, Olleie*	YS	1297	*Ullay*
YD	1253	*Ullay*	NV	1316	*Ullay*
KI	1285	*Ulley*	YD	1323	*Ulleye*

Dr Moorman gives a 13th cent. spelling *Ulflay* which may be interpreted 'wolf lea' and seems quite decisive. But doubts arise when we compare the early forms with those of Woolley, Wooldale, Woolrow, Woolthwaite, where the *f* of OE *wulf* or ON *ulfr* appears quite regularly down to the end of the 13th cent. Probably the correct interpretation is 'Ulla's lea.' In that case instead of an early form *Ulle·lei* we find *Ullei*, the *l* in *lei* having coalesced with that in *Ulle* at a very early date; compare Methley, Owston and Shafton.

UNDERBANK, UNDERCLIFFE, are examples of a small class of place-names formed by means of a preposition and a noun. The former occurs in Hunshelf, the latter in Bradford.

UPPERTHONG.—See Thong.

UPPERTHORPE is in Hallam.

UPTON, Badsworth, DB *Uptone*, KC 1218 *Opton*, NV 1316 *Uppeton*, PT 1379 *Vpton*, is derived from OE *up*, up, upwards, and *tūn*, an enclosure, farmstead. The meaning is simply 'high farm or enclosure.'

UTLEY, Keighley, DB *Utelai*, KI 1285 *Utteley*, PT 1379 *Uttelay*, *Vtlay*, appears to be 'the lea of Uta or Utta,' both forms of the personal name being recorded by Searle.

VISET, Hemsworth, DN 1555 *Biset*, is recorded by Clarke in 1828 as *Visit*, and the transformation in the name appears to be due to the influence of the common word 'visit.' The original name probably meant 'the seat of Bisi,' from the personal name recorded by Searle, and OE *set*, a seat, entrenchment, camp. Viset was for a time the home of Roger Dodsworth.

WADSLEY, WADSWORTH, WADWORTH, situate respectively near Sheffield, Hebden Bridge, and Doncaster, have the following early records :

DB 1086 *Wadesleia*	DB 1086 *Wadesuurde*	DB 1086 *Wadeuurde*
HR 1276 *Waddesley*	HR 1276 *Wadewyrth*	PR 1190 *Wadewurde*
YS 1297 *Wadeslay*	WCR 1307 *Waddeswrth*	PF 1202 *Waddewurth*
CR 1311 *Waddesley*	PT 1379 *Waddesworth*	KI 1285 *Waddeworth*

In the first element of Wadsley and Wadsworth we have the strong form Wade, genitive Wades, and in that of Wadworth the weak form Wada, genitive Wadan, both recorded by Searle. The meaning of Wadsley and Wadsworth, is, therefore, 'the lea of Wade' and 'the farmstead of Wade,' from OE *lēah* and *weorth*, while the meaning of Wadworth is 'the farmstead of Wada.'

WAKEFIELD.—There is no lack of post-Conquest records, of which the following is a typical selection :

DB	1086	*Wachefeld, Wachefelt*	WCR	1286	*Wakefeud*
PR	1103	*Wakfeld*	WCR	1298	*Wakefeud*
YR	1270	*Wakefeld*	NV	1316	*Wakefeld*
WCR	1274	*Wakefeud*	PT	1379	*Wakefeld*

Wakeley in Hertfordshire, DB *Wachelie*, is explained by Dr Skeat as 'the lea in which wakes were formerly held,' from OE *wacu*, a wake, vigil, an annual village merry-making ; and Wakefield may certainly have a similar meaning. On the other hand the correct interpretation may be 'Waca's field,' where the personal name is a weak form corresponding to Uach recorded in LV, and Vakr given by Naumann. Other place-names with the same prefix are Wakeham in Dorset and Wakehurst in Sussex.

On every side of Wakefield there is marked evidence of Danish occupation and settlement. No other town in South-west Yorkshire shows in its vicinity so large a number of 'thorpes.' Though some of these, like Chapelthorpe, are certainly post-Conquest, others very probably go back to the 10th century. Among the names of Scandinavian origin we may enumerate Ackton, Agbrigg, Altofts, Alverthorpe, Blacker, Carlton, Carr Gate, Cluntergate, Dirtcar, Flanshaw, Foulby, Gawthorpe, Gill, Hesketh, Hollingthorpe, Kettlethorpe, Kirkthorpe, Laithes, Lofthouse, Milnthorpe, Nooking, Normanton, Ouchthorpe, Painthorpe, Skitterick, Snydale, Snapethorpe, Thorpe, Woodthorpe, and Wragby.

Within a radius of about ten miles the meeting-places of five Wapentakes are clustered together. The sites of four—Agbrigg, Staincross, Morley, Skyrack—are known, and that of the fifth, though uncertain, must have been in the neighbourhood of Castleford and Pontefract, and so within the radius mentioned. Thus in every case the meeting-place must have stood at the extremity of the Wapentake nearest Wakefield.

Remembering that the Ridings are of Scandinavian origin, that Wakefield is the traditional capital of the West Riding, and that it still possesses the Registry of Deeds for the Riding, and recalling the points mentioned in the note on Tingley as well as those recounted in the paragraphs immediately preceding this, we shall come to the conclusion that Wakefield was most probably the Viking capital of the West Riding, and that, therefore, it was also a place of importance long before the Viking Age.

'The manor of Wakefield is very extensive, possessing a jurisdiction stretching from Normanton to the edge of Lancashire, and including the lordship of Halifax ; it is more than 30 miles in length from east to west, and comprises 118 towns, villages, and hamlets' (Clarke, 1828). From east to west the diocese of Wakefield, formed in 1888, has almost exactly the same extent ; it includes, however, many townships not in the ancient manor.

WALDERSHELF, WALDERSHAIGH, WALDERSLOW, north-west of Sheffield.—For the first DB gives *Sceuelt*, BD 1290 *Waldershelfe*, YD 1302 *Walderschelf*, YD 1307 *Walderschelf*. The first element in each is the OE name Wealdhere, army-wielder, and the endings come from OE *scylf*, a shelf or ledge, OE *haga*, an enclosure, homestead, OE *hlǣw*, a cairn or burial-mound.

WALES, WALESWOOD, WALSH, WALSHAW, WALTON.—Early spellings of Wales, Walton, and Walshaw are as follows :

DB 1086 *Wales, Walise*	DB 1086 *Waleton*	WCR 1277 *Wallesheyes*
HR 1276 *Wales*	KI 1285 *Walton*	PT 1379 *Walschagh*
KI 1285 *Weles*	NV 1316 *Walton*	HW 1543 *Walshaye*
PT 1379 *Wales*	PT 1379 *Walton*	HW 1549 *Walshay*

These names possess peculiar interest; they refer to the presence of Britons living side by side with the Anglian settlers. The OE word *wealh*, meant a foreigner, a Briton ; and in the nom. pl. its form was *wealas* or *walas*, the gen. pl. being *weala* or *wala*.

WALES, Rotherham, means 'the Britons,' from OE *wealas*. Its origin is exactly the same as the name of the country, which, like Norfolk and Suffolk, first referred to the people, and afterwards to the place where they dwelt.

WALESWOOD, Rotherham, YD 1311 *Walaswod*, YD 1326 *Waliswode*, is formed from the previous name, Wales, and the

OE *wudu*, a wood. Its meaning is simply 'the wood near Wales.'

WALSH, a group of cottages in Gomersal, probably comes from the OE adjective *Wælsc*, foreign, British, Welsh.

WALSHAW, Hebden Bridge, may fairly be explained as 'the copse of the Britons,' from OE *sceaga*, a small wood, and *weala*, gen. pl. of *wealh*.

WALTON, Wakefield, appears to represent OE *Weala·tūn*, 'the farmstead of the Britons,' from OE *tūn*, an enclosure or farmstead. Of Walden in Herts., DB *Waldene*, HR *Waledene*, Dr Skeat says 'The spelling with -le- is to be noted, as it shows that the name begins neither with AS *weald*, a wood, nor with *weall*, a wall. In fact, it precisely agrees with AS *Wealadene*, dative case of *Wealadenu*.' After explaining Walden as 'the valley of strangers,' Dr Skeat concludes by saying 'we here find a trace of the Celts.'

WALTON CROSS, Liversedge, where there is the base of an ancient cross, possibly of the 8th century, has the same origin and meaning.

WALKLEY, Sheffield, 1270 *Walkeley*, 1285 *Walkeleye*, HH 1366 *Walkelay*, PT 1379 *Walkmylne*. It seems clear that *Walkeley* has for its first element a weak personal name. The patronymic of corresponding form is found in the West Riding Walkingham, DB *Walchingeha'*, and in the East Riding Walkington, DB *Walchinton*, NY *Walkyngton*. Hence Walkley may be explained as 'the lea of *Wealca'; compare the Frisian name Walke recorded by Brons. The name *Walkmylne* on the other hand means 'fulling mill,' from OE *wealcan*, ME *walke*, to roll, revolve. From this OE word we get OE *wealcere*, ME *walker*, a fuller of cloth ; hence the personal name Walker. Perhaps *Wealca is of cognate origin.

WALL, WELL.—These words may fairly be taken together because of the instances where variation between one and the other is to be found, as, for example, in the field-name White Walls or White Wells, which occurs in Austonley, Dinnington, Ovenden, Silkstone, and elsewhere. In Lancashire there are several ancient names which present this phenomenon :

Aspinall,   1244 *Aspiwalle*   1247 *Aspenewell*
Childwall,   1224 *Childewal*,   11th c. *Cheldewell*
Halliwell,   1292 *Haliwall*,   1246 *Haliwell*
Thingwall,   1346 *Thingewall*, 1228 *Thingwell*

When the word is Scandinavian, variation of this kind can be fully accounted for, ON *vǫllr*, a field or plain, having a stem of the form *vall-* and a dat. sing. *velli*. But as this variation sometimes occurs where the first element is obviously Anglian—as in the case of Churwell, 1226 *Cherlewall*, 1296 *Chorelwell*—we find ourselves beset with difficulties. Perhaps (1) the common word ' well ' has been influenced by ON *vǫllr*; perhaps (2) it has been influenced by OE *weall*; perhaps (3) there is a variant of ' well ' having the form ' wall '—compare OFris. *walla*, a spring, and Dan. *væld* (for *væll*).

South-west Yorkshire has the following names where the source seems clearly OE *well*, *welle*, *wiell*, a well, spring, fountain : Birdwell, Churwell, Dudwell, Hollingwell, Ludwell, Mapplewell, Oakwell, Ouzlewell, and Spinkwell. Names probably Scandinavian are Braithwell, Heliwell, and Purlwell.

WALSH, WALSHAW, WALTON.—See Wales.

WARBURTON, WARLAND, WARLOW PIKE, WARSIDE.—Warburton occurs in Emley, Warland near Todmorden, Warlow Pike in Saddleworth, Warside in Ovenden. Without early forms of these names it is quite impossible to give definite explanations. Yet among the various sources from which the first element may come there are two much more likely than any other, namely, OE *weard*, ME *ward*, a guard or watchman, and ON *varða*, a beacon, a pile of stones, or cairn. The former occurs in Warborough, Oxfordshire, formerly *Weard·burg*, and may well occur in Warburton, the combination *rdb* becoming *rb* quite regularly. But the ON *varða*, later *warthe*, might occur in any of the four names, for the *th* would readily disappear.

WARDSEND, Ecclesfield, HH 1235 *Wereldsend*, YD 1323 *Werldishende*, HH 1366 *Werlsend*, PT 1379 *Werdeshend*, *Wardeshend*, ' world's end,' from OE *weoruld* or ON *vereld*, world, and OE *ende* or Dan. *ende*, end, quarter, district.

WARLEY, Halifax, provides an excellent example of the weakening of the unaccented syllable and its final loss.

DB	1086 *Werlafeslei*		WCR	1342 *Warleley*
WCR	1274 *Werloweley*		WCR	1345 *Warlilley*
WCR	1286 *Werloley*		WCR	1372 *Warlullay*
WCR	1309 *Werlouleye*		WCR	1374 *Wherolay*
WCR	1326 *Warouley*		WCR	1442 *Warley*

Spellings strikingly different from the above are

NV	1316 *Warlowby*		PT	1379 *Warbillay*

The Domesday form has been the subject of much discussion. In the enumeration of the lands of the King, although the berewicks of Wakefield are said to be nine, the Domesday record names only eight : *Sandala, Sorebi, Werlafeslei, Micleie, Wadeswurde, Cru'betonestun, Langefelt, Stanesfelt*. In order to make up the nine *Werlafeslei* has been divided into *Werla* and *Feslei* ; but obviously WCR 1274 *Werloweley* could not come from *Werla*. Further, copying from an early document Watson gives the following : '*Manerium de Wachfielde et ville de Sandala, Warlefester, Medene, Wadesworth, Crigestone, Bretone, Orberie, Oslesett, Stanleie, Scelfetone, Amelie, Seppleie, Scelveleye, Cumbreword, Crosland, Holme, Halifaxleie, et Thoac.*' In the form *Warlefester* we have the best possible support for DB *Werlafeslei*, even though the ending is different, and we may safely interpret Warley as 'the lea of Wærlaf,' while *Warlowby* is 'the farmstead of Wærlaf,' and *Warlefester* is 'the place of Wærlaf,' from ON *bȳr* and *staðr*. The form given in PT, *Warbillay*, appears to be merely a scribal error. Note the loss of the sign of the genitive, the loss of the first *l* through dissimilation in WCR 1326 and 1374, and the change from four syllables to three and then from three to two.

WARMFIELD, WARMSWORTH, Wakefield and Doncaster.—Early spellings of these names are plentiful, and tell their tale with sufficient clearness.

DB	1086 *Warnesfeld*		DB	1086 *Wermesford, Wemesforde*
RC	1215 *Warnefeld*		HR	1276 *Wermesworth*
YR	1252 *Warnefeld*		YS	1297 *Wermesworth*
NV	1316 *Warnefeld*		NV	1316 *Wermesworth*
VE	1535 *Warmefeld*		VE	1535 *Warmesworth*

WARMFIELD is 'the field of Wærn,' where the strong form
*Wærn corresponds to the weak form Wærna (Searle); compare
BCS *Wærnan·hyll* and *Wernan·broc*.
WARMSWORTH on the other hand is 'the farmstead of
*Werm,' such a personal name being assured by the patronymic
in Warmingham, Warminghurst, Warmington.

WATH-ON-DEARNE, DB *Wade, Wate, Wat*, YR 1234
Wath, KI 1285 *Wath*, NV 1316 *Wath*. This is from ON *vað*,
a wading place, ford. The word is found elsewhere qualified
by various prefixes; there are, for example, Sandwath and
Langwath, sand-ford and long-ford.

WELBECK, Stanley, WRM 1391 *Wilbyght, Wilbytht*,
appears to mean 'willow bend,' from OE *wilig*, ME *wilwe*,
willow, and OE *byht*, a bend, an angle. The name refers to
a great bend in the Calder opposite Kirkthorpe.

WELL.—See Wall.

WELLINGLEY, Tickhill, RC 1231 *Wellingleye*, YD 1374
Welyngley, YF 1494 *Wellyngley*, is 'the lea of Welling or the
Wellings.' This OE patronymic appears in Welling, the name
of a village in Kent; it also appears in Wellingham, Welling-
borough, and the four Wellingtons.

WENT, WENTBRIDGE.—See Chevet.

WENTWORTH, Sheffield, DB *Winteuuord, Winteuuorde,
Wintreuuorde*, YR 1234 *Wintewrth*, YI 1252 *Wintewrde*, YI
1308 *Wynteworthe Wodehous*. In Cambridge there is a second
Wentworth, DB *Wintewrde*, derived according to Professor
Skeat from the OE personal name Winta and OE *weorth* and
explained as 'Winta's farmstead.' This may well be the interpretation of our Yorkshire Wentworth; but see Went.

WESTERTON, Ardsley near Wakefield, PT 1379 *Westerton*, is probably 'the farm more to the west.'

WESTFIELD ROAD, Wakefield.—Although this name shows no sign of antiquity and awakens no desire to probe its history, it carries us back a thousand years and more. The district to which Westfield Road leads was in olden days the 'common field' of Wakefield. According to custom this common field was divided into three divisions to agree with a threefold rotation of crops. The names of the divisions were Cross Field, Middle Field, and West Field, and it is from the last of these that the modern road obtains its name.

The position of the three fields is shown in a map of Wakefield dated 1728. In this map we can see something of the larger divisions of the open field, something of the isolation of strip from strip in the possessions of one individual, and also something of the coalescing which gradually took place.

WESTNAL.—Bradfield had formerly four divisions, Waldershelf, Dungworth, Bradfield, and Westmonhalgh or Westnal. In YAS we find 1329 *Westmundhalgh*, 1335 *Westenhalgh*, 1380 *Westmundhalch*, 1398 *Westmonhall*, and YF 1560 has *Westmanhaugh*. As PF 1166 records the name Westmund we may explain Westnal as 'Westmund's corner,' from OE *healh*, a corner or meadow; see Hale.

WHAM.—In the Colne Valley there are Broad Wham, Cabe Wham, and Fore Wham; near Holmfirth, the Wham and Boshaw Whams; near Hebden Bridge, Whams Wood; and the name is also found in Fulstone, Thurlstone, Erringden and Golcar. EDD explains the dialect-word 'wham' which occurs in the Northern counties as a swamp, a marshy hollow, a dale among the hills, a hollow in a hill or mountain. According to the same authority the source of the word is ON *hvammr*, a grassy slope or vale.

WHARNCLIFFE, Sheffield, is probably 'mill cliff,' from OE *cweorn* or ON *kvern*, a mill, and OE *clif* or ON *klif*, a cliff; see Quarmby.

WHEATCROFT, WHEATLEY.—The first occurs in Ecclesfield, the second in Ovenden, WCR 1307 *Queteleyhirst*, and near Doncaster. Early spellings of the last are DB *Watelage*, YI 1279 *Waitele*, CR 1280 *Whetelagh*, YD 1394 *Qwhatelay*. The meaning is 'wheat lea,' from OE *hwǣte* and *lēah*.

WHELDALE, Pontefract.—One naturally divides the word thus, Whel-dale ; but such a division is topographically unlikely, and raises up difficulties in regard to the prefix. Early spellings are DB *Queldale*, PC 1240 *Queldale*, IN 1252 *Weldale*, NV 1316 *Queldale*, and the meaning is 'Cweld's corner' from OE *healh*, and the known personal name Cweld or Kveld. Compare the names Beal and Roall found in the immediate neighbourhood.

WHIRLOW, Sheffield, is recorded in 1501 as *Hurlowe*. There is conflict between the spellings, but the termination is certainly from OE *hlāw*, *hlǣw*, a burial mound or hill.

WHISTON, Rotherham, appears on the one hand as DB 1086 *Witestan*, YR 1280 *Wytstan*, KI 1285 *Wytstan*, YD 1342 *Whitstan*, 'white stone,' from OE *hwīt*, white, and *stān*, stone, and on the other hand as DB 1086 *Widestan*, PF 1196 *Wiᵹestan*, YR 1270 *Withstan*, where the first element may perhaps represent the gen. of the personal name Viði recorded by Naumann. Other forms like YD 1306 *Wystan*, YD 1314 *Wistan*, YD 1377 *Whystan*, spring naturally from either of the forms before mentioned.

WHITCLIFFE, WHITGIFT, WHITLEY, WHIT-WELL, WHITWOOD, WHITECHAPEL, WHITE-HAUGHS, WHITE LEE, WHITELEY.—Just as the OE *blǣc* often means dark and dull rather than black, so the OE *hwīt* frequently denotes bright and fair rather than white. This is the meaning in place-names. The change of vowel-length—*hwīt* becoming *whit* instead of *white*—corresponds to that already noted in *āc*, an oak-tree, *brād*, broad, *stān*, a stone, and *tūn*, an enclosure, words which as prefixes become quite regularly *ack*, *brad*, *stan*, and *tun*. The corresponding word in ON is *hvītr*, white.

WHITCLIFFE, Cleckheaton, 'fair cliff,' may be either Anglian or Scandinavian in both its elements.

WHITGIFT, Goole, SC 1154 *Witegift*, PF 1198 *Witegift*, CR 1203 *Wytegift*, is 'fair portion,' from OE *gift*, a portion or dowry, and the weak form of OE *hwīt*.

WHITELEY, Ecclesall, †1280 *Wyteleye*, 1366 *Whitley*, is the 'fair lea,' from OE *lēah*, a lea or meadow.

WHITELEY, Hebden Bridge, WCR 1308 *Wyteleye*, has the same origin and meaning.

WHITLEY, Knottingley, DB *Witelai*, PF 1202 *Witelay*, NV 1316 *Whitley*, comes from the same source.

WHITLEY BEAUMONT, Kirkheaton, DB *Witelei*, CR 1247 *Wyttelegh*, NV 1316 *Whiteley*, has the distinctive appellation *Bellomonte* in early documents, later forms being *Beumont* and *Beamont*, the 'fine mount.'

WHITWELL, Stocksbridge, YD 1302 *Whitewell*, YD 1307 *Wytewell*, is 'the clear spring,' from OE *well*.

WHITWOOD, Normanton, DB *Witeuude*, PC †1090 *Witewde*, is 'the fair wood,' from OE *wudu*.

WHITECHAPEL.—See Chapelthorpe.

WHITEHAUGHS, Fixby, WH *Wytehalge*, is 'the fair corner,' from OE *healh*, a corner or meadow.

WIBSEY, Bradford.—Early forms are DB *Wibetese*, CR 1283 *Wybecey*, CR 1311 *Wibbeseye*, PT 1379 *Wybsay*. The name is of the same type as Arksey and Pudsey, and we expect its first element to be a personal name. In DB we find such names as Bar and Baret, Eli and Eliet, Leue and Leuet, Tor and Toret; and, as the names Wibba and Wibbo are on record, we are justified in postulating the forms Wibo and Wibet. The latter would agree with the DB spelling *Wibetese*, and would warrant the explanation 'Wibet's island,' from OE *ēg*, an island. A perusal of the lists of Frisian names given by Brons shows the actual existence of the name Wibet, as well as Wibba, Wibbe, and Wibbo.

WICK, WICKEN, WICKER, WICKING, WICKINS, WYKE.—OE *wīc* meant a dwelling, an abode, a village, and ON *vīk* a creek, inlet, bay. It would seem impossible to make use of the latter for inland places; yet in Cumberland the form 'wike' is used to designate 'a narrow opening between rising grounds,' the maritime word being apparently converted to inland uses.

The terminal -wick occurs in the two Adwicks and the two Hardwicks, as well as in Cowick, Creswick, Fenwick, Huntwick, Pledwick, and Wilsick, all words of Anglian origin; but the terminal -wike is found only once, namely, in Heckmondwike.

WICKEN, Scholes, and WICKINS, Upperthong, may mean simply 'mountain-ash,' for that according to EDD is the meaning of the dialect-word Wicken or Quicken. But compare the Norw. place-name Viken, formerly *Wickenn*, from ON *vík*.

WICKER, Sheffield, HS 1637 *Whicker*, may be the dialect-word Wicker or Quicker, a quick-set hedge (EDD). But compare the Norw. place-name Viker pl. of ON *vík*.

WICKING, found in Wicking Lane in Sowerby, Wicking Slack in Widdop, and Wicking Green in Marsden, is perhaps derived from ON *vík*, and ON *eng*, a meadow.

WYKE, Bradford, DB *Wich, Wiche*, HR 1276 *Wyk*, PT 1379 *Wyke*, is interesting because the township contains just such a 'narrow opening between rising grounds' as is alluded to above. It seems very probable that we owe the name to ON *vík*.

THE WYKE, Horbury, is a tract of lowlying land alongside the Calder. The name is most probably from ON *vík*.

WICKERSLEY, Rotherham, DB *Wicresleia, Wincreslei*, RC 1186 *Wikerslai*, KI 1285 *Wykerslegh*, is 'the lea of Wikær,' from OE *lēah*, and the ODan. name recorded by Nielsen.

WIDDOP, on the Lancashire border north of Todmorden, is recorded in HW 1440 as *Wedehope* and HW 1548 as *Widope*. The meaning is the 'wide secluded valley,' from OE *wīd*, wide, and OE *hop*, a secluded valley.

WIGFALL, Worsborough, CH †1250 *Wigfall*, PT 1379 *Wigfall*, appears to be 'the sloping horse-pasture,' from OE *wicg*, ME *wig*, a horse.

WIGHTWIZZLE, Bradfield, CH †1280 *Wygestwysell*, *Wigestwysell*, 1311 *Wigtuisil*, 1335 *Wiggetwisell*, YF 1573 *Wyghtwysill*, is 'Wig's watersmeet,' from OE *twisla*, a confluence, and the recorded name Wig; see Briestwistle.

WILBERLEE, Slaithwaite, YS 1297 *Wildeborleye*, WCR 1308 *Wildborleyes*, may be 'the lea of the wild boar,' but is more probably 'Wildbore's lea,' the sign of the genitive being omitted as in Alverley and Alverthorpe. The personal name *Wildebore* occurs in DN 1355.

WILBY, WILTHORPE, Doncaster and Barnsley.—The latter may be the place referred to as *Wilthorp* in PF 1202. Both names are Scandinavian, and the first element in both is most probably the personal name Wili recorded by Naumann. Thus Wilby may be explained as 'Wili's farm,' from ON *bȳr*, and Wilthorpe as 'Wili's thorp,' from ON *thorp*.

WILSDEN, WILSICK, Bradford and Doncaster.—Early records of these names are as follows :

DB	1086	*Wilsedene*	DB	1086	*Wilseuuice*
PC	†1246	*Wilsyndene*	PR	1190	*Willesich*
NV	1316	*Wylseden*	KF	1303	*Wylsyk*
YF	1558	*Wylsden*	PT	1379	*Wilsewyke*

Among ON personal names several have the ending *-si*, e.g. Elfsi, Grimsi, Hugsi (Naumann); and among Frisian names many have the ending *-se*, e.g. Bense, Gatse, Inse (Brons); while to-day the name Wilse is found in Christiania. Hence we may explain Wilsden as 'the valley of Wilsa or Wilsi,' and Wilsick as 'the habitation of Wilsa or Wilsi,' from OE *denu* and *wīc*.

WILSHAW, Meltham, is probably 'the willow copse,' from OE *wilig*, a willow, and *sceaga*, a copse or wood.

WINCOBANK, Sheffield is the site of an ancient camp. The earliest available records are *Wyncobanke* in YF 1573, *Wincowbanke* in the Ecclesfield Registers of 1597 and 1600, and *Wincowbanke* in HS 1637; compare also HS 1637 *Wincowe Wood*. These are sufficient to warrant us in deriving the second syllable from ON *haugr*, a mound, hill. The first element is doubtless a personal name; and Searle gives Winco, which would account for the prefix in Winksley near Ripon. For Wincow- we require a weak form and must postulate such a name as Winca.

WINDHILL, WINDYBANK.—The former name occurs
(1) near Bradford, PT 1379 *Wyndehill*, YF 1578 *Wyndhyll*, and
(2) near Sheffield, 1307 *Wyndehullefall*. It goes back of course
to OE *wind*, ME *wind*, *wynd*, wind.

The latter name is also found twice, namely, in Southowram,
YD 1277 *Wyndibankes*, and in Liversedge, and derives its first
element from OE *windig*, windy.

WINTERSETT, Wakefield, PR 1190 *Winterseta*, CR 1215
Wintersete, CR 1280 *Wyntressete*, NV 1316 and PT 1379
Wynterset, is probably 'the seat of Winter.' The name Wintra
is recorded by Searle, and we may postulate the corresponding
strong form Winter; indeed Falkman records a Dan. personal
name Vinter. The suffix is from OE *set*, a seat, entrenchment,
camp, or OE *ge·set*, a dwelling, habitation.

WIRRAL, Sheffield.—In quite modern times an alternative
spelling, Worrall, has arisen. Early records are DB *Wihala*,
Wihale, HH 1350 *Wirall*, *Wyrall*, PT 1379 *Wirall*, *Wyrhall*,
YD 1432 *Wyrehall*. The first instance of the alternative form
is in YF 1562 *Worrall als Wyrrall*. The signification appears
to be similar to that of Wirrall in Cheshire which was *Wirhalum*
in 1002, namely, 'the corner of the wild-myrtle,' from OE *wīr*,
the wild-myrtle, and *healh*, a corner or meadow.

-WITH.—Derived from ON *viðr*, a wood, this termination
is found in Cupwith Hill, Slaithwaite; in Stockwith Lane,
Hoyland Nether; and in Bubwith, Pontefract.

WITHENS, WITHINS.—In the neighbourhood of
Halifax this name is of frequent occurrence. We find it in
Southowram, Ovenden, Luddenden, Heptonstall, Cragg Vale,
and Rishworth. There is also Withins Moor west of Penistone,
and DN 1362 has a *Within* in Fixby. Rygh records the name
* *Viðin*, now Vien, and derives it from ON *viðr*, wide, and *vin*, a
meadow, but more probably our words are connected with ON
viðir, a willow, for EDD explains 'withen' as a name given to
various species of willow, or to a piece of wet land where willows
grow. See Lund.

WOMBWELL, Barnsley, has a name of much interest, which is recorded in the following forms:

DB 1086 *Wanbella, Wanbuella*   YI 1307 *Wambewelle*
HE 1276 *Wambwell*   NV 1316 *Wambewell*
KI 1285 *Wambewell*   PT 1379 *Wombewell*

The substitution of *n* for *m* in the Domesday spellings is due to the Norman scribes ; but the change from 'wamb' to 'womb' is quite regular, and corresponds to the change from 'lang' to 'long' and 'strang' to 'strong.' Though the meaning is almost certainly 'the well in the hollow,' the origin is doubtful, as the first element may be either OE *wamb*, ME *womb*, or ON *vömb* (stem *vamb*), words used doubtless in the sense of a hollow place. We find in Icelandic such names as *Vambar·holmr* and *Vambar·dalr* (Vigfusson), and on the other hand we find in Staffordshire the name Wombourne, DB *Wamburne*, later *Wombeburne*, 'the brook in the hollow' (Duignan). See Thong and Wall.

WOMERSLEY, Pontefract.—Without early records it would be impossible to find the true explanation. DB gives *Wilmereslege, Wlmeresleia*, YI 1286 *Wilmeresley*, YD 1318 *Wylmersley*, PT 1379 *Wilmerslay*. Wilmær is a well-known personal name given by Searle, and the place-name may safely be interpreted as 'Wilmær's lea,' from OE *lēah*.

WOOD, WOODALL, WOODHALL, WOODHEAD, WOODHOUSE, WOODKIRK, WOODLANDS, WOODROW, WOODSETTS, WOODSOME, WOODTHORPE.— The word 'wood' is from OE *wudu*, ME *wode*, wood, timber, or a wood, a forest. The following names have this word for their termination : Blackwood, Eastwood, Ewood, Greenwood, Littlewood, Lockwood, Longwood, Middlewood, Morwood, Norwood, Outwood, Pickwood, Sowood, Waleswood, Westwood, Whitwood. The corresponding Scandinavian word, found in such names as Askwith and Birkwith, comes from ON *við̄r*, a wood.

WOODALL, Harthill, YD 1536 *Wodehill*, appears to have suffered a change in its termination.

WOODHALL, Darfield, correctly represents the early form *Wudehall* given in YS 1297.

WOODHEAD, Huddersfield, is given in YD 1369 as *Wodeheued*.

WOODHEAD occurs also near Penistone.

WOODHOUSE and WOODSOME form an interesting pair, being related to one another as singular and plural; the ending of the former is from the OE dative singular *hūse*, of the latter from the dative plural *hūsum*. The only example of the plural form occurs in Woodsome Hall, near Huddersfield, of which curiously enough the earliest record is in the singular, DN 1236 *Wodehuse*, though later spellings, DN 1373 *Wodsom*, CH 1375 *Wodhusum*, DN 1383 *Wodsum*, DN 1393 *Wodesom*, YF 1561 *Wodosom*, are obviously plural and signify 'wood houses.' Of the name Woodhouse eight examples have come to notice; they are situated at Ardsley, Cartworth, Emley, Normanton, Handsworth 1297 *Wodehouses*, Huddersfield DN 1383 *Wodehous*, Rastrick 1314 *Wodehowses*, and Shelley WCR 1275 *Wodehuses*.

WOODKIRK, Dewsbury, BM 1196 *Wodekirk*, CR 1215 *Wdekirka*, HR 1276 *Wodekirke*, is interesting because of the form 'kirk' and its association with the Anglian 'wood.' The connection of Woodkirk with the ancient Mystery Plays is well known, and the annual horse fairs held close at hand are no less famous, though after another fashion. An entry in WCR 1306 tells us of John, servant of the late Henry de Swynlington, that he 'stole a hide worth 15*d.* from Wodekirk Fair,' and concludes 'He is to be arrested.'

WOODLANDS occurs in Adwick-le-Street.

WOODROW, Methley, KC 1332 *Woderoue*, MPR 1612 *Wood-rowe*, is probably 'the row beside the wood,' from ME *rowe*, OE *rāw*, a row, line.

WOODSETTS, on the Nottinghamshire border, spelt *Wodesete* in 1324 and *Wodesetes* in 1354, seems to be 'the seat in the wood,' from OE *set*, a seat, entrenchment, camp.

WOODTHORPE, Wakefield, is mentioned in WCR under the forms *Wodethorp* in 1279 and *Wodethorpe* in 1286.

WOODTHORPE, Handsworth, was *Wodetorp* in †1277 and *Wodethorp* in †1300. The termination is derived from ON *thorp*, a village.

WOODLESFORD, on the Aire near Leeds, is recorded in PF 1170 as *Wridelesford*, in PF 1202 as *Wriddlesford*, CR 1250

Wudelesford, DN 1251 *Wodelesford,* PM 1258 *Wridelesford,* LC 1296 *Wridelesforde.* Apparently there were two forms struggling for the mastery, and a third form 1327 *Wriglesford,* RPR 1671 *Wriglesforth,* RPR 1670 *Wriglesworth,* is also to be found. The first element is clearly a personal name, and as Searle gives Wodel, we may explain the place-name as ' Wodel's ford.'

WOOLDALE, WOOLGREAVES, WOOLLEY, WOOL-ROW, WOOLTHWAITE.—In no single instance is there any connection with sheep. The prefixes are, in fact, 'wolves masquerading in sheep's clothing,' for the origin is either (1) OE *wulf,* a wolf, (2) ON *ūlfr,* a wolf, or (3) corresponding weak personal names Wulfa and Ulfi. We take first those names which are Anglian.

WOOLGREAVES, Cawthorne and Sandal, is 'wolf-thicket,' from OE *grǣfa,* a bush, thicket, grove.

WOOLLEY, which occurs three times, is either ' lea of the wolves,' from OE *wulfa,* gen. pl. of *wulf,* or ' lea of Wulfa,' witness the following early forms : (1) Woolley near Wakefield, DB *Wiluelai,* PC 1192 *Wlveleia,* YI 1297 *Wolvelay,* NV 1316 *Wolfelay* ; (2) Woolley, Shire Green, spelt *Wolleghes* about 1325 according to Eastwood ; (3) Woolley Head, Hipperholme, WCR 1297 *Wlveley heud.*

The following names are Scandinavian, and have for their first element either ON *ūlfa,* the gen. pl. of *ūlfr,* or Ulfa the gen. of a personal name Ulfi.

WOOLDALE, Holmfirth, commonly pronounced Oodle (*ūdl*), DB 1086 *Vluedel,* WCR 1274 and 1297 *Wlvedale,* WCR 1286 *Wolvedale,* gets its terminal from ON *dalr,* a valley.

WOOLROW, Shelley, YI 1266 *Wolewra,* WCR 1275 *Wlvewro,* gets its terminal from ON *vrá,* a nook or corner.

WOOLROW, Brighouse, WCR 1308 *Wollewro, Wolwro,* HW 1554 *Wolrawe,* has the same origin and meaning.

WOOLTHWAITE, Tickhill, BM *Wolvethwaite,* RC 1241 *Wlve-thwait,* comes from ON *thveit,* a clearing.

WORMALD occurs both in Barkisland and Rishworth.

Burton gives the early forms *Wlfrunwell* and *Wulfrunwall*, and other early forms are as follows :

WCR	1286 *Walronwalle*	PT	1379 *Wornewall*
WCR	1308 *Wollerenwalle*	HW	1402 *Wormewall*
WCR	1326 *Wolronwal*	DN	1632 *Hye Wormall*

The final *d* has been added in more recent times, as in the case of Backhold. The name is possibly Scandinavian, 'the field of Ulfrun,' from ON *vollr*, a field ; but it is possibly Anglian, its meaning 'Wulfrun's well.' Ulfrun is the Scandinavian form of the personal name—which is feminine—and Wulfrun is the Anglian form. See Wall, Well.

WORMLEY, Thorne, PT 1379 *Wormelay*, may perhaps be of the same origin as Wormley, Herts., DB *Wermelai*. In that case it means ' Wurma's lea,' being equivalent to OE *Wurman·lēah*, where Wurma is a short form of some such name as Wurm·beorht or Wurm·here. But see Wormald.

WORSBOROUGH, WORTLEY.—These place-names provide examples of a personal name in its strong and weak forms. They also provide examples where different ancient names have produced the same result.

Worsborough, Barnsley.	Wortley, Leeds.	Wortley, Sheffield.
DB 1086 *Wircesburg*	KC 1189 *Wirkeleia*	DB 1086 *Wirtleie, Wirlei*
CR 1249 *Wyrkesburc*	CC 1200 *Wirkelaia*	YS 1297 *Wortelay*
NV 1316 *Wyrkesburgh*	KI 1285 *Wirkelay*	NV 1316 *Wortelai*
PT 1379 *Wyrkesburgh*	KF 1303 *Wirkeley*	PT 1379 *Wortelay*

Worsborough has for its first element a personal name, doubtless the strong form Wyrc equivalent to the OE Weorc found in BCS *Weorces·mere*; compare also the Frisian name Wirke (Brons). We may explain Worsborough as ' Weorc's strong place,' from OE *burg*, a fortified post.

The early forms of the Leeds Wortley differ from those of Worsborough in omitting the final *s* from the personal name ; we are therefore dealing with a weak form of the name such as Weorca, and hence the meaning is ' Weorca's lea,' from OE *lēah*.

WORTLEY, Sheffield, shows a prefix of quite another character, derived from OE *wyrt*, a herb, vegetable. Old English had several compounds in which *wyrt* was the first element. The ancient word for garden was *wyrt·geard*, wort-yard, or *wyrt·tūn*, wort-enclosure; the gardener was *wyrt weard*, wort-ward; and physic was *wyrt drenc*, wort-drink. Perhaps Wortley, *wyrt·lēah*, was noted for its productiveness.

WORTH.—The OE *worth, weorth, wyrth*, was applied to a homestead or farm. According to Professor Skeat it is closely allied to the word 'worth' meaning 'value' and it may be explained as 'property' or 'holding.' OE had two derivatives *worthine* and *worthig*; these have given us the terminations in the Shropshire names Shrawardine and Cheswardine, and the Devon or Somerset names Bradworthy, Holsworthy, Selworthy.

Examples in South-west Yorkshire include Ackworth, Badsworth, Cudworth, Cullingworth, Cumberworth, Cusworth, Dodworth, Fallingworth, Hainworth, Handsworth, Haworth, Holdsworth, Holdworth, Ingbirchworth, Kimberworth, Oakworth, Rishworth, Roughbirchworth, Saddleworth, Tudworth, Wadsworth, Wadworth, Warmsworth, Wentworth.

WRAGBY, Wakefield, WCR 1308 *Wraggeby*, WCR 1326 *Wraggebi*, IN 1332 *Wragheby*, should be compared with the Lincolnshire Wragby, which appears in DB as *Waragebi* (for *Wragebi*), and later as *Wraggeby* and *Wragheby*. Nielsen records an ODan. personal name Wraghi, which appears in *Wragathorp*, now Vragerup, in Skane. Hence we may explain the two Wragbys as 'Wragi's farm,' from ON *bȳr*, and a personal name *Wragi. See Hagg, Haigh.

WRAITH HOUSE, Oxspring.—In EDD a dialect-word 'wreath' or 'wraith' is explained as a wattle, underwood, brushwood. Another dialect-word 'wreath' or 'wread' is described as an enclosure for cattle. The latter is doubtless from the OE *wrǣth*, which according to Professor Skeat is found in the Cambridgeshire names Shepreth and Meldreth.

WRANGBROOK, Pontefract, KC †1153 *Wrangebroc*, YR 1230 *Wrangbrok*, HR 1276 *Wrangbroc*, PT 1379 *Wraynebrok*, derives its first element from OE *wrang*, twisted, crooked, and its termination from OE *brōc*, a stream; compare BCS 944 *Wrangan·hylle*. It should be noted, however, that there is a stream in South Wales called Afon Wrangon; and Mr Henry Bradley suggests that Wrangon was the name of the Warwickshire Avon.

WRENTHORPE, Wakefield, provides an excellent example of the 'rounding' which the lapse of time tends to produce. Early records include HR 1276 *Wyverinthorp*, WCR 1298 *Wyverumthorpe*, WCR 1307 *Wyveromthorpe*, 1348 *Wyrenthorp*, 1425 *Wyrnethorp*. The first element is doubtless a personal name, and Searle has Wifrun, a name which neither Naumann nor Nielsen records, although they give Dagrun, Guthrun, Oddrun, and others. Wrenthorpe is 'the thorpe of Wifrun,' from ON *thorp*.

WROSE, Shipley, PT 1379 *Wrose*, YF 1547 *Wrose*, YF 1550 *Wrosse*, appears to be connected with the OE *wrāsan*, which means a knot or lumps.

WYKE.—See Wick.

YATEHOLME, Holmfirth.—See Holme.

SUPPLEMENTARY LIST OF NAMES

WITH ADDITIONAL NOTES AND CORRECTIONS

APPERLEY, p. 59.—In the curious Nottinghamshire name Styrrup there is probably support for the suggestion that *apa*, water, occurs in English place-names. Early forms of the name are:

DB	1086 *Estirape*		IL	1348 *Stirap*
HR	1278 *Stirap*		IL	1414 *Sterap*
IL	†1300 *Styrap*		IN	†1500 *Sterop*

This can scarcely be the common word 'stirrup,' which meant literally *sty·rope*, and which comes from OE *stīrāp* (for *stigrāp*), ME *stirop*. Rather, the first element is a mutated form of *stūr*, which is itself an early form of the common river-name Stour (McClure); compare the Westphalian place-name Stirpe and the Dutch stream-name Stierop, both of which according to Jellinghaus involve the word *apa*.

BRIDGE, BRIGG.—While 'brigg' and 'rigg' come from Scandinavian sources, viz. ON *bryggja* and ON *hryggr*, 'bridge' and 'ridge' are English in origin and come from OE *brycg* and OE *hrycg*.

FULNECK, Pudsey.—A settlement of the Moravian Brethren was established here in 1744. The name is derived from Fulneck in Moravia, which was one of the principal seats of the Community.

HEELEY.—See Healey, p. 162.

MAGDALE.—For the pronunciation compare Haigh. It is of course possible in such names as Mag Field and Mag Wood that the source of the first element is the dialect-word *mag*, a magpie.

NORWOOD.—See Norland, p. 219.

NOSTELL.—See Brierley, p. 72.

POTTER.—The suggestions in the note on Pott, Potter, must not be held to preclude an etymology from OE *pott*, ON *pottr*, a pot, ME *potter*, a potter.

PRIESTLEY, p. 232, is 'the lea of the priests,' from *prēosta*, the gen. pl. of OE *prēost*, a priest.

QUARMBY, p. 234, has Domesday forms which are in conflict with later forms and with one another. Perhaps DB *Cornelbi* is a scribal error; but DB *Cornebi* means 'Korni's farm,' and DB *Cornesbi* means 'Korn's farm,' the personal name in the former being weak and in the latter strong.

SKELDERGATE, p. 258, means 'shield-maker's road,' from ON *skjaldari*, a shield-maker, and ON *gata*, a road (Lindkvist). It is therefore connected with ON *skjold*, a shield, but not in the way previously suggested.

SKELLOW.—See Skelbrook, p. 256.

SKYRE.—See Skiers, p. 258.

SNAPETHORPE, p. 262, has for its first element a word connected with ON *sneypa*.

SOOTHILL, p. 263.—Forms like DC 1225 *Sotehill*, DC 1349 *Sotehull*, YF 1504 *Sotehill*, agree with a derivation from the ON personal name Sōti, gen. Sōta.

SPROTBOROUGH, p. 269, is 'Sprota's fortified place,' where *Sprota is a weak form corresponding to the recorded strong form Sprot.

TOPCLIFFE.—See Toft, p. 286.

TRIPPEY, p. 288.—The first element can scarcely come from ON *threp*, *threpi*; more probably it is to be connected with the dialect-word *trip*, a flock of sheep (EDD).

THE COMMON FIELD SYSTEM.—In connection with the note on page 10 it should be noted that while certain communities had *three* common fields, others had only *two*.

THE GENITIVE INFLECTION.—There are several points of interest in regard to the genitives dealt with in the foregoing pages.

THE GENITIVE SINGULAR.—(1) Many strong personal names have lost the -*s*- they once possessed, witness the early forms of Adlingfleet, Alverley, Armley, Armthorpe, Auckley, Austonley, Chellow, Dodworth, Heckmondwike, Kettlethorpe, Osgoldcross, Painthorpe, Rainborough, Streetthorpe, Warley, Warmfield, and Wightwizzle; compare also Dransfield, Kerisforth, and Tankersley.

(2) Other strong personal names have in their known history never possessed this -*s*-, witness the early forms of Alverthorpe, Edderthorpe, Herringthorpe, Osgathorpe, Renathorpe, Rodley, Skelmanthorpe, Thurstanland, and perhaps also Attercliffe, Chickenley, and Normanton. Of these Attercliffe, Skelmanthorpe, and Thurstanland are found in the Domesday record.

(3) There is no assured example of a weak genitive in -*en* representing OE -*an*; but see Rossington.

THE GENITIVE PLURAL.—(1) Several names appear to have possessed as their first element a genitive plural in -*a*, viz. Bramley, Brierley, Briestwistle, Churwell, Farnley, Farsley, Priestley, Rotherham, Shepley, Shipley, and possibly also Woolley and Normanton.

(2) A small number of names show in their first element the representative of a weak genitive plural in -*ena*, viz. Hewenden, Oxenhope, *Carlentone*, and *Rodenham*.

THE SUFFIXED ARTICLE.—The map of Norway shows large numbers of names ending with -*en* or -*et*, where -*en* is the Masc. or Fem. form of the suffixed article, and -*et* the Neut., older forms being -*in* and -*it*. From ON *ōss*, the mouth or outlet of a river or lake, we get the Norwegian place-names Os and Osen, the latter with and the former without the suffixed article. Other names of the same kind are Lunden, from ON *lundr*, a grove; Viken, from ON *vīk*, an inlet; Dalen, from ON *dalr*, a dale; Holtet, from ON *holt*, a wood. See Lund.

This suffixed article did not come fully into use until about the year 1200, and it has been stated that there is no trace of it in English; but Björkman reminds us that Jakobsen gives instances of its retention in the Shetlands and asks whether the ending -*īn* in *Orrmīn* in the *Ormulum* may not have the same origin[1]. An examination of the place-names in South-west Yorkshire reveals a number of instances which can scarcely be accounted for in any other way, early forms being 1318 *le Swythen* and 1362 *Within*. The list includes the following names: Collin, Collon, London, Magdalen, Stubbin, Swithen, Swithens, Withens, Withins, some of them several times repeated.

THE FIELD OF BRUNANBURH.—At Brunanburh in the year 937 was fought one of the most memorable of early battles—one which was known for many a day as 'the great battle.' In this historic fight the forces of a great confederation—Picts and Scots, and Strathclyde Britons, and Vikings from the West and North—were met by Athelstan and utterly defeated. The fight began with the dawn, and the long and fierce pursuit which followed was only ended by the darkness of night. There was terrible slaughter, and among the slain were five kings and seven earls; but the two leaders, Constantine and Anlaf, made good their escape, the former by land and the latter by water.

The scene of the struggle is still uncertain, and among the places suggested one, Burnswark, is as far north as Dumfries, and a second, Brunedown, is as far south as Devon, while others are Bourn and Brumby in Lincolnshire, Boroughbridge in Yorkshire, Bromborough in Cheshire, Burnley in Lancashire, and Bromfield in Cumberland.

In early Chronicles the place where the fight took place is called *Brunandune, Brunanwerc, Brunefeld, Bruneford, Brumesburh, Brunesburh*, and *Bruneswerc*, as well as *Brunanburh* and *Brunanbyrig*; but it is also called *Brune* by the Welsh Chronicle, *Othlyn* by the Annals of Clonmacnois, and *Wendune* by Symeon

---

[1] Björkman, *Scandinavian Loan-words in Middle English*, p. 21.

of Durham. Obviously therefore the site showed a *dun*, that is, a hill; a *burh* or *werc*, that is, a fortified place; a *lyn*, that is, a pool; and also a *ford*. Further, according to Florence of Worcester and Symeon of Durham, Anlaf brought his Viking fleet up the Humber, and according to Ingulf the battle was fought in Northumbria. Still further, an army marching from Wessex towards York would probably follow the course of Riknild Street, which after passing Derby and Chesterfield entered Yorkshire and crossed the Don near Rotherham.

Under these circumstances it seems proper to draw attention to certain facts connected with our South-west Yorkshire place-names, leaving any further discussion to others.

1. BRINSWORTH, Rotherham, DB *Brinesford*, may possibly be derived from an earlier **Brunesford*; compare Crigglestone and Crimbles. At Brinsworth there is an ancient ford over the Don, and close beside the ford an extensive rectangular earthwork believed to be of Roman origin and now called Templeborough.

2. WENT, formerly *Wenet*, occurs as the name of a stream six miles south of Castleford and fifteen north of Rotherham; and the name Wentworth, DB *Winteuuorde*, occurs three or four miles north-east of Rotherham. These may possibly have a link with *Wendune*.

3. MORTHEN, five miles south-east of Rotherham, formerly *Morhtheng*, is of Scandinavian origin, and appears to be the site of an ancient battle, its meaning being 'slaughter meadow.'

Milton Keynes UK
Ingram Content Group UK Ltd.
UKHW041521181024
449640UK00009B/123